Aquaculture Ecosystems:
Adaptability and Sustainability

Aquaculture Ecosystems: Adaptability and Sustainability

Editor: Douglas Rodriquez

RCALLISTO
REFERENCE

www.callistoreference.com

Callisto Reference,
118-35 Queens Blvd., Suite 400,
Forest Hills, NY 11375, USA

Visit us on the World Wide Web at:
www.callistoreference.com

ISBN: 978-1-64116-130-5 (Hardback)

Cataloging-in-Publication Data

Aquaculture ecosystems : adaptability and sustainability / edited by Douglas Rodriquez.
   p. cm.
Includes bibliographical references and index.
ISBN 978-1-64116-130-5
1. Sustainable aquaculture. 2. Aquaculture--Environmental aspects.
3. Aquaculture. I. Rodriquez, Douglas.
SH136.S88 A68 2019
639.8--dc23

Table of Contents

Preface

The world is advancing at a fast pace like never before. Therefore, the need is to keep up with the latest developments. This book was an idea that came to fruition when the specialists in the area realized the need to coordinate together and document essential themes in the subject. That's when I was requested to be the editor. Editing this book has been an honour as it brings together diverse authors researching on different streams of the field. The book collates essential materials contributed by veterans in the area which can be utilized by students and researchers alike.

Aquaculture involves the farming of freshwater and saltwater organisms under controlled conditions. Various interventions in the rearing process are implemented to enhance production. Some of these include regular stocking, feeding, protection from predators, etc. Aquaculture is classified according to the species of organisms harvested, into fish farming, shrimp farming, oyster farming, etc. Aquaculture also focuses on animal welfare in terms of fitness, physical and mental well being. Modern techniques for sustainable aquaculture include the application of Integrated Pest Management, use of vaccines, adoption of recirculating aquaculture systems, etc. This book is compiled in such a manner, that it will provide in-depth knowledge about the theory and practice of aquaculture. It also discusses all relevant issues of adaptability and sustainability of aquaculture ecosystems. It aims to serve as a resource guide for students and experts alike and contribute to the growth of the field.

Each chapter is a sole-standing publication that reflects each author's interpretation. Thus, the book displays a multi-facetted picture of our current understanding of application, resources and aspects of the field. I would like to thank the contributors of this book and my family for their endless support.

Editor

Redox stratification drives enhanced growth in a deposit-feeding invertebrate: implications for aquaculture bioremediation

Georgina Robinson[1,2,*], Gary S. Caldwell[1], Clifford L. W. Jones[2],
Matthew J. Slater[1,3], Selina M. Stead[1]

[1]School of Marine Science and Technology, Newcastle University, Newcastle NE1 7RU, UK
[2]Department of Ichthyology and Fisheries Science, Rhodes University, Grahamstown 6140, South Africa
[3]Alfred-Wegener-Institute, Helmholtz Center for Polar and Marine Research, Am Handelshafen 12, 27570 Bremerhaven, Germany

ABSTRACT: Effective and affordable treatment of waste solids is a key sustainability challenge for the aquaculture industry. Here, we investigated the potential for a deposit-feeding sea cucumber, *Holothuria scabra*, to provide a remediation service whilst concurrently yielding a high-value secondary product in a land-based recirculating aquaculture system (RAS). The effect of sediment depth, particle size and redox regime were examined in relation to changes in the behaviour, growth and biochemical composition of juvenile sea cucumbers cultured for 81 d in manipulated sediment systems, describing either fully oxic or stratified (oxic–anoxic) redox regimes. The redox regime was the principal factor affecting growth, biochemical composition and behaviour, while substrate depth and particle size did not significantly affect growth rate or biomass production. Animals cultured under fully oxic conditions exhibited negative growth and had higher lipid and carbohydrate contents, potentially due to compensatory feeding in response to higher microphytobenthic production. In contrast, animals in the stratified treatments spent more time feeding, generated faster growth and produced significantly higher biomass yields (626.89 ± 35.44 g m^{-2} versus 449.22 ± 14.24 g m^{-2}; mean ± SE). Further, unlike in oxic treatments, growth in the stratified treatments did not reach maximum biomass carrying capacity, indicating that stratified sediment is more suitable for culturing sea cucumbers. However, the stratified sediments may exhibit reduced bioremediation ability relative to the oxic sediment, signifying a trade-off between remediation efficiency and exploitable biomass yield.

KEY WORDS: Sea cucumber · *Holothuria scabra* · Sandfish · Value-added aquaculture · Recirculating aquaculture · Bioturbation · Compensatory feeding

INTRODUCTION

Bioremediation — the biological treatment of waste streams and pollutants — is a widely established process, increasingly applied by a broad range of industries operating within varied environmental and ecological settings and constraints (Alexander 1999). In general, bioremediation technologies are bacteria-driven, with selective stimulation of the degrading activities of endogenous microbial populations, a fundamental concept underpinning the approach (Colleran 1997). Increasingly, bioremediation operations are seeking to add value to their processes by converting the derived biomass into exploitable products, e.g. feeds, fuels and high-value goods (Gifford et al. 2004, Muradov et al. 2014). Bioremediation strategies are being developed for and deployed in the aquaculture industry, primarily as a means to

*Corresponding author: g.robinson3@ncl.ac.uk

capture nutrient leachates and suspended solids from effluent and sediments (Chávez-Crooker & Obreque-Contreras 2010, Kim et al. 2013), but also to reduce pathogen discharge to the environment (Zhang et al. 2010).

The effective and affordable treatment of sediments and suspended solid wastes from intensive aquaculture operations remains a key sustainability challenge facing the industry. Land-based aquaculture operations, in particular recirculating aquaculture systems (RAS), offer the greatest potential to separate waste streams for downstream treatment. Despite the fact that concentrated solid wastes from closed recirculating aquaculture systems provide for a number of alternative waste management strategies, proportionally, research efforts have placed greater emphasis on the treatment of dissolved inorganic nutrients. The processing and/or removal of particulate organic wastes, either *in situ* or *ex situ*, poses an environmental threat and remains a significant cost to the industry. If added economic value is secured as part of the process, it will provide further incentives for the operators to adopt best environmental farming practices.

Deposit-feeding polychaetes and sea cucumbers are prime candidates for value-added bioremediation of aquaculture waste solids due to their ability to assimilate particulate organic wastes (comprising decaying feed and faeces) derived from intensive land-based aquaculture (Palmer 2010, Watanabe et al. 2012). Furthermore, deposit-feeders accelerate the depletion of organic matter reservoirs through bioturbation (Yingst 1976, Moriarty 1982, Baskar 1994, Mercier et al. 1999); thereby improving sediment quality and contributing to nutrient remineralisation (Aller 1994, Aller & Aller 1998, Kristensen 2000, Uthicke 2001, Mermillod-Blondin & Rosenberg 2006). The commercially valuable sea cucumber, *Holothuria scabra* (Echinodermata: Holothuroidae), commonly referred to as sandfish, is a promising candidate for cultivation as a high-value secondary product in effluent treatment systems, due to its natural ecological affinity for organically rich sediments in tropical environments. It is currently the most valuable tropical sea cucumber species, commanding high prices (up to US$ 1668 kg^{-1}) and a strong market demand (Purcell 2014); however, wild stocks are over-exploited, and the species is registered as endangered (Purcell et al. 2014). Aquaculture production is considered the only viable means to fulfil market demand.

Sandfish require a substrate for optimal growth in culture tanks; therefore, determination of the opti-

mum physicochemical substrate parameters is a prerequisite for the successful design of land-based bioremediation applications (Robinson et al. 2013). A number of factors affect the rate of organic matter mineralisation in sediments, including organic matter characteristics, temperature, pH, redox potential, bioturbation, and sediment depth and grain size (Stahlberg et al. 2006). The dynamics between sediment mineralisation and bioturbation processes are characterised by strong feedback loops between deposit-feeders, their food, and their microbial and chemical environment (Herman et al. 1999). Deposit-feeder–microbial (i.e. detrital) food chains offer a great potential for manipulation to improve their overall ecological and commercial efficiency in terms of organic loading and biomass production (Moriarty 1987). However, it is essential to improve understanding of the physicochemical and biological processes that govern the degradation kinetics of organic matter in detrital food chains and sediment-based aquaculture effluent treatment systems integrating deposit-feeders (Pullin 1987, Stahlberg et al. 2006).

The supplemental delivery of external electron acceptors is one method used for *in situ* bioremediation technologies to alleviate the constraints imposed by the naturally slow mineralisation process. Increased oxidant supply via the percolation of oxygen-rich water is commonly used to stimulate aerobic decomposition (Colleran 1997); however, it remains unclear how oxidant supply and sediment manipulation affects waste availability and remediation efficiency in such systems. Furthermore, it is anticipated that this approach may have negative consequences for the nutritional environment of deposit-feeders, which inhabit redox-stratified sediments in the wild (Lopez & Levinton 1987).

This study examines the effect of manipulated surficial sediment systems describing fully oxic versus stratified oxic–anoxic conditions, sediment particle size and depth on the growth and carrying capacity — defined here as the maximum sustainable population density supported by a given set of environmental factors — of an invertebrate (sea cucumber) deposit-feeder (Grassle & Grassle 1974, Tenore 1981). A number of studies have examined the effects of substrate particle size and depth on the growth and survival of sandfish (Battaglene et al. 1999, Mercier et al. 1999, Pitt et al. 2001); however, the effects of other physical parameters or factor interactions have not been investigated. From the empirical evidence, inferences on whether *H. scabra* can increase the overall assimilative capacity of an

Table 1. Description of the experimental treatments. Fine particle size: 125–250 µm; medium particle size: 250–500 µm

Treatment no.	Tank structure	Aeration	Factor 1: redox regime	Factor 2: sediment depth (cm)	Factor 3: particle size
1	Plenum	Airlift pump	Oxic	2	Fine
2	Plenum	Airlift pump	Oxic	2	Medium
3	Plenum	Airlift pump	Oxic	4	Fine
4	Plenum	Airlift pump	Oxic	4	Medium
5	No plenum	Airstone	Oxic–anoxic	2	Fine
6	No plenum	Airstone	Oxic–anoxic	2	Medium
7	No plenum	Airstone	Oxic–anoxic	4	Fine
8	No plenum	Airstone	Oxic–anoxic	4	Medium

aquaculture effluent treatment system, either directly via the consumption and assimilation of particulate organic matter or indirectly through the stimulation of benthic microbial metabolism, can be determined. This information is important for understanding the benefits deposit-feeding invertebrates like the sea cucumber can have on mitigating the environmental impacts of aquaculture on its aquatic environment.

MATERIALS AND METHODS

Animal husbandry and experimental conditions

The study was conducted at HIK Abalone Farm (Pty) Ltd in Hermanus, on the southwest coast of South Africa (34° 26′ 04.35″ S; 19° 13′ 12.51″ E) between 20 March and 9 June 2012. A total of 2000 hatchery-reared juvenile *Holothuria scabra* weighing 2 g each were imported from a commercial hatchery (Madagascar Holothurie S.A., Madagascar) on 3 November 2011 and quarantined in a biosecure facility for 6 wk in accordance with South African importation and scientific investigations licences. Following the quarantine period and prior to experimentation, the animals were held in a recirculating aquaculture system in tanks filled with 4.0 cm of calcium carbonate sand sediment and were fed a 34% protein commercial abalone weaning diet (S34 Abfeed®, 1.0 mm sugar grain pellet; Marifeed).

A 2 × 2 × 2 factorial design was used to investigate the response of *H. scabra* to:

(1) the sediment redox regime (oxic versus oxic–anoxic),

(2) sediment depth (2 and 4 cm) and

(3) sediment particle size (fine: 125–250 µm; medium: 250–500 µm; Table 1). Redox stratified (oxic–anoxic) sediments were intended to mimic the redox

state of sediments in the natural habitat of *H. scabra* which exhibit a shallow oxic–anoxic interface below which the sediment remains anoxic (Michio et al. 2003, Wolkenhauer et al. 2010). Substrate particle size and depth were included as experimental factors, as they can affect organic matter content and the distribution of microbial communities that mediate organic matter decomposition and biogeochemical transformations. Eight experimental treatments were allocated to 32 polyethylene tanks (455 × 328 × 175 mm) using a randomised block design of 8 tanks distributed in 4 blocks with 1 replicate per block.

Tanks were supplied with heated coastal seawater at a flow rate of 0.75 l min⁻¹ tank⁻¹ filtered through a recirculating system comprising a composite sand filter, protein skimmer and biofilter. Aeration was supplied continuously, except during feeding when the air and water supplies were interrupted for 15 min to allow the feed to settle. Sea cucumbers were fed a 34% protein commercial abalone weaning diet (S34 Abfeed®, 1.0 mm sugar grain pellet; Marifeed) once per day at 16:00 h. This reference diet was used to provide a reproducible baseline against which to compare any subsequent feeding trials using aquaculture wastes as feeds. Feeding was standardised across experimental treatments: daily feed rations were calculated at 1% of the total tank biomass and adjusted every 2 wk based on predicted biomass gains in between weight assessments (Battaglene et al. 1999). Decaying uneaten food and any arising white bacterial patches were removed by siphoning every 48 h (present in oxic–anoxic treatments only). All tanks were cleaned once per month; tank walls were manually scrubbed to remove the biofilm and any epiphytic algae or cyanobacteria. Experimental tanks were subjected to a natural photoperiod of 10 h light:14 h dark (07:40 to 17:40 h light, local time).

The sediment consisted of calcium carbonate 'builders sand' sourced from a commercial dune quarry (SSB Mining). The sediment was sieved to achieve the requisite particle sizes using a series of decreasing nylon mesh sizes (500, 250 and 125 µm.) An internal tank liner made from 95% shade cloth was used to contain the sediment in all tanks (Fig. 1). Oxic treatment tanks (n = 16) were fitted with a plastic grid partition with perforations of 2 cm² supported 4.5 cm above the tank base to create a sediment-free plenum (false bottom) (Jaubert 2008) directly under the liner.

A) Fully oxic sediment tank

Water inflow

Water surface

Tank

Substrate

Plenum

Plenum support

Up-stand outflow

Airline tube

Tank liner

Airlift pump

Direction of water flow

Approximate scale:
50 mm

B) Redox stratified (oxic-anoxic) sediment tank

Water inflow

Water surface

Tank

Substrate

Up-stand outflow

Airline tube

Tank liner

Air stone

Approximate scale:
50 mm

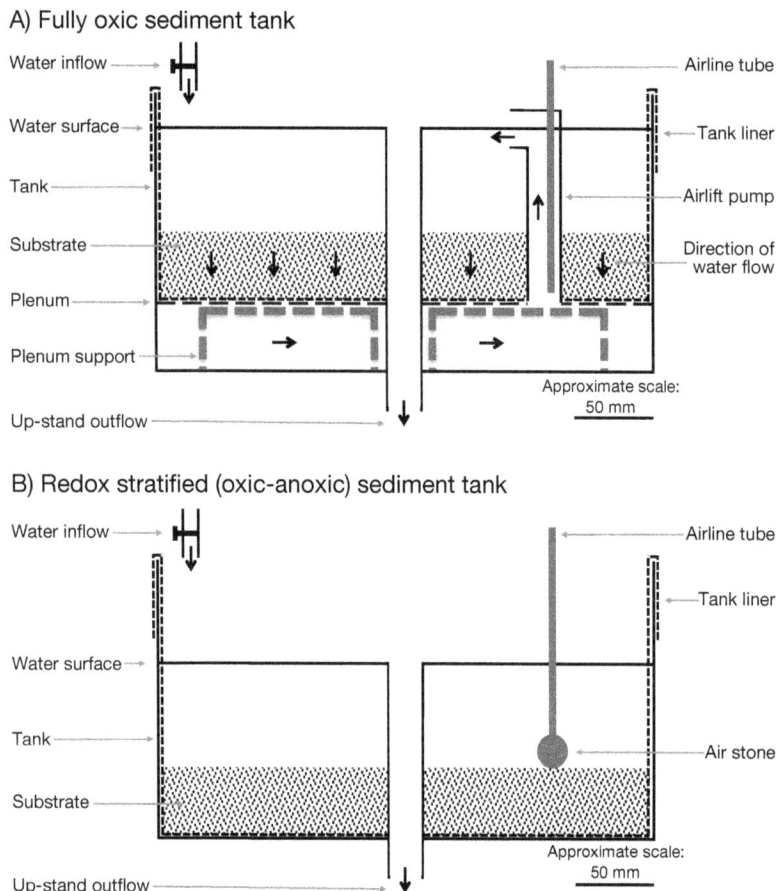

Fig. 1. Schematic diagram of the tank design used to create the 2 different sediment redox regimes. (A) The oxic treatment tanks were fitted with a plenum, and an airlift pump was used to circulate oxygenated water downwards through the sediment to maintain the sediment under a fully oxic regime. (B) In the oxic–anoxic treatment tanks, there was no plenum or water movement through the sediment, so a naturally stratified oxic–anoxic sediment developed; dissolved oxygen levels were maintained with an airstone

a damp cloth for 1 min, weighed to the nearest 0.01 g and photographed for individual photo-identification to permit monitoring of individual growth rates (Raj 1998). Animals with a mean ($\pm$SE) weight of 7.3 ± 0.07 g ind.$^{-1}$ were allocated randomly to 32 groups of 4 individuals per group. Each individual was gut-evacuated for 24 h and reweighed every 27 d over the 81 d experimental period. Wet weight data were used to calculated specific growth rate (SGR), growth rate and coefficient of variation (CV) as follows:

$$\text{SGR} \ (\% \ d^{-1}) = 100 \ (\ln W_2 - \ln W_1) \ / \ T \ (1)$$

$$\text{Growth rate} \ (g \ d^{-1}) = (W_2 - W_1) \ / \ T \quad (2)$$

$$\text{CV} \ (\%) = 100 \times (\text{SD} \ / \ \bar{x}) \qquad (3)$$

where W_1 and W_2 are initial and final wet body weight (g) of sea cucumbers in each experimental tank, T is the duration of the experiment (days); SD is the standard deviation in body weight and $\bar{x}$ is the mean wet weight (g) of sea cucumbers in each experimental tank for a particular sampling period.

Behavioural observations

Behavioural observations were carried out over 3 consecutive 24 h periods in the penultimate week of the study. Observations were made at 4 h intervals, commencing at noon. Red light was used to facilitate night time observations. During each observation period, each tank was observed for 2 min and the number of animals in each burial state and their levels of activity were recorded (Table 2).

Proximate composition analysis

At the end of the trial all animals from 3 of the 4 replicate tanks for each treatment were pooled and homogenised, thereby creating 1 composite sample per tank, which was frozen at −20°C. The composite samples were then lyophilised at −80°C, ground to a fine powder (~50 μm) with a pestle and mortar, and their proximate composition was analysed according to the Association of Official Analytical Chemists' (AOAC) official methods (AOAC 2010). Moisture

An airlift pump was used to circulate oxygenated water downwards through the sediment, maintain the sediment under a fully oxic regime and maintain dissolved oxygen levels within the water column. The water movement was sufficiently gentle as to cause no particle movement within the sand. In the stratified oxic–anoxic treatment tanks, the sediment was directly exposed to the base of the tanks (i.e. no plenum and no water movement through the sediment) so that naturally stratified oxic–anoxic sediment developed. Aeration was provided to the oxic–anoxic tanks using airstones to maintain dissolved oxygen levels within the water column (Table 1).

Prior to stocking into experimental tanks, the juvenile sea cucumbers (n = 128) were suspended in mesh bags for 24 h to ensure gut contents were evacuated prior to weighing. They were then drained on

Table 2. *Holothuria scabra*. Definition of sea cucumber behaviour categorised into burial state and activity according to Wolkenhauer (2008)

Behaviour	Definition
Burial state	
Buried	Entire body under the sediment surface
Semi-buried	>50% of the body under sediment surface
On the surface	Entire body on sediment surface
Activity	
Resting	Animal is inactive with no movement observed
Feeding	Animal is actively feeding on sediment or walls; tentacles are exposed, and head performs sweeping movements

was determined by weight loss after drying at 95°C for 72 h (AOAC Method 934.01), while ash was determined by weight loss on combustion after ashing in a furnace for 4 h at 550°C (AOAC Method 942.05). Crude protein was analysed in a LECO Truspec nitrogen analyser using the Dumas combustion method (AOAC Method 990.03). Crude fibre was analysed using a Dosi-Fibre machine (AOAC Method 978.10). Gross energy was determined using a LECO AC500 automatic bomb calorimeter (LECO Corporation). Carbohydrate was calculated indirectly by adding the percentage values determined for crude protein, lipid, crude fibre and ash, and subtracting the total from 100.

Water and sediment analysis

Water-quality parameters were recorded weekly from water sampled adjacent to the outflow of each tank during mid-morning. Temperature and pH were measured using a pH meter (YSI Inc. Model No. 60/10 FT). Dissolved oxygen concentration was measured using an oxygen meter (YSI Inc. Model No. 55D). Total ammonia nitrogen (NH_4-N; TAN) was determined using the method of Solorzano (1969). Nitrite concentration was measured using a commercially available test kit (Merck nitrite test kit, Cat. No. 1.14776.0001) with colour absorbance read by a spectrophotometer (Prim Light, Secomam). Absorbance was converted into the concentration of total ammonia nitrogen or nitrite using the coefficients derived from standard curves of least-squared linear regression.

At the end of the trial on Day 81, the sediment reduction–oxidation (redox) potential was measured

in millivolts by inserting a redox probe (Eutech Instruments pH 6+ portable meter) to the base of the sand sediment. Readings were taken following stabilisation (after approximately 5 min). Since technological limitations did not allow for a full vertical profile measurement of the redox potential of the sediment, in addition, 4 replicate cores were taken from different positions within each tank using a 10 × 1 cm (length × diameter) Perspex coring device and the depth from the sediment surface to the oxic–anoxic interface was recorded. The mean depth (cm) of anoxic sediment in each tank was converted to the percentage of anoxic sediment in the core to allow direct comparisons between treatments, since sediment depth varied. Sediment samples were collected from the upper 3 mm of 3 replicate tanks and dried to a constant weight at 50°C for 48 h. Samples were analysed for organic carbon and total nitrogen content using a LECO TruSpec micro elemental analyser prior to and after carbonate removal. Carbonates were removed by fuming with 2 M HCl for 48 h, after which the samples were rinsed 3 times with distilled water, dried to constant weight and re-analysed for total organic carbon. Carbon to nitrogen ratios were then calculated for each replicate sample. Total chlorophyll concentration was determined using 90% acetone extraction before the first spectrophotometric step, and the concentrations of chlorophyll *a*, *b* and *c* were calculated using the trichromatic equation of Jeffrey & Welschmeyer (1997).

Statistical analyses

Mean biomass of individual *H. scabra* (per replicate tank) and mean (per replicate tank) water and sediment characteristics were tested for normality using Shapiro-Wilk's test and for homogeneity of variance using Levene's test. Data that met the test assumptions were compared across the 8 experimental treatments using multifactor analysis of variance (ANOVA), and Duncan's multiple range tests were used to compare differences among means of dependent variables (Quinn & Keough 2012). Data that did not meet the test assumptions were log transformed before analysis. If log-transformed data did not meet the assumption of homogeneity of variance either, a Kruskal-Wallis 1-way ANOVA was used to test for significant differences in the means between treatments. Differences were considered significant at $p < 0.05$. Mean values are given ± SE.

The numbers of animals engaging in each specific behaviour were averaged to give the mean number

Table 3. Summary of sediment characteristics subjected to 2 contrasting redox regimes, each with different depths and particle sizes (fine particles: 125–250 µm; medium particles: 250–500 µm). Data are presented as means ± SE. Different superscripts indicate significant differences (multifactor ANOVA, p < 0.05) identified by Duncan's multiple range tests. –: not applicable

Redox regime	Depth (cm)	Particle size	Total chlorophyll (µg g^{-1})	Redox potential (mV)	Anoxic sediment (%)	Organic carbon (%)	Total nitrogen (%)	C/N ratio
Oxic	2	Fine	15.58 ± 5.05	169.33 ± 40.92	–[a]	0.26 ± 0.01[c]	0.040 ± 0.000[c]	6.49 ± 0.25[c]
Oxic	2	Medium	7.44 ± 0.97	158.08 ± 58.18	–[a]	0.69 ± 0.06[a]	0.030 ± 0.003[a]	20.99 ± 2.54[ab]
Oxic	4	Fine	9.83 ± 3.62	137.44 ± 41.80	–[a]	0.39 ± 0.01[cd]	0.030 ± 0.000[a]	13.03 ± 0.49[d]
Oxic	4	Medium	8.67 ± 1.65	121.88 ± 41.13	–[a]	0.56 ± 0.00[ab]	0.033 ± 0.000[a]	18.52 ± 0.16[a]
Oxic–anoxic	2	Fine	5.72 ± 0.15	95.83 ± 13.60	81.97 ± 8.02[b]	0.45 ± 0.03[bd]	0.027 ± 0.003[a]	17.39 ± 1.78[a]
Oxic–anoxic	2	Medium	6.23 ± 3.06	101.25 ± 57.81	76.98 ± 9.26[b]	0.69 ± 0.05[a]	0.030 ± 0.000[a]	23.09 ± 1.80[b]
Oxic–anoxic	4	Fine	2.24 ± 0.22	41.81 ± 42.40	93.34 ± 1.00[b]	0.59 ± 0.09[ab]	0.027 ± 0.003[a]	21.74 ± 0.83[ab]
Oxic–anoxic	4	Medium	2.93 ± 0.51	24.38 ± 46.38	79.74 ± 9.26[b]	0.58 ± 0.02[ab]	0.020 ± 0.000[b]	29.15 ± 0.91[e]

of animals per replicate tank in each burial state or activity at 6 different time intervals and analysed using repeated measures ANOVA. A multivariate approach was used with redox regime, sediment depth and particle size as categorical predictors and the mean number of animals engaged in a specific behaviour at each time period as the dependent variable (within effects). Although the assumptions for normality and homogeneity of variance were not met using the Shapiro-Wilk's and Levene's tests, even with transformed data, repeated measures ANOVA was still deemed sufficiently robust to compare treatment means over time (Moser & Stevens 1992). A Mauchly's test examined sphericity of the variance–covariance matrix. As sphericity was violated in the majority of cases, a Greenhouse-Geisser epsilon correction was used to adjust F statistics conservatively. Significant differences among treatment means were identified using a Tukey's honestly significant difference (HSD) post hoc test.

RESULTS

Water quality

The mean water temperature was 25.71 ± 0.05°C and varied between 25.18 and 26.40°C over the experimental period. pH ranged from 8.01 to 8.43, and salinity was constant at 35 g l^{-1}. Dissolved oxygen concentrations varied between 5.77 and 6.97 mg l^{-1} (mean: 6.6 ± 0.04 mg l^{-1}). Total ammonia concentrations varied between 17.06 and 64.21 µg l^{-1} (mean: 29.00 ± 1.74 µg l^{-1}) and nitrite concentrations ranged from 17.68 to 37.00 µg l^{-1} (mean: 26.01 ± 0.78 µg l^{-1}). There were no significant differences in water-quality parameters between treatments (p > 0.05), with the exception of Treatment 4 that exhibited a significantly lower nitrite concentration over

the experimental period (multifactor ANOVA, $F_{1, 24}$ = 6.32, p = 0.019), which was explained by the interaction between sediment depth and the manipulated redox regime; sediment particle size made no significant contribution.

Sediment characteristics

The sediment core profiles confirmed that the incorporation of a plenum into the tank design maintained fully oxic conditions (Treatments 1–4); hereafter referred to as oxic. In contrast, in tanks without a plenum, 83.01 ± 3.80% of the sediment profile was anoxic (Treatments 5–8); hereafter referred to as oxic–anoxic (multifactor ANOVA, $F_{1, 24}$ = 465.53, p < 0.001). The redox potential of the oxic treatments was significantly higher (147.98 ± 21.00 mV) compared with the oxic–anoxic treatments (67.27 ± 21.18 mV; multifactor ANOVA, $F_{1, 24}$ = 5.81, p = 0.028; Table 3).

The sediment redox regime significantly affected the total chlorophyll concentration on the sediment surface (multifactor ANOVA, $F_{1, 21}$ = 13.42, p = 0.001; Table 3), with higher concentrations in the oxic treatments (10.03 ± 1.51 µg g^{-1}) compared with oxic-anoxic treatments (4.29 ± 0.74 µg g^{-1}), indicating greater microphytobenthic production. Observations of the sediment surface between the contrasting redox regimes also supported this quantitative data: oxic treatments were characterised by markedly thicker growth of benthic diatoms (*Nitzschia* and *Navicula* species) and *Oscillatoria* cyanobacteria.

Levels of organic carbon in the sediment were generally low, ranging from 0.26 ± 0.01 to 0.69 ± 0.06%, although the percentage of organic carbon was significantly higher in sediments with medium particle sizes (0.63 ± 0.03%) than in those with fine particles (0.42 ± 0.04%; multifactor ANOVA, $F_{1, 16}$ = 42.15, p < 0.001; Table 3). Sediment redox regime, depth and

Table 4. *Holothuria scabra*. Summary of growth performance of sea cucumbers subjected to 2 substrates with contrasting redox regimes, each with different depths and particle sizes (fine particles: 125–250 µm; medium particles: 250–500 µm). Data are presented as means± SE. Different superscripts indicate significant differences (multifactor ANOVA, p < 0.05) identified by Duncan's multiple range tests. SGR: specific growth rate; CV: coefficient of variation

Redox regime	Depth (cm)	Particle size	Survival (%)	Initial weight (g)	Final weight (g)	Growth rate (g d^{-1})	SGR (% d^{-1})	Biomass density (g m^{-1})	CV (%)
Oxic	2	Fine	100	7.16 ± 0.22	17.20 ± 0.67^a	0.12 ± 0.01^{ab}	1.03 ± 0.06^a	461.68 ± 18.01^a	23.11 ± 4.34^{ab}
Oxic	2	Medium	100	7.38 ± 0.10	16.09 ± 1.04^a	0.11 ± 0.01^a	0.95 ± 0.08^a	432.00 ± 27.85^a	11.58 ± 1.90^a
Oxic	4	Fine	100	7.05 ± 0.17	16.90 ± 1.35^a	0.12 ± 0.01^a	1.00 ± 0.05^a	453.67 ± 36.37^a	21.28 ± 7.61^{ab}
Oxic	4	Medium	100	7.18 ± 0.10	17.97 ± 1.36^{ab}	0.13 ± 0.02^{ab}	1.10 ± 0.09^{ab}	449.53 ± 38.38^a	19.03 ± 4.72^{ab}
Oxic–anoxic	2	Fine	100	7.60 ± 0.06	23.97 ± 2.72^{bc}	0.20 ± 0.03^{ab}	1.36 ± 0.13^{ab}	643.56 ± 72.99^b	27.90 ± 7.58^{ab}
Oxic–anoxic	2	Medium	100	7.09 ± 0.09	22.35 ± 2.07^{abc}	0.20 ± 0.03^{bc}	1.30 ± 0.15^{ab}	600.03 ± 55.60^{ab}	51.82 ± 7.34^c
Oxic–anoxic	4	Fine	100	7.16 ± 0.33	26.66 ± 2.83^c	0.24 ± 0.04^c	1.54 ± 0.23^b	715.76 ± 75.97^b	34.00 ± 5.64^b
Oxic–anoxic	4	Medium	100	7.80 ± 0.25	20.42 ± 2.86^{abc}	0.16 ± 0.04^{bc}	1.10 ± 0.20^{ab}	548.20 ± 76.84^{ab}	23.36 ± 6.73^{ab}

particle size interacted to affect total nitrogen content (multifactor ANOVA, $F_{1, 16} = 8.33$, p = 0.011) and the carbon to nitrogen ratio (multifactor ANOVA, $F_{1, 16} = 7.85$, p = 0.013; Table 3). The redox regime explained the oxic–anoxic treatments' higher organic carbon content (0.58 ± 0.04 % versus 0.47 ± 0.05 %; multifactor ANOVA, $F_{1, 16} = 10.72$, p = 0.005), while oxic treatments had a higher total nitrogen content (0.033 ± 0.001 % versus 0.026 ± 0.001 %; multifactor ANOVA, $F_{1, 16} = 27.00$, p < 0.001) and resulted in higher carbon to nitrogen ratios in the oxic–anoxic treatments (22.84 ± 1.40 % versus 14.76 ± 1.77 %; multifactor ANOVA, $F_{1, 16} = 71.29$, p < 0.001).

Growth and survival

There were no significant differences in mean sea cucumber biomass (29.20 ± 2.29 g) between treatments at the start of the experiment (Kruskal-Wallis, $H_{7, 32} = 11.17$, p = 0.13). Survival was 100 % in all treatments (Table 4). The final mean growth rate of animals reared in the oxic–anoxic treatments was significantly higher at 0.20 ± 0.02 g d^{-1} compared with 0.12 ± 0.01 g d^{-1} for animals in oxic treatments (multifactor ANOVA, $F_{1, 24} = 19.59$, p < 0.001). The growth rate of juveniles in oxic treatments decreased over time from 0.23 ± 0.02 g d^{-1} between Days 0 and 27 to a negative growth rate of –0.01 ± 0.02 g d^{-1} between Days 54 and 81 (Fig. 2). Conversely, animals reared in oxic–anoxic treatments maintained an initial growth rate >0.2 g d^{-1}, with a slight reduction to 0.14 ± 0.04 g d^{-1} between Days 54 and 81 (Fig. 2). Biomass from oxic sediments reached a maximum mean density of 454.84 ± 14.30 g m^{-2} on Day 54, after which no further increase in biomass was observed (Fig. 3). The final biomass density of animals in the oxic–anoxic treatments was significantly higher at

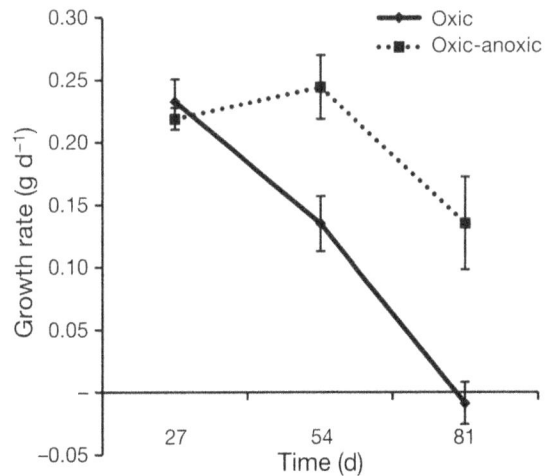

Fig. 2. Mean (±SE) growth rate per sampling period of *Holothuria scabra* (n = 4) reared under 2 contrasting sediment redox regimes: fully aerobic (oxic) and stratified (oxic–anoxic)

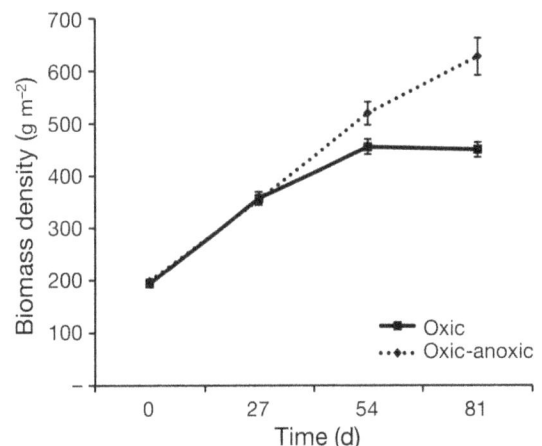

Fig. 3. Mean (±SE) cumulative biomass density of *Holothuria scabra* (n = 4) reared for 81 d under 2 contrasting sediment redox regimes: fully aerobic (oxic) and stratified (oxic–anoxic)

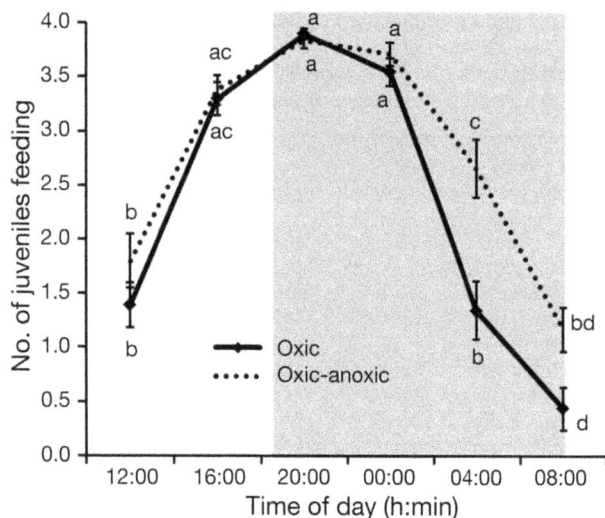

Fig. 4. Mean (±SE) number of juvenile *Holothuria scabra* (n = 4) observed to be feeding on the sediment surface and the tank walls when reared under 2 contrasting sediment redox regimes: fully aerobic (oxic) and stratified (oxic–anoxic). Different lower case letters indicate significant differences between the number of juveniles feeding over time for each redox regime (Tukey's HSD post hoc test). The shaded area represents dark hours (17:40–07:40 h, local time)

626.89 ± 35.44 g m^{-2} at the end of the trial compared with 449.22 ± 14.24 g m^{-2} for oxic treatments (multifactor ANOVA, $F_{1,24} = 21.05$, $p < 0.001$; Fig. 3). Significant differences in growth rates were principally related to the depth of the oxic–anoxic interface (multiple regression, $r^2 = 0.40$; $\beta = 0.63$; $p < 0.001$) and the sediment redox potential (multiple regression, $r^2 = 0.26$; $\beta = -0.51$; $p < 0.001$). The manipulated redox regime was the foremost factor responsible for this difference in growth rate and final biomass production, as sediment depth and particle size had no significant effect on growth rate or biomass density.

Behaviour

Time and the interaction between sediment redox regime and time significantly affected all categories of observed burial behaviour and activity level defined in Table 2 ($p < 0.05$). Animals reared in oxic–anoxic treatments displayed a shorter burial cycle, remaining on the surface (exposed) for longer in the early morning before burying (repeated measures ANOVA, $F_{3.43, 82.40} = 7.27$, $p < 0.001$) and spent significantly more time feeding (repeated measures ANOVA, $F_{2.59, 62.21} = 4.85$, $p = 0.006$; Fig. 4). The most notable differences in burial cycle and feeding activity occurred in the early morning between midnight

and 08:00 h, when animals in oxic treatments ceased feeding and began to bury earlier compared with the animals in oxic–anoxic treatments; for example, at 04:00 h 2.66 ± 0.27 animals were still feeding on the oxic–anoxic sediment compared with 1.34 ± 0.27 feeding on the oxic sediment (Fig. 4).

Proximate composition

The sediment redox regime significantly affected the sea cucumber biochemical composition (Table 5), with crude lipid and carbohydrate content significantly higher in animals reared in oxic treatments (multifactor ANOVA, $F_{1,16} = 7.72$, $p = 0.013$ and $F_{1,16} = 8.16$, $p = 0.011$ for lipid and carbohydrate, respectively). In contrast, ash content was lower in animals reared in oxic treatments; $62.82 \pm 1.59\%$ compared with $67.03 \pm 1.10\%$ in oxic–anoxic treatments (multifactor ANOVA, $F_{1,16} = 5.01$, $p = 0.04$). Redox regime and substrate depth interacted to account for a significant difference in crude fibre (multifactor ANOVA, $F_{1,16} = 7.34$, $p = 0.015$; Table 5); however, there were no significant differences in protein, moisture, or gross energy content (multifactor ANOVA, $p > 0.05$). The sediment redox regime was again the causative factor for ash, lipid and carbohydrate content.

DISCUSSION

This study investigated the growth responses of juvenile *Holothuria scabra*, a commercially and ecologically important sea cucumber species, cultured with surficial sediment of manipulated physicochemical characteristics, namely, sediment depth, particle size and redox regime. The redox regime (either: oxic–anoxic or fully oxic) was the principal factor affecting sea cucumber growth, biomass production, behaviour and biochemical composition, in addition to driving differences in sediment quality (e.g. C/N ratio, chlorophyll *a* concentration). The oxic–anoxic treatments supported an approximately 30% increase in growth rate relative to the fully oxic sediments that had, by the end of the trial, translated into a substantial increase in sea cucumber biomass production (626.89 ± 35.44 versus 454.84 ± 14.29 g m^{-2}). In the oxic sediment system, sea cucumbers exhibited negative allometric growth, with individual growth rates decreased to zero or becoming regressive once the carrying capacity was exceeded (454.84 ± 14.29 g m^{-2} at Day 54). Indeed, in contrast to the oxic treatments, the biomass carrying capacity

Table 5. *Holothuria scabra*. Mean (±SE) proximate composition of sea cucumbers reared on substrates with 2 contrasting redox regimes, each with different depths and particle sizes (fine particles: 125–250 µm; medium particles: 250–500 µm). Different super-scripts indicate significant differences (multifactor ANOVA, $p < 0.05$) identified by Duncan's multiple range tests

Redox regime	Depth (cm)	Particle size	Crude protein (%)	Crude lipid (%)	Crude fibre (%)	Carbo-hydrate (%)	Ash (%)	Moisture (%)	Gross energy (%)
Oxic	2	Fine	21.40 ± 3.21	0.89 ± 0.15[a]	0.40 ± 0.08[ab]	13.91 ± 1.41[ab]	63.39 ± 3.89[ab]	87.83 ± 1.66	4.89 ± 0.88
Oxic	2	Medium	23.83 ± 0.11	0.84 ± 0.04[a]	0.50 ± 0.01[b]	16.39 ± 1.36[b]	58.45 ± 1.43[b]	88.82 ± 0.58	5.76 ± 0.50
Oxic	4	Fine	21.91 ± 1.92	0.87 ± 0.15[a]	0.45 ± 0.17[ab]	13.82 ± 0.41[ab]	62.95 ± 1.72[ab]	88.64 ± 0.53	5.09 ± 0.23
Oxic	4	Medium	18.73 ± 2.61	0.74 ± 0.04[ab]	0.23 ± 0.04[a]	13.78 ± 2.10[ab]	66.51 ± 4.34[ab]	87.07 ± 1.69	4.07 ± 0.65
Oxic–anoxic	2	Fine	18.35 ± 2.11	0.65 ± 0.08[ab]	0.22 ± 0.05[a]	11.47 ± 0.70[a]	69.32 ± 2.84[a]	87.61 ± 0.74	4.14 ± 0.59
Oxic–anoxic	2	Medium	21.26 ± 1.61	0.77 ± 0.01[ab]	0.34 ± 0.03[ab]	13.26 ± 1.30[ab]	64.37 ± 1.81[ab]	90.13 ± 0.17	4.85 ± 0.34
Oxic–anoxic	4	Fine	18.69 ± 1.70	0.51 ± 0.03[b]	0.39 ± 0.05[ab]	11.54 ± 0.46[a]	68.87 ± 1.82[a]	88.51 ± 0.65	4.00 ± 0.44
Oxic–anoxic	4	Medium	20.96 ± 2.00	0.72 ± 0.07[ab]	0.54 ± 0.06[b]	12.24 ± 0.32[a]	65.55 ± 1.67[ab]	88.92 ± 0.60	4.89 ± 0.56

of the oxic–anoxic sediments had not been reached by the trial end and growth rates remained positive. Populations of soft sediment deposit-feeders such as sea cucumbers are commonly food limited (Lopez & Levinton 1987, Josefson 1998); however, the growth limitation experienced in the oxic system cannot be explained by differences in the feeding regimen since rations were standardised across all treatments and adjusted on a 2-weekly basis in the same way for each treatment.

All stages in organic matter degradation involve the interaction of organisms, resource quality, physical and chemical environmental conditions (Anderson 1987). Consequently, deposit-feeder nutrition is fundamentally related to the decomposition of particulate organic matter (detritus) (Tenore et al. 1982, Rice & Rhoads 1989), with strong feedback loops existing between deposit-feeders, their food and their chemical environment (Herman et al. 1999). The oxygenation status of marine sediments is the main factor influencing the rate and intensity of organic matter decomposition, since the presence or absence of oxygen influences microbial metabolism and enzymatic capacity. In oxidised sediment, organic matter is rapidly decomposed by aerobic heterotrophs which have the enzymatic capacity to perform complete oxidation of complex organic substrates via the tricarboxylic acid cycle (Kristensen 2001). In anoxic sediments, organic matter is degraded more slowly by an anaerobic food chain involving the sequential and/or simultaneous activities of a mutualistic consortium of anaerobic bacteria, since no single type of anaerobic bacterium seems capable of complete mineralisation (Middelburg et al. 1993, Fenchel et al. 1998). While oxygen can be substituted by other forms of chemically bound oxygen in anaerobic respiration, there is no equivalent of molecular oxygen that can fulfil its function in the primary

rate-limiting hydrolytic step in organic matter decomposition (Kristensen et al. 1995).

In the oxic treatments all sources of autochthonous and allochthonous organic matter would have been rapidly respired and lost as CO_2. The higher C/N ratios in the oxic–anoxic treatments indicate a greater pool of refractory organic matter from which bacteria have preferentially hydrolysed nitrogen-rich molecules, i.e. nucleic acids and proteins. The predominately anaerobic conditions of the oxic–anoxic sediments, which mirror the natural habitat of *H. scabra*, are likely to have provided a steady release of bioavailable food resources, thereby supporting a higher secondary biomass. Further, sea cucumbers are known to absorb dissolved organics across the epithelium (Jangoux & Lawerence 1982) or through the respiratory trees as part of a bipolar feeding strategy (Jaeckle & Strathmann 2013). Therefore, the more refractory organic matter and energy-rich fermentation products may have functioned as a more durable and persistent food source (i.e. an overall increase in food quantity) in the oxic–anoxic treatments, contributing to long-term productivity despite the apparent lower nutritive quality (Schroeder 1987, Karlsen 2010).

The influence of microbial communities on the rate and pathways of nitrogen cycling is also sensitive to the presence or absence of oxygen (Aller & Aller 1998). In the oxic sediment layer, ammonium is oxidised in a step-wise reaction to nitrite and subsequently to nitrate, which is then reduced to free nitrogen gas by denitrification in the anoxic layer (Vanderborght & Billen 1975). Nitrification rate is regulated by the availability of NH_4^+ and oxygen (Fenchel et al. 1998). The nitrification efficiency would therefore have been high in the oxic sediment system where oxygen penetration was maximised and NH_4^+ production rapid. The significantly lower nitrite concentration in water from the 4 cm oxic treatment tanks indicates that nitrite, as the intermediate product in nitrification, may have been oxidised more efficiently to nitrate by nitrite-oxidising bacteria that proliferated in the sediment under increased oxygen concentrations.

The greater microphytobenthic standing stock in the oxic treatments is likely to have resulted from enhanced nitrate remineralisation. In shallow euphotic benthic systems there is tight coupling between autotrophic and heterotrophic processes, and in well-mixed systems this coupling can be intense (Fenchel et al. 1998, Herman et al. 1999). Algal mats are efficient at assimilating nutrients from the sediment surface, and an increased nutrient supply may affect the dominance of bottom-up forces leading to overgrowth by algal mats (Levinton & Kelaher 2004). While nitrate availability is likely to have had the most impact on microphytobenthic production, increased concentrations of dissolved inorganic carbon resulting from the rapid, aerobic oxidation of organic matter may have also played a role (Ludden et al. 1985).

Deposit-feeding holothurians are broadly adapted to process large quantities of sediments of low nutritional quality (Lopez & Levinton 1987, Roberts et al. 2000); however, they exhibit a general plasticity of behavioural and feeding strategies in response to variations in resource availability (Roberts et al. 2000). In the oxic sediment, animals spent less time feeding and spent longer periods buried. In a more detailed behavioural study, Yu (2012) made similar observations, in that juvenile *H. scabra* reared on oxic sediments exhibited much lower feeding and activity levels. Sea cucumbers have a behavioural capacity to alter their ingestion rate in response to food quality. The 'optimal ingestion rate' predicts that the ingestion rate increases with increasing food value (Taghon 1981, Taghon & Jumars 1984), while 'compensatory feeding' predicts that the ingestion rate decreases with increasing food value (Calow 1977, Cammen 1979). Apparent compensatory slowdowns in ingestion rates in response to rich food sources such as benthic diatoms have been observed for deposit-feeders including sea cucumbers (Cammen 1979, Lopez & Levinton 1987, Zamora & Jeffs 2011). As their net gain of energy and nutrients is determined by foraging and digestion, sea cucumbers in the oxic treatments may have decreased their feeding rate and activity levels due to the limited need to forage in response to the abundant microphytobenthic production. The demonstration by Phillips (1984) that compensatory regulation of energy intake can be consistent with an optimal foraging model may explain the behaviour and growth limitation of sea cucumbers in the oxic treatments.

The higher lipid and carbohydrate contents of animals reared in the oxic sediment system may be a reflection of the higher quality of food resources, such as benthic diatoms, which have a relatively high intrinsic food value and assimilation efficiencies (Yingst 1976, Watanabe et al. 2012). As lipids and carbohydrates are believed to be the primary nutrient reserves in sea cucumbers (Krishnan 1968, Féral 1985), it would appear that animals were allocating the products of digestion to storage as opposed to using them for maintenance and growth. This type of compensatory feeding is typical in some deposit-feeders in the face of abundant food reserves and allows them to regulate their intake of energy and store excess nutrients for later use (Calow 1977).

The current study highlights a facet of sea cucumber feeding ecology that has profound, and arguably conflicting, repercussions for their application as bioremediation organisms and for improvements in farming practices. Achieving good oxygenation status is a common and inexpensive stimulatory strategy to accelerate the aerobic microbial decomposition of organic-rich sediments; yet based on the current results this approach (driven by oxic sediment redox regimes) limits and retards the growth and eventual biomass yield of sea cucumbers by rapid mineralisation of nutrients, which would otherwise be available for assimilation into sea cucumber biomass. Conversely, where sediment is managed to maximise sea cucumber biomass yield (employing a stratified redox regime), the rate of organic matter mineralisation is considerably slowed (Torres-Beristain et al. 2006). Thus, from a commercial perspective maximising the system's bioremediating capacity occurs at the expense of a valuable cash crop, while compromising environmental remediation enhances economic return. In reality, this need not be an either/or scenario. A 'hybrid' sediment management system could be conceived wherein oxygenation strategies could be selectively optimised over the times and spatial scales that may provide an acceptable compromise. Alternatively, the organic loading of the oxic system (i.e. the feeding regime) may have been overly conservative. In a commercial situation, the sediment-based effluent treatment system would receive frequent waste addition, comprising principally waste feed and faeces. In this scenario, wherein the rate of organic loading exceeds the rate of carbon loss, oxic sediment systems would have an increased and hitherto untapped capacity to support significantly greater sea cucumber growth and biomass production.

Oxygenation of subsurface sediments and the removal of metabolites are considered the 2 most important underlying mechanisms for stimulating carbon oxidation in controlled sediment regimes (Kristensen 2001). In the current study oxic sediment systems experienced increased rates of organic matter decomposition, including the more refractory organic matter pools. This will have enhanced porewater column solute exchange, thereby contributing towards an overall reduction in food resource availability to deposit-feeding sea cucumbers and providing a proximal cause for limited growth. Strictly aerobic detrital systems in nature are rare (Plante et al. 1990) and are unlikely to be a suitable medium for deposit-feeder growth in the long term, unless the rate of organic loading matches the rate of carbon loss (respiration and organism growth). In contrast, the oxic–anoxic system, which resembles the natural habitat of *H. scabra* and supported both aerobic and anaerobic mineralisation pathways, supported strong and sustained sea cucumber growth, presumably due to the presence of more refractory organic matter. Consequently, we conclude that the observed divergence in biomass carrying capacity was driven by differences in inorganic nutrient mineralisation and cycling between the contrasting redox regimes with the oxic–anoxic system supporting a more durable food resource. Future research should focus on determining the quality, quantity and frequency of the addition of waste effluents to support the optimal production of deposit-feeders in land-based bioremediation systems in tandem with a more varied approach to oxygenation integrated over space and time.

Acknowledgements. This research was funded by a Biotechnology and Biological Sciences Research Council (BBSRC) Industrial CASE Studentship to G.R. (Grant Code BB/J01141X/1) with HIK Abalone Farm Pty as the CASE partner, and with additional contributions from the THRIP program of the National Research Foundation, South Africa (Grant Number TP2011070800007). The work was conceptualised and funding was secured by G.R., C.L.W.J., M.J.S. and S.M.S. Experiments were performed by G.R. Data were analysed by G.R. and C.L.W.J., the manuscript was written by G.R. and G.S.C. and edited by M.J.S., C.L.W.J. and S.M.S.

LITERATURE CITED

Alexander M (1999) Biodegradation and bioremediation, 2nd edn. Academic Press, New York, NY

Aller RC (1994) Bioturbation and remineralization of sedimentary organic-matter — effects of redox oscillation. Chem Geol 114:331–345

Aller RC, Aller JY (1998) The effect of biogenic irrigation intensity and solute exchange on diagenetic reaction rates in marine sediments. J Mar Res 56:905–936

Anderson JM (1987) Production and decomposition in aquatic ecosystems and implications for aquaculture. In: Moriarty DJW, Pullin RSV (eds) Detritus and microbial ecology in aquaculture. ICLARM, Manila, p 123–145

AOAC (Association of Official Analytical Chemists) (2010) Official methods of analysis of AOAC International, 18th edn. AOAC, Washington, DC

Baskar BK (1994) Some observations on biology of the holothurian *Holothuria* (*Metriatyla*) *scabra* (Jaeger). CMFRI Bull 46:39–43

Battaglene SC, Seymour JE, Rajagopalan M, Ramofafia C (1999) Culture of tropical sea cucumbers for stock restoration and enhancement. Aquaculture 178:293–322

Calow P (1977) Ecology, evolution and energetics: a study in metabolic adaptation. Adv Ecol Res 10:1–62

Cammen LM (1979) Ingestion rate: an empirical model for aquatic deposit feeders and detritivores. Oecologia 44: 303–310

Chávez-Crooker P, Obreque-Contreras J (2010) Bioremediation of aquaculture wastes. Curr Opin Biotechnol 21: 313–317

Colleran E (1997) Uses of bacteria in bioremediation. In: Sheehan D (ed) Methods in biotechnology, bioremediation protocols, Vol 2. Humana Press, Clifton, NJ, p 3–22

Fenchel T, King GM, Blackburn TH (1998) Bacterial biogeochemistry. The ecophysiology of mineral cycling. Academic Press, San Diego, CA

Féral JP (1985) Effect of short-term starvation on the biochemical composition of the apodous holothurian *Leptosynapta galliennei* (Echinodermata): possible role of dissolved organic material as an energy source. Mar Biol 86: 297–306

Gifford S, Dunstan RH, O'Connor W, Roberts T, Toia R (2004) Pearl aquaculture — profitable environmental remediation? Sci Total Environ 319:27–37

Grassle JF, Grassle JP (1974) Opportunistic life histories and genetic systems in marine benthic polychaetes. J Mar Res 32:253–284

Herman PMJ, Middelburg JJ, Van de Koppel J, Heip CHR (1999) Ecology of estuarine macrobenthos. In: Nedwell DB, Raffaelli DG (eds) Advances in ecological research, Vol 29. Academic press, New York, NY, p 195–240

Jaeckle WB, Strathmann RR (2013) The anus as a second mouth: anal suspension feeding by an oral deposit-feeding sea cucumber. Invertebr Biol 132:62–68

Jangoux M, Lawerence JM (1982) Echinoderm nutrition. Balkema, Rotterdam

Jaubert JM (2008) Scientific considerations on a technique of ecological purification that made possible the cultivation of reef-building corals in Monaco. In: Leewis RJ, Janse M (eds) Advances in coral husbandry in public aquariums. Burgers' Zoo, Arnhem, p 155–126

Jeffrey SW, Welshmeyer NA (1997) Spectrophotometric and fluorometric equations in common use in oceanography. In: Jeffrey SW, Mantoura RFC, Wright SW (eds) Phytoplankton pigments in oceanography, monographs on oceanographic methodology No. 10. UNESCO Publishing, Paris, p 597–615

Josefson AB (1998) Resource limitation in marine soft sediments — differential effects of food and space in the association between the brittle-star *Amphiura filiformis* and the bivalve *Mysella bidentata*? Hydrobiologia 375/376: 297–305

Karlsen A (2010) Benthic use of phytoplankton blooms: uptake, burial and biodiversity effects in a species-poor system. Stockholm University

Kim JK, Duston J, Corey P, Garbary DJ (2013) Marine finfish effluent bioremediation: effects of stocking density and temperature on nitrogen removal capacity of *Chondrus crispus* and *Palmaria palmata* (Rhodophyta). Aquaculture 414/415:210–216

Krishnan S (1968) Histochemical studies on reproductive and nutritional cycles of the holothurian, *Holothuria scabra*. Mar Biol 2:54–65

Kristensen E (2000) Organic matter diagenesis at the oxic/anoxic interface in coastal marine sediments, with emphasis on the role of burrowing animals. Hydrobiologia 426:1–24

Kristensen E (2001) Impact of polychaetes (*Nereis* spp. and *Arenicola marina*) on carbon biogeochemistry in coastal marine sediments. Geochem Trans 2:92

Kristensen E, Ahmed SI, Devol AH (1995) Aerobic and anaerobic decomposition of organic matter in marine sediment: Which is fastest? Limnol Oceanogr 40:1430–1437

Levinton J, Kelaher B (2004) Opposing organizing forces of deposit-feeding marine communities. J Exp Mar Biol Ecol 300:65–82

Lopez GR, Levinton JS (1987) Ecology of deposit-feeding animals in marine sediments. Q Rev Biol 62:235–260

Ludden E, Admiraal W, Colijn F (1985) Cycling of carbon and oxygen in layers of marine microphytes—a simulation model and its ecophysiological implications. Oecologia 66:50–59

Mercier A, Battaglene SC, Hamel JF (1999) Daily burrowing cycle and feeding activity of juvenile sea cucumbers *Holothuria scabra* in response to environmental factors. J Exp Mar Biol Ecol 239:125–156

Mermillod-Blondin F, Rosenberg R (2006) Ecosystem engineering: the impact of bioturbation on biogeochemical processes in marine and freshwater benthic habitats. Aquat Sci 68:434–442

Michio K, Kengo K, Yasunori K, Hitoshi M, Takayuki Y, Hideaki Y, Hiroshi S (2003) Effects of deposit feeder *Stichopus japonicus* on algal bloom and organic matter contents of bottom sediments of the enclosed sea. Mar Pollut Bull 47:118–125

Middelburg JJ, Vlug T, Vandernat F (1993) Organic matter mineralization in marine systems. Global Planet Change 8:47–58

Moriarty DJW (1982) Feeding of *Holothuria atra* and *Stichopus chloronotus* on bacteria, organic carbon and organic nitrogen in sediments of the Great Barrier Reef. Aust J Mar Freshw Res 33:255–263

Moriarty DJW (1987) Methodology for determining biomass and productivity of microorganisms in detrital food webs. In: Moriarty DJW, Pullin RSV (eds) Detritus and microbial ecology in aquaculture. International Center for Living Aquatic Resources Management, Bellagio, Como, p 4–31

Moser BK, Stevens GR (1992) Homogeneity of variance in the two-sample means test. Am Stat 46:19–21

Muradov N, Taha M, Miranda AF, Kadali K and others (2014) Dual application of duckweed and azolla plants for wastewater treatment and renewable fuels and petrochemicals production. Biotechnol Biofuels 7:30

Palmer PJ (2010) Polychaete-assisted sand filters. Aquaculture 306:369–377

Phillips NW (1984) Compensatory intake can be consistent with an optimal foraging model. Am Nat 123:867–872

Pitt R, Thu NTX, Mihn MD, Phuc HN (2001) Preliminary sandfish growth trials in tanks, ponds and pens in Vietnam. SPC Beche-de-mer Inf Bull 15:17–27

Plante CJ, Jumars PA, Baross JA (1990) Digestive associations between marine detritivores and bacteria. Annu Rev Ecol Syst 21:93–127

Pullin RSV (1987) Session on manipulation of detrital systems for aquaculture, chairman's overview. In: Moriarty DJW, Pullin RSV (eds) Detritus and microbial ecology in aquaculture. International Center for Living Aquatic Resources Management, Bellagio, Como, p 420

Purcell SW (2014) Value, market preferences and trade of beche-de-mer from Pacific island sea cucumbers. PLoS ONE 9:e95075

Purcell SW, Polidoro BA, Hamel JF, Gamboa RU, Mercier A (2014) The cost of being valuable: predictors of extinction risk in marine invertebrates exploited as luxury seafood. Proc R Soc B 281:20133296

Quinn GP, Keough MJ (2012) Experimental design and data analysis for biologists. Cambridge University Press, Cambridge

Raj L (1998) Photo-identification of *Stichopus mollis*. SPC Beche-de mer Inf Bull 10:29–31

Rice DL, Rhoads DC (1989) Early diagenesis of organic matter and the nutritional value of sediment. Lect Notes Coast Estuar Stud 31:59–97

Roberts D, Gebruk A, Levin V, Manship BAD (2000) Feeding and digestive strategies in deposit-feeding holothurians. Oceanogr Mar Biol 38:257–310

Robinson G, Slater M, Jones CLW, Stead S (2013) Role of sand as substrate and dietary component in recirculating aquaculture systems for juvenile sea cucumber *Holothuria scabra*. Aquaculture 392–395:23–25

Schroeder GL (1987) Carbon pathways in aquatic detrital systems. In: Moriarty DJW, Pullin RSV (eds) Detritus and microbial ecology in aquaculture. ICLARM, Manila, p 420

Solorzano L (1969) Determination of ammonia in natural waters by phenol hypochlorite method. Limnol Oceanogr 14:799–801

Stahlberg C, Bastviken D, Svensson BH, Rahm L (2006) Mineralisation of organic matter in coastal sediments at different frequency and duration of resuspension. Estuar Coast Shelf Sci 70:317–325

Taghon GL (1981) Beyond selection: optimal ingestion rate as a function of food value. Am Nat 118:202–214

Taghon GL, Jumars PA (1984) Variable ingestion rate and its role in optimal foraging behavior of marine deposit feeders. Ecology 65:549–558

Tenore KR (1981) Organic nitrogen and caloric content of detritus. I. Utilization by the deposit-feeding polychaete, *Capitella capitata*. Estuar Coast Shelf Sci 12:39–47

Tenore KR, Cammen L, Findlay SEG, Phillips N (1982) Perspectives of research on detritus—Do factors controlling the availability of detritus to macro-consumers depend on its source? J Mar Res 40:473–490

Torres-Beristain B, Verdegem M, Kerepeczki E, Verreth J (2006) Decomposition of high protein aquaculture feed under variable oxic conditions. Water Res 40:1341–1350

Uthicke S (2001) Nutrient regeneration by abundant coral reef holothurians. J Exp Mar Biol Ecol 265:153–170

Vanderborght JP, Billen G (1975) Vertical distribution of nitrate concentration in interstitial water of marine sediments with nitrification and denitrification. Limnol Oceanogr 20:953–961

Watanabe S, Kodama M, Zarate JM, Lebata-Ramos MJH, Nievales MFJ (2012) Ability of sandfish (*Holothuria scabra*) to utilise organic matter in black tiger shrimp ponds. In: Hair CA, Pickering TD, Mills DJ (eds) Proceedings Asia–Pacific tropical sea cucumber aquaculture. Australian Centre for International Agricultural Research, Canberra, p 113–120

Wolkenhauer SM (2008) Burying and feeding activity of adult *Holothuria scabra* (Echinodermata: Holothuroidea) in a controlled environment. SPC Beche-de-mer Inf Bull 27:25–28

Wolkenhauer SM, Uthicke S, Burridge C, Skewes T, Pitcher R (2010) The ecological role of *Holothuria scabra* (Echinodermata: Holothuroidea) within subtropical sea-grass beds. J Mar Biol Assoc UK 90:215–223

Yingst JY (1976) The utilization of organic matter in shallow marine sediments by an epibenthic deposit-feeding holothurian. J Exp Mar Biol Ecol 23:55–69

Yu W (2012) The effect of a plenum on the growth and behaviour of cultured sandfish, *Holothuria scabra*. Rhodes University, Grahamstown

Zamora LN, Jeffs AG (2011) Feeding, selection, digestion and absorption of the organic matter from mussel waste by juveniles of the deposit-feeding sea cucumber, *Australostichopus mollis*. Aquaculture 317:223–228

Zhang X, Zhang W, Xue L, Zhang B, Jin M, Fu W (2010) Bioremediation of bacteria pollution using the marine sponge *Hymeniacidon perlevis* in the intensive mariculture water system of turbot *Scophthalmus maximus*. Biotechnol Bioeng 105:59–68

Assessment of spawning of Atlantic bluefin tuna farmed in the western Mediterranean Sea

Antonio Medina[1],[*], Guillermo Aranda[1], Silvia Gherardi[1], Agustín Santos[1], Begonya Mèlich[2], Manuel Lara[2]

[1]University of Cádiz, Department of Biology (Zoology), Faculty of Marine and Environmental Sciences, Campus de Excelencia Internacional del Mar (CEI·MAR), 11510 Puerto Real, Cádiz, Spain

[2]Grup Balfegó, Pol. Ind. edifici 'Balfegó', 43860 L'Ametlla de Mar, Tarragona, Spain

ABSTRACT: Mediterranean tuna farms account for >60% of the eastern Atlantic bluefin tuna (ABFT) catch quota. Besides the direct impact of purse seining on wild stocks, ABFT farming practices may have environmental implications that are still poorly known. An unexplored potential source of interactions of ABFT farms with wildlife is the release of eggs into the environment in places other than spawning grounds. Purse seine-caught ABFT schools are known to spawn in towing cages as they are transported to farms. We show here that farmed ABFT are also capable of spawning during at least 2 subsequent reproductive seasons following their capture. The reproductive potential of ABFT commercial stocks was investigated in a farm located in the western Mediterranean Sea from 2012 through 2014, using occurrence and number of postovulatory follicles as proxies of spawning fraction and realised batch fecundity, respectively. Although the spawning fraction among farmed fish was lower than that in the wild, the mean fecundity of captive spawners was similar to that of wild fish; consequently, the number of fertile eggs released from grow-out cages is thought to be significant. Larvae hatched from eggs spawned in farms are likely to grow and join wild-born ABFT juveniles that use nearshore areas of the western Mediterranean as foraging grounds. Depending on the volume of fish ranched for >1 yr and the larval survival rate in the region, the escape through spawning may have a significant impact on the ecosystem and could affect recruitment, thus influencing the population dynamics of ABFT in the Mediterranean Sea.

KEY WORDS: *Thunnus thynnus* · Bluefin tuna farms · Reproductive maturation · Spawning · Fecundity · Egg production

INTRODUCTION

Since the 1990s, farming of Atlantic bluefin tuna (ABFT) *Thunnus thynnus* (Linnaeus, 1758), in the Mediterranean Sea has become a profitable activity which relies on the capture of live fish from the wild (Ottolenghi 2008, Mylonas et al. 2010, Vitalini et al. 2010, Metian et al. 2014). ABFT are caught in spring and summer by purse-seine fleets and transported to offshore sea cages for fattening/farming over 4 mo to 1–2 yr. These capture-based aquaculture practices have received strong criticism, as they can negatively affect the natural resources (Sumaila & Huang 2012).

In addition to the direct impact of purse seining on wild populations, ABFT farming may have a suite of significant ecological implications that are still poorly understood. Most studies on environmental impacts of tuna farms have been focused on the effects of waste discharged from the cages (e.g. Vezzulli et al. 2008, Piedecausa et al. 2010, Sarà et al. 2011, Vizzini & Mazzola 2012, Moraitis et al. 2013, Mangion et al. 2014). However, very little data currently exist regarding other interactions between wildlife and ABFT farming activities. A widespread phenomenon that has been identified recently is the strong attracting effect of Mediterranean ABFT farms on wild individ-

*Corresponding author: antonio.medina@uca.es

uals, which could cause alterations in their schooling behaviour and migratory patterns (Arechavala-López et al. 2015).

Another source of potential interactions with wildlife is the production of eggs by farmed ABFT. Although spawning has been observed in commercial cages, no investigation has been conducted to monitor and assess the reproductive performance of ABFT in farms. However, given the substantial proportion (>60%) of the eastern ABFT total allowable catch eventually absorbed by purse-seine fleets that supply live fish to tuna farms (Ortiz 2015), it is likely that the number of eggs leaked from the cages to the environment is significant. Therefore, the assessment of the egg production capacity of farm stocks is worthy of consideration with a view to improving our perception of potential impacts of the ABFT fattening/farming industry. Bluefin tuna held in captivity undergo physiological impairment that results in significant reduction of their reproductive capacity (Mylonas et al. 2007). However, experiments carried out on cage-reared ABFT have shown that most of the individuals treated with implants loaded with gonadotropin releasing hormone agonist (GnRHa) recovered the capacity to mature and release fertile gametes (Corriero et al. 2007, 2009, Mylonas et al. 2007, de la Gándara et al. 2010, De Metrio et al. 2010, Aranda et al. 2011, Rosenfeld et al. 2012). A small proportion of the untreated fish in the experimental broodstock were likewise capable of spawning as early as 1 yr after their capture from the wild, showing fecundity rates that were similar to those of GnRHa-treated individuals (Corriero et al. 2007, Aranda et al. 2011). Uninduced spawning has also been observed in an experimental ABFT stock maintained in captivity for 1 yr and transported to the Balearic spawning grounds during the reproductive season (Gordoa & Carreras 2014).

In this study, we assessed the short-term reproductive performance of farmed ABFT by gonad histology analysis during the breeding season. The occurrence and number of postovulatory follicles (POFs) in the ovaries were estimated to determine the female spawning fraction and batch fecundity. Such estimates can be used to calculate the egg production capacity of commercial stocks.

MATERIALS AND METHODS

Animals and farming conditions

ABFT were caught by purse seining around the Balearic Islands (Spain) in June 2010, 2012 and 2013.

They were transported to grow-out floating cages in the farming facilities of Grup Balfegó, located 4 km off L'Ametlla de Mar (Tarragona, NE Spain; Fig. 1). The holding cages were circular (50 m in diameter and 30 m deep) or elliptical (120 m long, 60 m wide, 30 m deep) and were moored in water of a total depth of 50 m. The initial stocking density was ~3 kg m^{-3}. After an adaptation period of a few weeks, the fish were fed to satiation 5 d wk^{-1} with defrosted baitfish, mostly mackerel. The initial ratio of bait supplied per total ABFT weight was ~8% in summer and decreased gradually to ~2% in winter. The sea surface water temperature (SST) was recorded daily from each cage. SST ranged from 11.6°C (February 2013) to 28.3°C (August 2012).

Systematic experimental sampling was impracticable, as the harvesting dates were fully dependent on the market demand. Harvesting took place between 06:00 and 08:00 h (UTC). Total body weight (BW, to the nearest 0.01 kg) and straight fork length (FL, to the nearest 1 cm) were recorded for each fish as they were harvested. Gonads were sampled from late May to early August in order to span the natural reproductive season of ABFT in the western Mediterranean Sea (Heinisch et al. 2008). Gonad weight (GW) was recorded to the nearest 0.01 kg, and the gonadosomatic index (GSI) was calculated as GSI = 100 GW BW^{-1}. The ovarian volume (OV) was estimated from the ovarian mass according to the equation: OV = 0.9174 GW (Medina et al. 2007).

All fish used in this study far exceeded 135 cm in FL, which is the size at which 100% of eastern ABFT

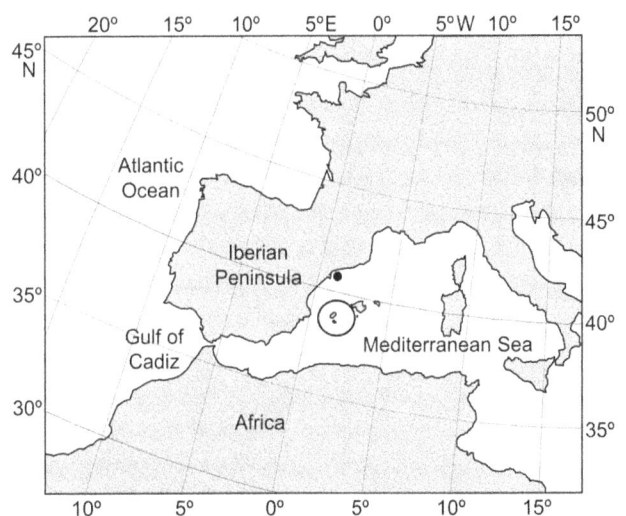

Fig. 1. Approximate location of the Atlantic bluefin tuna *Thunnus thynnus* farm where the study was conducted (black dot) and the fishing grounds where the fish were caught (circle)

are assumed to attain sexual maturity (Corriero et al. 2005). This, added to the fact that Mediterranean ABFT purse seiners primarily target schools of spawners, suggests that all the individuals examined were sexually mature.

Histology and functional classification of gonads

A piece of tissue was removed from the central part of one of the gonads and fixed in 10% phosphate-buffered formalin (4% formaldehyde in 0.1 M phosphate buffer, pH 7.2). The tissue samples were then washed in buffer, dehydrated in ethyl alcohol, cleared in xylene and embedded in paraffin. Serial 10 µm sections of the ovaries were stained with haematoxylin-VOF (Gutiérrez 1967) and photographed on a light microscope Nikon eclipse Ci® equipped with a Jenoptik ProgRes® CT5 digital camera. Histological sections of the testes were stained with haematoxylin-eosin.

Following Schaefer (1998), female ABTF were classified into 4 reproductive functional stages (Table 1), based on the most advanced group of ovarian follicles and the extent of atresia (for atretic states, see Hunter & Macewicz 1985a). The ovaries of resting (R) females contain only unyolked and/or early yolked oocytes and no sign of atresia. Active nonspawning (ANS) females show large yolked oocytes and minor (<50%), if any, α atresia. The active spawning (AS) condition is characterized by signs of either imminent spawning (presence of migratory-nucleus and/or hydrated oocytes) or recent spawning (evidenced by postovulatory follicles). Inactive mature (IM) females have entered regression following the end of reproductive activity, hence the ovaries contain either pre-vitellogenic or early yolked oocytes plus α and/or β atresia, or advanced yolked oocytes plus major (>50%) α atresia (Fig. 2). Among females, the spawning fraction was calculated as the proportion of

mature fish with POFs (Hunter & Macewicz 1985b).

Three or 4 distinct developmental stages have been identified from gonad histology in captive bluefin tuna males (e.g. Corriero et al. 2007, Sawada et al. 2007, Seoka et al. 2007). The male reproductive stages distinguished in our samples are shown in Table 2, which is based on Corriero et al. (2007). In fish at late spermatogenesis (LS), the germinal epithelium of the testes consists mainly of cysts composed of spermatids and spermatozoa, although spermatocyte and spermatid cysts are also present; the lumen of the seminiferous lobules and central ducts becomes filled with sperm. The testes of spent (S) males lack germinal cysts, and the lumen of the seminiferous lobules appears completely empty or shows scarce, loose residual sperm (Fig. 3).

Stereology

POFs were quantified by the physical disector method of Sterio (1984) adapted to fish ovarian samples (Aragón et al. 2010, Aranda et al. 2011, Ganias et al. 2014). The disector pairs consisted of 2 consecutive sections (referred to as reference and look-up sections) that were 40 or 60 µm apart (depending on the POF size in the histological sample). Three counting frames of 9.78 mm^2 were used per disector pair, and the total number of counting frames used per ovary was 18. POFs that appeared in the counting frame on the reference section, but not in the look-up section, were counted. When POFs touched the left or bottom lines of the frame, they were not counted. Counts were also made in the opposite direction. The volume number density of POFs was estimated according to the formula: $N_V = \Sigma Q^- (2 \Sigma P a/f h)^{-1}$, where ΣQ^- is the total number of POFs counted, ΣP is the number of disectors used per ovary (18), h is the disector thickness (40 or 60 µm), and a/f is the area of the working frame (9.78 mm^2). During histological processing, tissue

Table 1. Classification of Atlantic bluefin tuna *Thunnus thynnus* females in 4 different reproductive stages based on histological features according to Schaefer (1998). POF: postovulatory follicle

Stage	Histological features	Number of fish
Resting[a] (R)	Unyolked or early yolked oocytes and no atresia	8
Active nonspawning (ANS)	Advanced yolked oocytes and no atresia or minor (<50%) α atresia	28
Active spawning (AS)	Avanced yolked oocytes and no or minor α atresia plus POFs and/or migratory-nucleus oocytes	18 (1 with no POF)
Inactive mature (IM)	Previtellogenic or early yolked oocytes plus α and/or β atresia, or advanced yolked oocytes plus major (>50%) α atresia	4
[a]Referred to as 'immature' in the original classification of Schaefer (1998)		

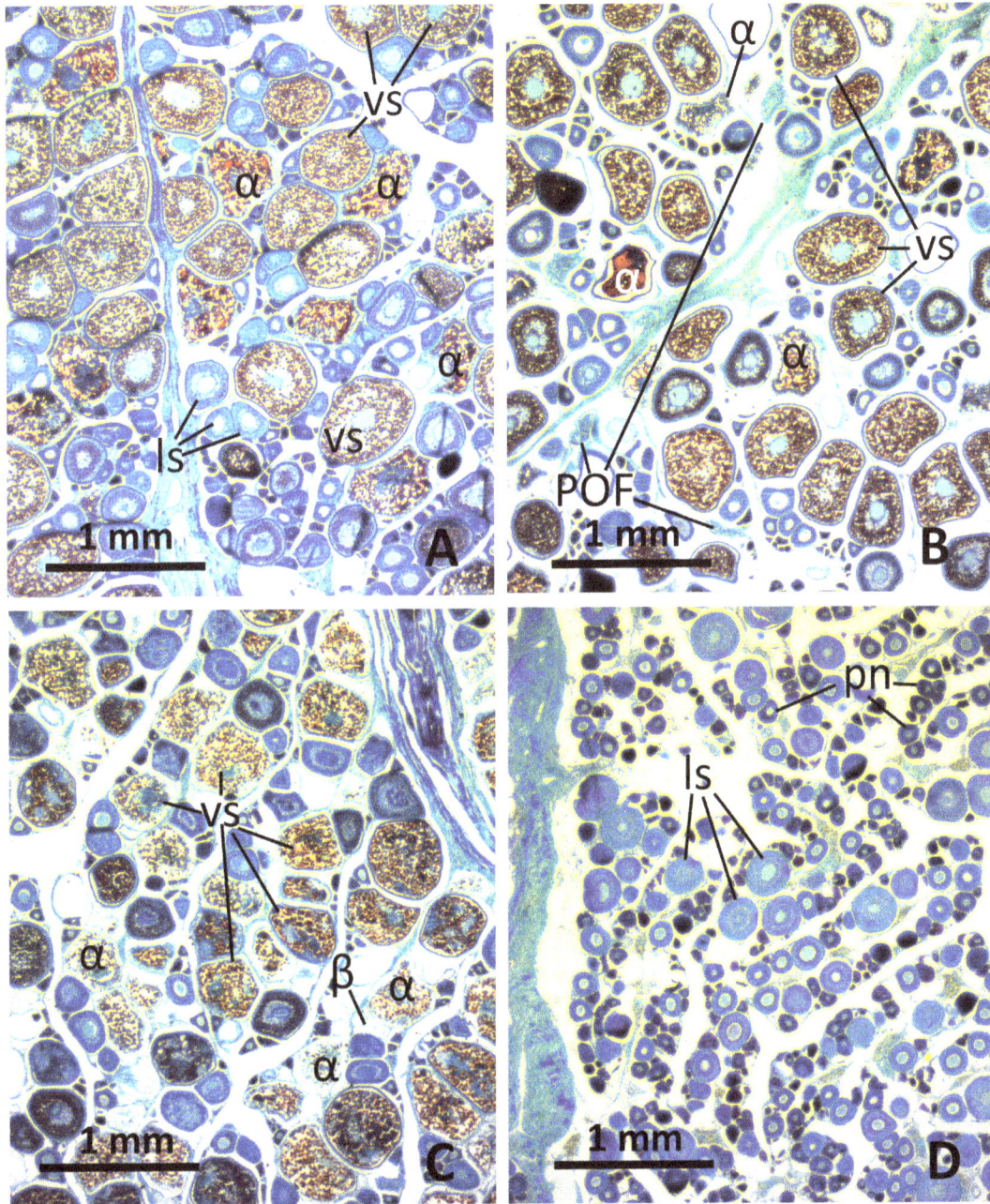

Fig. 2. Micrographs of Atlantic bluefin tuna *Thunnus thynnus* ovaries at the 4 developmental stages identified in this study: (A) active nonspawning (ANS); (B) active spawning (AS); (C) inactive mature (IM); (D) resting (R). ls: lipid-stage oocyte; pn: peri-nucleolar oocyte; POF: postovulatory follicle; vs: vitellogenic oocyte; α: α-atretic follicle; β: β-atretic follicle. Haematoxylin-VOF staining

samples experience a mean volume loss of 34.8% (Knapp et al. 2014). Therefore, a correction factor was applied to account for sample shrinkage. The total number of POFs (realised batch fecundity) was thus calculated as $N = 0.652 \cdot OV \cdot N_V$ (OV: ovarian volume), and the relative realised batch fecundity as the number of POFs g^{-1} BW.

Statistical analysis

Differences of means between groups of data were analysed using either Student's *t*-test or ANOVA followed by Tukey's HSD post hoc test. In order to integrate data of Gregorian dates into statistical analysis, they were converted to days of the year, rang-

Table 2. Classification of Atlantic bluefin tuna *Thunnus thynnus* males in the 2 different reproductive stages identified in this study on the basis of histological criteria according to Corriero et al. (2007)

Stage	Histological features	Number of fish
Late spermato-genesis (LS)	Germinal epithelium of the testicular lobes containing cysts of developing male germ cells where spermatids predominate; lumina of the seminiferous lobules and central system of ducts filled with spermatozoa	19
Spent (S)	Germinal epithelium devoid of germinal cysts, and consisting mostly of spermatogonia; lumina of seminiferous lobules completely empty or showing loose residual sperm	4

ing from 1 to 366 for 2012, and 1 to 365 for 2013 and 2014. Relationships between continuous variables were assessed by linear regression and bivariate Pearson's correlation analyses. A significance level of $\alpha = 0.05$ was considered in all tests. The statistical analyses were performed using SPSS version 15.0 and R Statistical Software version 3.2.0 (R Core Team 2015). Collective data are expressed as means ± SD.

RESULTS

Animals and morphometry

Eighteen fish (all females) were sampled in 2012, of which 17 had been captured in 2010 and 1 in 2012. In 2013, 22 females and 7 males sourced from the wild in 2012 and 2013 were examined. All tuna sampled in 2014 (18 females and 17 males) were captured in 2013, with the exception of 1 male which had been caught in the previous month in 2014 (see Table S1 in the Supplement at www.int-res.com/articles/suppl/q008p089_supp.pdf).

ABFT size ranged from 144 to 266 cm FL, and from 63 to 453 kg BW. The mean (±SD) size of the females analysed in 2012 (163.50 ± 13.46 cm) was significantly lower than that of the females sampled in the 2 subsequent years (211.45 ± 9.92 cm and 212.22 ± 15.05 cm; ANOVA, $F_{2,55} = 88.56$, p < 0.001, followed by Tukey's HSD test, p < 0.001). The mean size of the males was similar in 2013 (223.17 ± 13.47 cm) and 2014 (220.59 ± 22.94 cm; $t_{21} = 0.26$, p = 0.80).

Histology

The majority (12) of the 18 tuna sampled in 2012 (all females) were classified as ANS; 4 of these fish (ANS* in Table S1) were entering the regression phase as they showed advanced yolked oocytes and the proportion of α atretic oocytes was close to 50% (>40%). One of the fish was at the regenerating phase at the time of sampling (6 August); 2 individuals that were

sacrificed in late July were IM, and the remaining 3 fish, which were sampled between 11 June and 12 July, were at the AS stage (Fig. 4). Spawning occurred at SSTs between 21.5 and 24.3°C (Table S1). The estimated spawning fraction was 0.2 (Fig. 4).

Five of the 22 females analysed in 2013 were sampled only 1 mo after their capture from the wild. One of them, sacrificed on 10 July, was IM, whereas the 4 others, which were sampled in late July, had completed ovarian regression and were therefore in the R stage. Among the fish that had spent over 1 yr in captivity, 7 were AS females, which were harvested between 10 June and 23 July. One of them had no POFs but did contain migratory-nucleus stage oocytes. Spawning occurred at SSTs between 21.47 and 27.0°C, with the exception of 1 individual sampled on 10 June 2013 at 17.2°C SST. Nine females were found at the ANS stage, 7 of which were nearing the regression phase (close to 50% α atresia). The remaining 2 females held in captivity for over 1 yr (sampled in late July and early August) were R (Fig. 4, Table S1). The estimated spawning fraction was 0.4 (Fig. 4).

The males harvested on 5 June and 10 July 2013, 1 yr after their capture, were found to be at the LS stage. The 4 males sampled in late July had been caught during the year's fishing season 1 mo earlier, and were all considered S (Table S1).

In 2014, the 8 females classified as AS were sampled between 9 June and 14 July. Spawning occurred at SST between 20.3 and 22.5°C. Another 8 females were classified as ANS; 1 of them was apparently in the transition towards the IM stage. One IM female was harvested on 31 July, and the only R individual was sampled on 4 August (Fig. 4, Table S1). The estimated spawning fraction was 0.5 (Fig. 4).

All males appeared to be reproductively active (LS stage) throughout the sampling period (30 May to 4 August; Table S1).

We found significant differences in GSI values among different histological categories in both males and females (ANOVA, $F_{3,54} = 9.38$, p < 0.001). GSI was significantly lower in R females (mean ± SD:

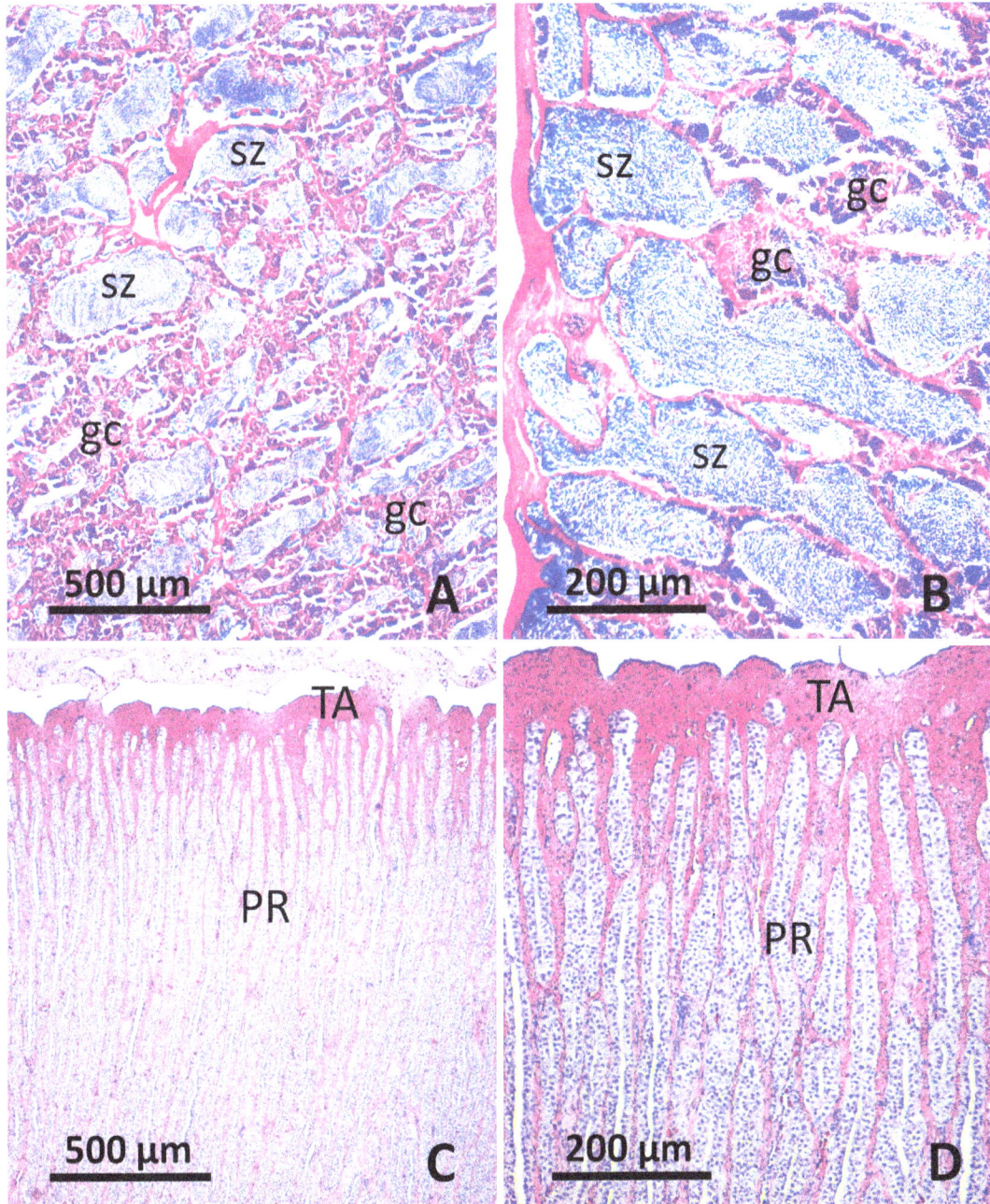

Fig. 3. Micrographs of Atlantic bluefin tuna *Thunnus thynnus* testes at the 2 maturation stages recognised in this study: (A) and (B) late spermatogenic stage (LS); (C) and (D) spent (S). gc: germinal cyst; PR: peripheral region of the testis showing lobules devoid of germinal cysts and spermatozoa; sz: spermatozoa; TA: tunica albuginea. Haematoxylin-eosin staining

1.06 ± 0.31) than in ANS (3.23 ± 1.06) and AS females (3.02 ± 1.01) (Tukey's HSD test, p < 0.001), which showed similar GSI values, whereas IM females (2.73 ± 1.74) had intermediate values that did not significantly differ from the other 3 categories. The mean GSI was significantly higher in LS males (1.99 ± 1.21) than in S males (0.31 ± 0.11) (*t*-test assuming unequal variances, $t_{16} = 5.45$, p < 0.001).

Stereology

The mean (±SD) realised batch fecundities estimated from the number of POFs (in millions of eggs) were 2.07 ± 1.69 (year 2012), 8.54 ± 5.28 (2013) and 12.71 ± 5.98 (2014). ANOVA indicated significant differences among the 3 groups of samples ($F_{2,14} = 4.481$, p = 0.03). Even when the number of spawned

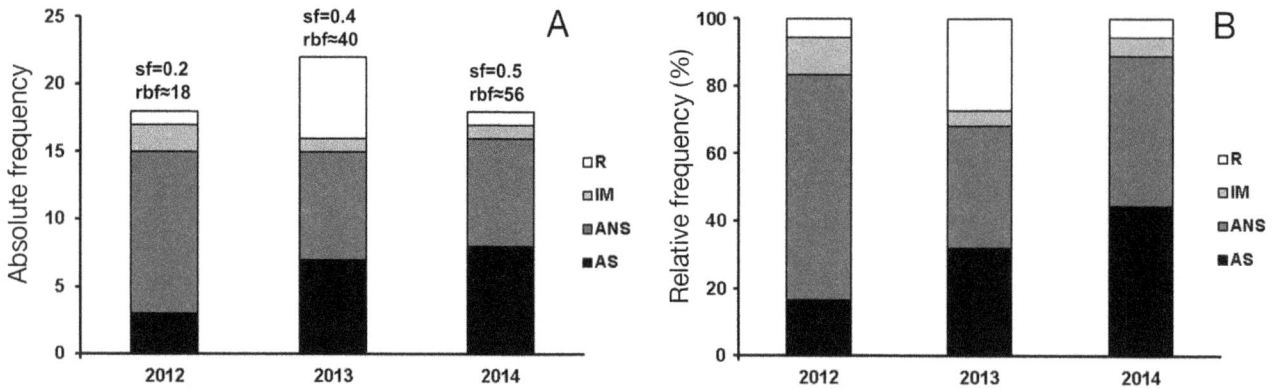

Fig. 4. (A) Absolute and (B) relative frequencies of the maturity stages found in Atlantic bluefin tuna *Thunnus thynnus* females sampled in 2012, 2013 and 2014. ANS: active nonspawning; AS: active spawning; IM: inactive mature; R: resting; rbf: relative batch fecundity (eggs g^{-1}); sf: spawning fraction

eggs was calculated relative to the fish weight (mean [±SD] relative batch fecundities of 17.89 ± 10.40 g^{-1}, 39.92 ± 22.41 g^{-1} and 56.06 ± 17.72 g^{-1}, respectively, Fig. 4), the statistical analysis indicated significant differences, although the result of the test approached the limit of significance ($F_{2,14}$ = 3.90, p = 0.045). For both absolute and relative fecundities, Tukey's HSD post hoc analysis showed that the fish sampled in 2012 were significantly less fecund than those sampled in 2014, whereas significant differences were not detected between the individuals from 2013 and the 2 other groups at a level of significance of 0.05.

Linear regressions revealed significant relationships between the realised batch fecundity and FL, BW and GW, but the predictive capacity of the models was relatively poor. The best predictor of the batch fecundity was BW through the equation: N POF (realised batch fecundity) = −4.51 × 10^6 + 67.52 × 10^3 BW ($F_{1,15}$ = 27.44, p = 0.0001, r^2 = 0.62). Table 3 shows the values of Pearson's correlation coefficients of pairwise comparisons between variables. As seen in the table, relative batch fecundity was not significantly correlated with any of the variables considered.

DISCUSSION

ABFT schools caught by purse seining in the Balearic Sea have been observed to spawn in towing cages during their transportation to commercial facilities (Gordoa & Carreras 2014). Our results show that farmed ABFT are also capable of spawning in commercial cages in subsequent years following their capture, even if the farm facility is relatively far from their natural spawning grounds. Histological evidence of spawning was observed between 9–11 June

and 12–23 July. This period matches the reproductive peak reported previously for wild spawners in the western Mediterranean Sea (Heinisch et al. 2008, Gordoa & Carreras 2014). Substantial oocyte resorption (IM reproductive stage) was identified from mid-July, leading to spent ovaries (R stage) in late July to August. This temporal pattern is also consistent with the migratory dynamics revealed by electronic tagging in the region (Aranda et al. 2013a). Spawning occurred at SSTs ranging between 20 and 27°C, but POFs were unexpectedly found in the ovaries of a fish sampled on 10 June 2013 at SST slightly over 17°C, although the batch fecundity of this specimen was the lowest of all females examined. These observations support earlier results indicating that spawning may take place at SSTs well below the currently assumed minimum spawning temperature of ~24°C (Gordoa & Carreras 2014).

The fraction of spawning females appears to be lower in the captive environment than in the wild. A proportion of spawning females over 80% has been reported for ABFT breeding schools in the western Mediterranean Sea (Medina et al. 2002, 2007, Aranda et al. 2013a,b). In contrast, only 3 of the captive females sampled in 2012 were classified as AS, representing 20% of the total number of individuals found in the spawning-capable phase sensu Brown-Peterson et al. (2011), i.e. ANS plus AS fish. The spawning fraction was substantially greater in the samples of the 2 other years, where the percentage of AS females related to the total spawning-capable fish was 40% in the sample of 2013 and 50% in 2014. Such differences could be due to the smaller size of the fish in the 2012 group, but the effect of size/age on spawning fraction requires further investigation to draw reliable conclusions.

Table 3. Output of bivariate correlation analysis between the continuous variables considered in the study showing values for Pearson's correlation coefficient r. BW: body weight; FL: straight fork length; GW: ovarian weight; GSI: gonadosomatic index; DOY: day of year; N POF: number of postovulatory follicles (proxy for realised batch fecundity); N POF g^{-1}: number of POFs per gram of total body weight (proxy for relative realised batch fecundity); SST: sea surface temperature. Correlations were significant at $^*p < 0.05$ and $^{**}p < 0.01$

	N POF g^{-1}	FL	BW	GW	GSI	DOY	SST
N POF	0.843**	0.662**	0.804**	0.746**	0.351	−0.087	−0.087
N POF g^{-1}		0.360	0.441	0.474	0.409	−0.053	0.015
FL			0.911**	0.865**	0.497*	−0.270	−0.266
BW				0.922**	0.383	−0.256	−0.311
GW					0.686**	−0.431	−0.324
GSI						−0.547*	−0.232
DOY							0.824**

The POF size and morphology varied among individuals, indicating they had different ages (see Aragón et al. 2010). Since all harvests were performed within a narrow time range early in the morning, distinct POF morphologies could reflect an irregular temporal spawning pattern in the cages. This is consistent with observations of spawning at different hours of the day (pers. comm. from the farm staff). In the wild, conversely, spawning appears to take place at nighttime within a narrow temporal window, approximately between 12:00 and 03:00 h UTC (Gordoa et al. 2009, Aranda et al. 2013a, Gordoa & Carreras 2014).

Preceding studies on experimental ABFT broodstocks (Corriero et al. 2007, Aranda et al. 2011) found that the spawning fraction in cages was low (similar to our 2012 data) even after a 3 yr period of adaptation to captivity. In ABFT, as in other teleosts, the stressful captive environment causes reproductive dysfunctions that primarily include oocyte maturation and ovulation failure in females, and reduced sperm quality/quantity in males (Mylonas et al. 2007). Although previous experimental trials have shown poor egg production in captive ABFT compared to wild populations, our observations suggest that a substantial proportion of ABFT held in farm grow-out cages are functional spawners during the species' natural reproductive season.

The increasing aquaculture of fish species, such as cod or sea bream, until the reproductive age results in escapes of large numbers of eggs into the environment (Uglem et al. 2012, Somarakis et al. 2013). The ABFT, like the Pacific bluefin tuna, is an example of those species where so-called 'escape through spawning' (Uglem et al. 2012) may be significant in commercial facilities, since >60% of the ABFT eastern stock total allowable catch ends up in Mediterranean tuna farms (Ortiz 2015). We know that ABFT

eggs collected from grow-out cages show a high hatching rate (75–87%) and produce apparently healthy yolk-sac larvae at temperatures between 21 and 26°C (Ortega 2015). However, the number and quality of the eggs produced in farms are difficult to determine, as it is virtually impossible to make sure that the entire spawns are collected from offshore cages.

Nevertheless, while reliable direct estimations of numbers of spawned eggs are unfeasible, the occurrence of POFs is useful to estimate the spawning fraction of females (Hunter & Macewicz 1985b, Hunter et al. 1986, Ganias 2012), and the quantification of these structures from histological samples provides accurate measures of fecundity. POFs are not detectable for more than 24 h in tuna species, such as the ABFT, that spawn in warm waters (Schaefer 2001). Thus, the number of eggs spawned within the past 24 h by ABFT can be estimated from unbiased stereological counts of POFs in ovarian samples (Aragón et al. 2010, Ganias et al. 2014). The average batch fecundity estimated for AS females in 2012 was ~2 million eggs (relative batch fecundity ~18 eggs g^{-1}), which is far below the estimates for the 2 other years: 8.5 and 12.7 million eggs (~40 and ~53 eggs g^{-1}, respectively). The 2 latter figures are similar to the fecundities calculated elsewhere for captive and wild ABFT using the same counting technique (Aranda et al. 2011). In agreement with a previous study conducted on wild ABFT (Aranda et al. 2013b), the absolute batch fecundity was strongly correlated with fish size, expressed as both FL and BW. However, the relative batch fecundity was not dependent on fish size, so that, generally speaking, all fish would contribute to the number of eggs produced by the broodstock in proportion to their weight. Consequently, the daily egg production of a stock could be easily estimated from data of the biomass stocked in the cages, the sex ratio of the stock and the estimates of spawning fraction and relative realised batch fecundity. For instance, the proportion of spawning females found in this study was ~0.3 (17 fish with POFs out of 58 females sampled in all 3 years, Table 1). Assuming a batch fecundity of 50 eggs g^{-1}, a sex ratio of 1:1 and equal sizes in males and females, a sea-cage holding 100 t of mature ABFT would produce as many as ~750 million eggs daily. Estimations of total annual egg production in tuna farms would need accurate data on the average duration of the spawning period, which still remains to be determined in captivity conditions. Current progress in genetic analysis (e.g. Nakadate et al. 2011, Gordoa et al. 2015), however, is promising for the estimation of key reproductive parameters in bluefin tunas, including individual spawning duration in captivity.

Jensen et al. (2013) showed that Atlantic salmon escaped from farms at the smolt stage or early in the post-smolt stage may grow, migrate and survive to adulthood as wild specimens do. A similar situation may occur with ABFT larvae hatched in farms. The productive coast off the Ebro River delta, where the studied farm is located, is used as a feeding habitat by early stages of ABFT (about 3–4 mo old) coming from the breeding grounds around the Balearic archipelago (Medina et al. 2015). Although the survival rate from hatching to age 3 mo in this area is unknown, some of the larvae hatched from eggs spawned in the farm are likely to grow and join wild-born young ABFT in the nearby feeding ground. The same may hold true for other Mediterranean farms located close to age-0 ABFT feeding areas. This could have an impact on the recruitment and hence on the population dynamics of ABFT in the Mediterranean Sea. In order to assess the survivorship of ABFT larvae, ongoing experimental work is being conducted to investigate the temperature tolerances of ABFT larval and postlarval stages.

Depending on the number of fish that are ranched for over 1 yr and the larval survival rate in the zone, the 'escape through spawning' phenomenon could have significant ecological implications and, hence, warrant further investigation regarding population dynamics and stock assessment.

Acknowledgements. This work was funded by the Spanish Ministry of Economy and Competitiveness (contract no. AGL2014-52003-C2-1-R). It would not have been possible without the collaboration and advice from the owners and staff of Grup Balfegó. This is contribution no. 113 of CEI·MAR.

LITERATURE CITED

ä Aragón L, Aranda G, Santos A, Medina A (2010) Quantification of ovarian follicles in bluefin tuna *Thunnus thynnus* by two stereological methods. J Fish Biol 77:719–730

ä Aranda G, Aragón L, Corriero A, Mylonas CC, de la Gándara F, Belmonte A, Medina A (2011) GnRHa-induced spawning in cage-reared Atlantic bluefin tuna: an evaluation using stereological quantification of ovarian postovulatory follicles. Aquaculture 317:255–259

ä Aranda G, Abascal FJ, Varela JL, Medina A (2013a) Spawning behaviour and post-spawning migration patterns of Atlantic bluefin tuna (*Thunnus thynnus*) ascertained from satellite archival tags. PLoS ONE 8:e76445

ä Aranda G, Medina A, Santos A, Abascal FJ, Galaz T (2013b) Evaluation of Atlantic bluefin tuna reproductive potential in the western Mediterranean Sea. J Sea Res 76:154–160

Arechavala-López P, Borg JA, Šegvić-Bubić T, Tomassetti P, Özgül A, Sánchez-Jerez P (2015) Aggregations of wild Atlantic bluefin tuna (*Thunnus thynnus* L.) at Mediterranean offshore fish farm sites: environmental and management considerations. Fish Res 164:178–184

Brown-Peterson NJ, Wyanski DM, Saborido-Rey F, Macewicz BJ, Lowerre-Barbieri SK (2011) A standardized terminology for describing reproductive development in fishes. Mar Coast Fish 3:52–70

ä Corriero A, Karakulak S, Santamaria N, Deflorio M and others (2005) Size and age at sexual maturity of female bluefin tuna (*Thunnus thynnus* L. 1758) from the Mediterranean Sea. J Appl Ichthyol 21:483–486

Corriero A, Medina A, Mylonas CC, Abascal FJ and others (2007) Histological study of the effects of treatment with gonadotropin-releasing hormone agonist (GnRHa) on the reproductive maturation of captive-reared Atlantic bluefin tuna (*Thunnus thynnus* L.). Aquaculture 272:675–686

ä Corriero A, Medina A, Mylonas CC, Bridges CR and others (2009) Proliferation and apoptosis of male germ cells in captive Atlantic bluefin tuna (*Thunnus thynnus* L.) treated with gonadotropin-releasing hormone agonist (GnRHa). Anim Reprod Sci 116:346–357

De la Gándara F, Mylonas CC, Coves D, Ortega A and others (2010) Seedling production of Atlantic bluefin tuna (ABFT) *Thunnus thynnus*, the SELFDOTT project. In: Miyashita S, Takii K, Sakamoto W, Biswas A (eds) Towards the sustainable aquaculture of bluefin tuna. Proc Int Symp 40th Anniv Pacific Bluefin Tuna Aquacult. Kinki University Press, Kinki, p 45–52

ä De Metrio G, Bridges CR, Mylonas CC, Caggiano M and others (2010) Spawning induction and large-scale collection of fertilized eggs in captive Atlantic bluefin tuna (*Thunnus thynnus* L.) and the first larval rearing efforts. J Appl Ichthyol 26:596–599

ä Ganias K (2012) Thirty years of using the postovulatory follicles method: overview, problems and alternatives. Fish Res 117-118:63–74

Ganias K, Murua H, Claramunt G, Dominguez-Petit R and others (2014) Egg production. In: Domínguez-Petit R, Murua H, Saborido-Rey F, Trippel E (eds) Handbook of applied fisheries reproductive biology for stock assessment and management. Northwest Atlantic Fisheries Organization, Vigo. Available at http://hdl.handle.net/10261/87768

ä Gordoa A, Carreras G (2014) Determination of temporal spawning patterns and hatching time in response to temperature of Atlantic bluefin tuna (*Thunnus thynnus*) in the western Mediterranean. PLoS ONE 9:e90691

Gordoa A, Olivar MP, Arévalo R, Viñas J, Molí B, Illas X (2009) Determination of Atlantic bluefin tuna (*Thunnus thynnus*) spawning time within a transport cage in the western Mediterranean. ICES J Mar Sci 66:2205–2210

► Gordoa A, Sanz N, Viñas J (2015) Individual spawning duration of captive Atlantic bluefin tuna (*Thunnus thynnus*) revealed by mitochondrial DNA analysis of eggs. PLoS ONE 10:e0136733

Gutiérrez M (1967) Coloración histológica para ovarios de peces, crustáceos y moluscos. Invest Pesq (Spain) 31: 265–271

► Heinisch G, Corriero A, Medina A, Abascal FJ and others (2008) Spatial-temporal pattern of bluefin tuna (*Thunnus thynnus* L. 1758) gonad maturation across the Mediterranean Sea. Mar Biol 154:623–630

Hunter JR, Macewicz BJ (1985a) Rates of atresia in the ovary of captive and wild northern anchovy, *Engraulis mordax.* Fish Bull 83:119–136

Hunter JR, Macewicz BJ (1985b) Measurement of spawning frequency in multiple spawning fishes. In: Lasker R (ed) An egg production method for estimating spawning biomass of pelagic fish: application to the northern anchovy, *Engraulis mordax.* NOAA Tech Rep NMFS 36:79–94

Hunter JR, Macewicz BJ, Sibert JR (1986) The spawning frequency of skipjack tuna, *Katsuwonus pelamis*, from the South Pacific. Fish Bull 84:895–903

► Jensen AJ, Karlsson S, Fiske P, Hansen LP, Hindar K, Østborg GM (2013) Escaped farmed Atlantic salmon grow, migrate and disperse throughout the Arctic Ocean like wild salmon. Aquacult Environ Interact 3:223–229

► Knapp JM, Aranda G, Medina A, Lutcavage M (2014) Comparative assessment of the reproductive status of female Atlantic bluefin tuna from the Gulf of Mexico and the Mediterranean Sea. PLoS ONE 9:e98233

► Mangion M, Borg JA, Thompson R, Schembri PJ (2014) Influence of tuna penning activities on soft bottom macrobenthic assemblages. Mar Pollut Bull 79:164–174

► Medina A, Abascal FJ, Megina C, García A (2002) Stereological assessment of the reproductive status of female Atlantic northern bluefin tuna during migration to Mediterranean spawning grounds through the Strait of Gibraltar. J Fish Biol 60:203–217

► Medina A, Abascal FJ, Aragón L, Mourente G and others (2007) Influence of sampling gear in assessment of reproductive parameters for bluefin tuna in the western Mediterranean. Mar Ecol Prog Ser 337:221–230

► Medina A, Goñi N, Arrizabalaga H, Varela JL (2015) Feeding patterns of age-0 bluefin tuna in the western Mediterranean inferred from stomach-content and isotope analyses. Mar Ecol Prog Ser 527:193–204

► Metian M, Pouil S, Boustany A, Troell M (2014) Farming of bluefin tuna –reconsidering global estimates and sustainability concerns. Rev Fish Sci Aquacult 22:184–192

► Moraitis M, Papageorgiou N, Dimitriou PD, Petrou A, Karakassis I (2013) Effects of offshore tuna farming on benthic assemblages in the Eastern Mediterranean. Aquacult Environ Interact 4:41–51

► Mylonas CC, Bridges CR, Gordin H, Belmonte Ríos A and others (2007) Preparation and administration of gonadotropin-releasing hormone agonist (GnRHa) implants for the artificial control of reproductive maturation in captive-reared Atlantic bluefin tuna (*Thunnus thynnus*). Rev Fish Sci 15:183–210

► Mylonas CC, de la Gándara F, Corriero A, Belmonte Ríos A (2010) Atlantic bluefin tuna (*Thunnus thynnus*) farming and fattening in the Mediterranean Sea. Rev Fish Sci 18:266–280

► Nakadate M, Kusano T, Fushimi H, Kondo H, Hirono I, Aoki T (2011) Multiple spawning of captive Pacific bluefin tuna (*Thunnus orientalis*) as revealed by mitochondrial

DNA analysis. Aquaculture 310:325–328

Ortega A (2015) Cultivo integral de dos especies de escómbridos: atún rojo del Atlántico (*Thunnus thynnus*, L. 1758) y bonito Atlántico (*Sarda sarda*, Bloch 1793). PhD dissertation, University of Murcia

Ortiz M (2015) Preliminary evaluations of potential growth of fattened/farmed eastern bluefin tuna (*Thunnus thynnus*) from ICCAT farm size database. Col Vol Sci Pap ICCAT 71:1505–1525

Ottolenghi F (2008) Capture-based aquaculture of bluefin tuna. In: Lovatelli A, Holthus PF (eds) Capture-based aquaculture, Vol 508. Food and Agriculture Organization of the United Nations, Rome, p 169–182

► Piedecausa MA, Aguado-Giménez F, Cerezo-Valverde J, Hernández-Llorente MD, García-García B (2010) Simulating the temporal pattern of waste production in farmed gilthead seabream (*Sparus aurata*), European seabass (*Dicentrarchus labrax*) and Atlantic bluefin tuna (*Thunnus thynnus*). Ecol Model 221:634–640

R Core Team (2015) R: a language and environment for statistical computing. R Foundation for Statistical Computing, Vienna. www.R-project.org/

► Rosenfeld H, Mylonas CC, Bridges CR, Heinisch G and others (2012) GnRHa-mediated stimulation of the reproductive endocrine axis in captive Atlantic bluefin tuna, *Thunnus thynnus.* Gen Comp Endocrinol 175:55–64

► Sarà G, Lo Martire M, Sanfilippo M, Pulicanò G and others (2011) Impacts of marine aquaculture at large spatial scales: evidences from N and P catchment loading and phytoplankton biomass. Mar Environ Res 71:317–324

► Sawada Y, Seoka M, Kato K, Tamura T and others (2007) Testes maturation of reared Pacific bluefin tuna *Thunnus orientalis* at two-plus years old. Fish Sci 73:1070–1077

Schaefer KM (1998) Reproductive biology of yellowfin tuna (*Thunnus albacares*) in the eastern Pacific Ocean. Inter-Am Trop Tuna Comm Bull 21:201–272

Schaefer KM (2001) Reproductive biology of tunas. In: Block BA, Stevens ED (eds) Tuna: physiology, ecology and evolution. Academic Press, London, p 225–270

Seoka M, Kato K, Kubo T, Mukai Y, Sakamoto W, Kumai H, Murata O (2007) Gonadal maturation of Pacific bluefin tuna *Thunnus orientalis* in captivity. Aquacult Sci 55:289–292

► Somarakis S, Pavlidis M, Saapoglou C, Tsigenopoulos CS, Dempster T (2013) Evidence for 'escape through spawning' in large gilthead sea bream *Sparus aurata* reared in commercial sea-cages. Aquacult Environ Interact 3:135–152

► Sterio DC (1984) The unbiased estimation of number and sizes of arbitrary particles using the disector. J Microsc 134:127–136

► Sumaila UR, Huang L (2012) Managing bluefin tuna in the Mediterranean Sea. Mar Policy 36:502–511

► Uglem I, Knutsen Ø, Kjesbu OS, Hansen ØJ and others (2012) Extent and ecological importance of escape through spawning in sea-cages for Atlantic cod. Aquacult Environ Interact 3:33–49

► Vezzulli L, Moreno M, Marin V, Pezzati E, Bartoli M, Fabiano M (2008) Organic waste impact of capture-based Atlantic bluefin tuna aquaculture at an exposed site in the Mediterranean Sea. Estuar Coast Shelf Sci 78: 369–384

Vitalini V, Benetti D, Caprioli R, Forrestal F (2010) Northern bluefin tuna (*Thunnus thynnus thynnus*) fattening in the Mediterranean Sea: status and perspectives. World Aquacult 41:30–36

► Vizzini S, Mazzola A (2012) Tracking multiple pathways of waste from a northern bluefin tuna farm in a marine-coastal area. Mar Environ Res 77:103–111

3

Salmon lice dispersion in a northern Norwegian fjord system and the impact of vertical movements

I. A. Johnsen*, L. C. Asplin, A. D. Sandvik, R. M. Serra-Llinares

Institute of Marine Research, PO Box 1870 Nordnes, 5817 Bergen, Norway

ABSTRACT: The abundance and distribution of salmon lice *Lepeophtheirus salmons* originating from fish farms in a northern Norwegian fjord during the summer of 2010 was investigated by means of a numerical model, underpinned by field observations. In order to evaluate the robustness of the simulated distribution of the lice, we re-ran the simulation several times, changing the vertical responses of the lice to environmental cues such as light and turbulence, in addition to altering their vertical swimming velocity. The model was able to realistically reproduce the observed currents and stratification in the region. The simulated distribution of lice was not sensitive to different implementations of surface light nor to the light sensitivity level of the lice. However, the vertical swimming velocity and a mixing parameter influenced both their vertical distribution and horizontal dispersion. The aggregation of lice along land was influenced by their response to turbulent water. The simulated infectious stages of the lice were transported on average 20 to 45 km from their release site. The simulated concentrations of infectious lice varied in synchronisation with lice infestations observed on wild fish in the area. Less than 1% of the simulated lice reached a farm site. The ratio between internal and external exposure ranged from 7 to 57%. Farms in the north of the fjord system were more exposed to lice released in the south than vice versa.

KEY WORDS: Salmon lice · *Lepeophtheirus salmonis* · Dispersion · Fjord · Aquaculture management · IBM · Model

INTRODUCTION

The salmon louse *Lepeophtheirus salmonis* is a naturally occurring parasite in the northern hemisphere that feeds on the skin, fat and mucus of salmonid fish (Pike 1989, Dawson et al. 1998, Pike & Wadsworth 1999). With the industrialization of aquaculture, the numbers of potential hosts have increased, leading to unnaturally high lice infestations on both farmed and wild fish (Finstad et al. 2000, Bjorn et al. 2001, Heuch & Mo 2001, Heuch et al. 2009, 2011). Lice levels are closely monitored at fish farms. At the time of this study, delousing was required if the level exceeded an average of 0.5 adult female lice per fish (Directorate of Fisheries 2013). On individual farms, different initiatives are being taken to minimize infestations, such as the use of cleaner fish or lice skirts. The combination of delousing and preventive actions ensures that the farms are capable of keeping the number of lice within the permitted limit; however, it is expensive, time consuming, stressful for the fish and can result in the release of harmful chemicals into the environment (McHenery et al. 1991, Burridge et al. 2010). Moreover, even with these actions in place, large amounts of salmon lice are regularly observed on wild fish in Norway, and these infestations frequently influence stock size in exposed areas (Bjorn et al. 2001). Today, Norway has the largest aquacultured production of salmonid fish in the world, exceeding 1.2 million t of fish sold annually. Both the aquaculture industry and the Norwegian government want a sustainable

*Corresponding author: ingrid.johnsen@imr.no

increase in production (Ministry of Trade Industry and Fisheries 2015); however, the potential negative effects of salmon lice on wild fish populations (salmon, sea trout and Arctic char) is regarded as one of the most serious limitations to sustainable growth (Taranger et al. 2014a).

Salmon lice carry eggs in 2 eggstrings at the rear of the body. When these eggs hatch, the lice are released in the water, where they go through 2 non-infectious nauplii stages before reaching the copepodid stage. At this stage they are able to—and are crucially dependent on—infecting salmonid fish. During these stages, lice are planktonic and drift with the currents. Due to their relatively long planktonic phase (Johnson & Albright 1991, Stien et al. 2005), information about the movement of the currents is crucial in order to determine the transport routes and residence areas of the lice. Temporal and spatial variability are high, thus the best estimates of 3-dimentional environmental variables (i.e. temperature, salinity and currents) can be achieved from numerical models, validated through field observations.

Salmon lice have the ability to swim vertically in the water column and are observed to reside close to the surface (Heuch et al. 1995, Gravil 1996, Penston et al. 2004, 2008). The motivation for these vertical movements is probably to avoid predators and to seek host-rich environments (Bron et al. 1993). Nauplii reside deeper in the water column than the copepodids, and recent studies have suggested that nauplii may reside even deeper during the winter than summer (Gravil 1996, Penston et al. 2008, Johnsen et al. 2014, á Norði et al. 2015). Norwegian fjords generally exhibit a large vertical gradient in temperature, salinity and currents in the upper 20 m of the water column—exactly where we assume the planktonic salmon lice to reside. Hence knowing the vertical position of the lice is important in order to simulate realistic lice dispersion (Johnsen et al. 2014).

As shown in previous studies, we have developed a numerical modelling system to estimate the dispersion of salmon lice hatched on fish farms (Asplin et al. 2004, 2011, 2014, Johnsen et al. 2014). The system consists of several separate models that compute forcing and boundary conditions for the overall fjord model, estimating current and hydrography with high resolution both in time (h) and space (~200 m) (Albretsen 2011). These variables are used to drive the salmon lice growth and transportation model, providing hourly density fields of infective lice.

Dispersion modelling for salmon lice has seen recent, rapid proliferation (Asplin et al. 2004, 2011, 2014, Murray & Gillibrand 2006, Gillibrandt & Willis

2007, Amundrud & Murray 2009, Murray et al. 2011, Stucchi et al. 2011, Adams et al. 2012, Salama et al. 2013, 2015, Stormoen et al. 2013, Johnsen et al. 2014). The models are widely used to calculate dispersion distance and the area of influence of salmon lice from single farms, to map high concentration areas in larger fjord or loch systems and to identify the exposure connectivity between fish farms. With increasing computer resources, most models have become more sophisticated with increasing quality in terms of the forcing of wind and freshwater and boundary conditions towards the coast.

A general challenge with salmon lice dispersion models is that they calculate the density of salmon lice. This represents the exposure or the infectious dose, whereas what usually is needed for management is the actual number of lice on individual fish (i.e. the response of this dose). The infectivity of salmon lice depends on salinity, temperature, the age of the copepodid and the swimming speed of the host (Bron et al. 1993, Tucker et al. 2000, Bricknell et al. 2006, Samsing et al. 2015). Thus, it is not necessarily simply a linear relationship between the number of copepodids in a region and the infection of salmonid fish.

In this paper, we present a novel approach to downscale the simulated lice density and relate it to infections found on wild fish. Our objective was to test the applicability of our salmon lice dispersion modelling system in a northern Norwegian fjord system. We have found good agreement between similar models and lice infection on fish in the southern part of Norway (A. D. Sandvik et al. unpubl data), but there are a few potentially important differences in conditions between the southern and northern regions. There is a different light regime in northern Norway, with midnight sun occurring during the summer months. In addition, water temperatures are normally lower there compared to further south. There are generally more lice in the south; during 2010 and 2012 the observed prevalence (the proportion of the fish infested with lice) of salmon lice infestation was on average the same in Folda as in the Hardangerfjord further south (~60°N); however, the intensity (the number of lice on the infested fish) was almost 3 times higher in the Hardangerfjord than in Folda (Taranger et al. 2014b). Topographically, there are differences between the north and south, with smaller and wider fjords to the north. These factors were all included in our circulation model.

To determine the ability of the model to predict lice densities in a northern Norwegian fjord we first compared the model-estimated circulation with available observations. This was done to ensure high quality forcing for the lice model. We then examined the sensitivity of the salmon lice model to light response and vertical swimming velocity as well as how turbulence in the water masses affected the dispersion of the salmon lice. The simulated distribution of the salmon lice was evaluated by comparing it to observed infestations on wild fish caught in the area. Finally, we discuss how the different farms in the fjord were exposed to salmon lice.

MATERIALS AND METHODS

Area description

The Folda fjord system is located in the northern region of Norway at ~67° N and stretches in from the large oceanic bay Vestfjorden east of the Lofoten archipelago (Fig. 1). The fjord has 2 distinct branches: Sørfolda to the south and Nordfolda to the north.

From the mouth to the fjord head, the distance is ~60 km for both branches. The fjord width varies from 1 km at the inner parts to several km at the outer parts. Folda has basins exceeding 500 m in depth and a sill depth of around 250 m, preventing the horizontal exchange of water between the basins and the coastal ocean. As is typical of Norwegian fjords, the terrain surrounding Folda consists of relatively steep mountains.

Freshwater runoff from rivers, heat transfer from the atmosphere, wind, tides and internal waves control the movement of the water close to the surface. The fjord system receives 160 m^3 s^{-1} (mean amount calculated between 1961 and 1990) of freshwater from rivers; most of which is released into Sørfolda (Myksvoll et al. 2011). During the simulated period for this study, the fjord received close to the 30 yr mean runoff. The tidal amplitude is relatively large with a difference of ~1–2 m between high and low tides. The water masses in the Vestfjorden outside the mouth of Folda are variable and dominated by eddies (Mitchelson-Jacob & Sundby 2001).

There are 10 approved fish farm locations in the fjord. Two of the fish farms were not active during the study period, but were included in the simulations in order to evaluate the potential dispersion of salmon lice even from these locations. The fjord system contains a number of wild salmonid fish; salmon *Salmo salar*, sea trout *Salmo trutta* and Arctic charr *Salvelinus alpinus*, all of which could potentially be exposed to salmon lice infections.

The fjord current model

The hydrography of the fjord (i.e. current, temperature and salinity) was calculated using the Regional Ocean Modeling System (ROMS; www.myroms.org). The horizontal resolution was set to 200 m and the vertical resolution to 35 terrain-following coordinates. Results from the coastal ocean current model NorKyst800 (Albretsen 2011) were used along the open boundaries of the fjord model, updating values of salinity, temperature, currents and

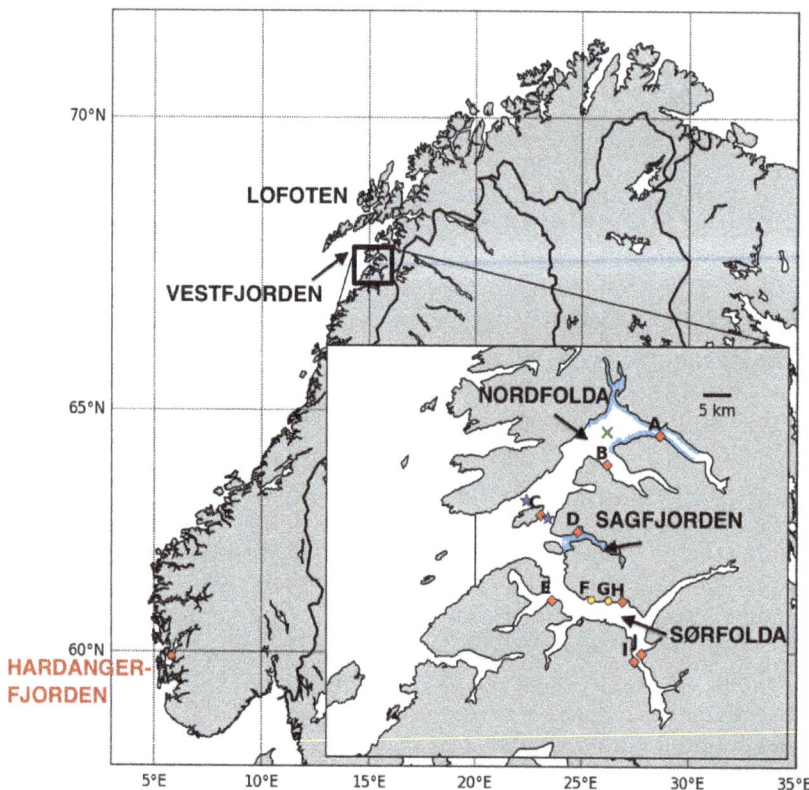

Fig. 1. The Folda fjord system on the northern coast of Norway. Red marks: active salmon farms; orange marks: fish farms not in production during the simulated period. Vertical profiles of temperature and salinity were measured at the locations indicated by the blue stars (7 March and 21 July 2010). Blue areas indicate the assumed residence areas for wild sea trout. The black line at the fjord mouth indicate the outer limit of calculated salmon lice retention. The green cross shows the position of buoy measuring current and salinity at 3 m depth and temperature at 4 m depth (2 July to 30 August 2010)

water levels hourly. The Norwegian Water Resources and Energy Directorate (NVE) provided the freshwater input for the model for 58 rivers as a daily freshwater volume flux. Atmospheric forcing consisting of surface wind and radiation was calculated by the Weather Research and Forecasting model (WRF; www.wrf-model.org) with 3 km horizontal resolution, updated every 3 h. Results of the fjord current model consisted of the 3-dimensional fields of currents, salinity, temperature and turbulence stored as hourly instantaneous values. The simulation period was from 1 April to 31 August 2010.

The salmon lice dispersion model

The Lagrangian Diffusion Model (LADIM) was used to simulate the dispersion of salmon lice from the 10 farms in the fjord system. Every hour, particles were released from these 10 sources, representing salmon lice nauplii produced at those farm locations at the given times. The particles were horizontally transported by the ambient currents, but were given the ability to swim towards the surface or sink as a response to environmental factors. Age was counted in degree-days to determine the infectious stage of each particle, which was assumed to be between 50 and 150 degree-days (Asplin et al. 2004, Stien et al. 2005). The environmental factors influencing the particles' vertical positioning were light, salinity and vertical mixing. The particles' responses to light and low salinity were as given in Johnsen et al. (2014), but whereas vertical mixing was held constant in earlier models (Asplin et al. 2011, 2014, Johnsen et al. 2014), we included vertical mixing of salmon lice controlled by the simulated level of turbulence to

determine if it influenced horizontal dispersion. The turbulent vertical mixing in Norwegian fjords depends on the wind-driven current, tides, stratification and the shape of the fjord (Stigebrandt & Aure 1989), and generally decreases with distance to the source (typically the surface or bottom). The implementation of dynamic turbulence was assumed to give a more realistic dispersion of salmon lice. The simulated diffusion coefficient from the fjord current model was used for calculating the turbulent vertical movement according to Visser (1997). We assumed that salmon lice stopped swimming upwards when their turbulent movements exceeded a given swimming velocity. Therefore, simulated salmon lice exposed to turbulent vertical displacement greater than 5×10^{-4} m s^{-1} were set to sink with a vertical velocity of 10^{-3} m s^{-1} (Bricknell et al. 2006) until less turbulent conditions were reached. Turbulence levels exceeding 0.1 m^2 s^{-1} were assumed to be numerical miscalculations and were set to 0.1 m^2 s^{-1}.

Sensitivity parameters tested

To test the sensitivity of the vertical distribution (and hence horizontal dispersion) of salmon lice to environmental factors we performed 6 simulations, varying the following parameters: (1) the light-sensing level of the lice, (2) level of surface light, (3) vertical mixing of the lice, and (4) swimming velocity of the lice. Various model parameterizations for the sensitivity experiment are provided in Table 1. Salmon lice are known to be positively phototactic (Bron et al. 1993, Heuch et al. 1995, Gravil 1996, Flamarique et al. 2000). The light levels assumed to trigger lice to swim vertically in the simulations

Table 1. Parameter settings used for triggering vertical swimming during the salmon lice *Lepeophtheirus salmonis* simulation

Parameter	Light sensitivity (µmol photons s^{-1} m^{-2})	Surface light (µmol photons s^{-1} m^{-2})	Vertical mixing (m^2 s^{-1})	Vertical swimming velocity (m s^{-1})
Random walk	Nauplii: 0.3 Copepodids: 2.06×10^{-5}	Calculated at actual position	Constant: $D = 10^{-3}$	5×10^{-4}
Dynamic mixing	Nauplii: 0.3 Copepodids: 2.06×10^{-5}	Calculated at actual position	Dynamic: from fjord model	5×10^{-4}
Less light sensitive	All stages: 0.1	Calculated at actual position	Dynamic: from fjord model	5×10^{-4}
60° N-light	Nauplii: 0.3 Copepodids: 2.06×10^{-5}	Calculated for 60° N	Dynamic: from fjord model	5×10^{-4}
Slow swimmers	Nauplii: 0.3 Copepodids: 2.06×10^{-5}	Calculated at actual position	Dynamic: from fjord model	10^{-4}
Fast swimmers	Nauplii: 0.3 Copepodids: 2.06×10^{-5}	Calculated at actual position	Dynamic: from fjord model	10^{-3}

'random walk', 'dynamic mixing', '60° N-light', 'slow swimmers' and 'fast swimmers' were selected based on Flamarique et al. (2000). Salmon lice are also known to exhibit burst swimming (Gravil 1996, Heuch & Karlsen 1997). As the swimming bursts occur on a smaller time scale than the temporal resolution of the model, we assumed that burst swimming episodes could be represented by a mean swimming velocity of ca. 1 body length s^{-1}. Surface light was calculated by latitude and time of day according to Skartveit & Olseth (1988); see Johnsen et al. (2014) for details. In addition, the simulated distribution was plotted with and without calculated mortality, to investigate the impact of mortality on the distribution of the lice.

Simulations were run with an equal release of salmon lice from all farms to test potential connectivity between the farms, but also with a realistic release of salmon lice calculated from observed infestation levels at the farm sites. All fish farms in Norway are obligated to monitor the amount of salmon lice on their fish. In 2010, the number of lice per fish was counted monthly on all farms in the area and reported to the authorities. The number of adult female lice per fish was made available to us, along with the number of fish and the measured seawater temperature at 3 m depth from all locations. We assumed that the lice counting was performed on the first day every month, and that the adult females were continuously producing eggs in 2 egg strings, with each string containing 150 eggs (Jackson & Minchin 1993). The estimated release of salmon lice nauplii from each location was based on the formula described in Stien et al. (2005), i.e. a daily production of 25 ind. female^{-1} d^{-1} at 10°C. The number of eggs was linearly interpolated to daily values.

Observed hydrography and currents

Vertical salinity and temperature profiles were obtained on 7 March and 21 July 2010 for the positions marked by blue stars in Fig. 1 and compared to the model results as validation. We used the CTD-sonde SAIV SD204 (www.saivas.no) with a 1 s sampling interval.

An observation buoy located in Nordfolda (Fig. 1, green cross) measured salinity and temperature at 4 and 10 m depths, and current at 3 m depth from 1 July to 30 August 2010. Salinity and temperature were measured using conductivity sensor number 3019 from Aanderaa Instruments (www.aadi.no) with a 2 s sampling interval. Current measurements were made with an Aanderaa Instruments Doppler current sensor DCS4100 (www.aadi.no), measuring flow close to the sensor with a 1 s sampling interval.

Salmon lice infestation on wild fish

Sea trout were caught at 2 locations (Sagfjorden and Nordfolda; Fig. 1) using floating gill nets following the methods described in Serra-Llinares et al. (2014). Lice stages (copepodid/chalimi, pre-adult, adult) were identified on a morphological basis according to Johnson & Albright (1991) and Schram (1993). Only newly attached salmon lice (copepodid and chalimus stages) were used in this paper. This data was previously published in a MSc thesis (Svedberg 2011).

Comparing simulated densities of salmon lice to observed salmon lice infections

In order to compare the simulated salmon lice density with observations on wild fish, we defined a potential infection area around the observation site as being within 7 km of all river mouths and 200 m from land, indicated by the blue areas in Fig. 1. We assumed that the infestations of lice occurred within 20 d prior fishing, which is the time period that is required for development to the pre-adult stage at 10–12°C (Stien et al. 2005). The swimming routes of the fish were not known, but we assumed that each fish would have been exposed to a fraction of the water volume within the infection area. We calculated the water volume that a fish would be exposed to during 20 d of swimming at a velocity of 0.15 m s^{-1} (Thorstad et al. 2004, 2014), assuming a lice detection range of 0.03 m (Heuch et al. 2007; a cylinder with 0.03 m radius as long as the travelled distance). Furthermore, we assumed that 10% of the lice within this volume were able to successfully attach to the fish (Tucker et al. 2000).

As the number of captured fish was low (14–24 fish), no statistical analysis was conducted.

RESULTS

Fjord hydrography

The currents and hydrography of the Folda are typical of fjords with a freshwater supply, with a mean surface current out of the fjord in the brackish

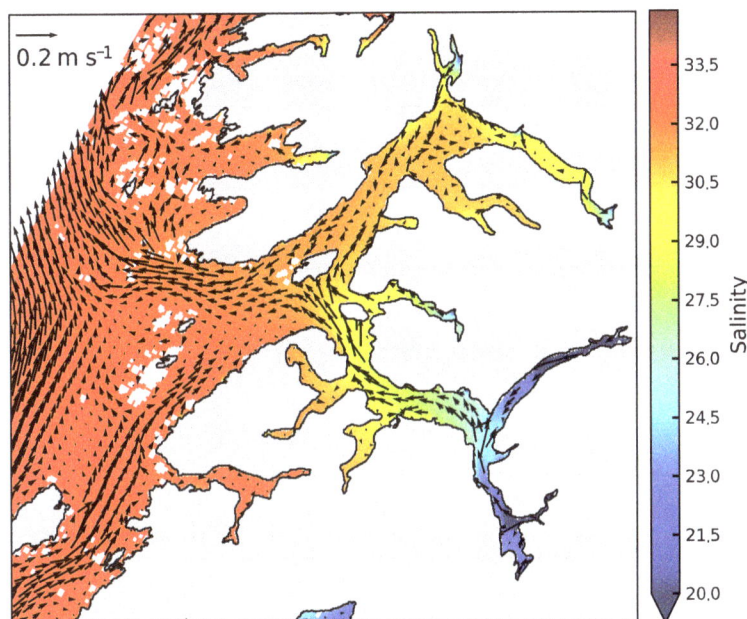

Fig. 2. Surface salinity and current vectors showing average values for the simulation period (1 April to 31 August 2010)

surface layer. In Sørfolda, where the river discharge is largest, the mean surface current was directed straight out of the fjord. Nordfolda, being a wider fjord with a lower freshwater supply, has a more complex surface circulation. On the southern side of Nordfolda the mean current is directed into the fjord, whilst a stronger outflow occurs on the western side (Fig. 2). The salmon lice in the fjord were influenced by the freshwater supply in terms of the general circulation, but also because the river discharge caused areas of low salinity water (<20) which the salmon lice avoided. These low salinity areas were close to the surface (seldom lower than 2 m depth), but fluctuated over time. Normally the low salinity surface was within 5 km of the inner fjord heads, but on 25 May low salinity water was observed over the whole surface of Sørfolda. There were northwesterly winds at the coast prior to this event, which created an aggregation of low salinity water within the fjord. The low salinity surface water extended deeper than normal in the inner part of the fjord days prior to 25 May, and when the coastal wind lessened in magnitude, the low salinity water that had collected at the fjord head flushed out of Sørfolda. The situation lasted less than 1 d.

The mean surface circulation pattern shown in Fig. 2 extended as deep as 5 m, and decreased in magnitude with depth. At 10–30 m depth, the mean circulation was in a counter-current direction into Sørfolda, with a magnitude of 0.05 m s^{-1}. A relatively

frequent exchange of water through propagation of long internal waves between Folda and the coast with currents of 0.1–0.3 m s^{-1} occurred irregularly (about every 14 d), and lasted 3–7 d. The water exchange between the fjord system and the coastal water correlates with the wind at the coast (Fig. 3). The currents outside the fjord were mainly directed parallel to the coast (~45° compass direction). The southwesterly wind transported water into the fjord while the northeasterly wind transported water out of the fjord. Currents directed out of the fjord affected a shallower part of the surface layer than the inflowing current (Fig. 3), indicating that water transport out of the fjord is generated by wind stress at the surface and is restricted to the low-salinity water close to the surface, whereas inflow occurs as a result of density-driven waves propagating into the fjord as described in Asplin et al. (1999). The inflowing water passed the fjord mouth and entered both Sørfolda and Nordfolda. In Nordfolda, the inflowing water followed the east side of the fjord, and subsequently generated an outflow at the western side of the fjord (Fig. 3). The tides have an amplitude of 1 to 2 m in the area, but since the fjord is both relatively wide and deep, the resulting tidal current speed was limited (0.05 m s^{-1}).

Fjord model validation

The observed salinity and temperature profiles showed increased stratification in the fjord from 7 May to 21 June. In the beginning of May, the water was well-mixed at 5°C and a salinity of 32–34. In June, the temperature had increased to 11–12°C at the surface, decreasing with depth to 9°C at 30 m. This stratification was a result of absorbed solar radiation, heat flux from the atmosphere and river runoff. The model recreated the stratification well in the top 10 m of the water column, but at 10–30 m depth the modelled water temperature was up to 1°C colder than the observed temperature. At the buoy position in Nordfolda, the simulated and observed temperatures followed each other in time, with the model on average underestimating the temperature at 4 m depth by 0.5°C, and overestimating the temperature by <0.1°C at 10 m depth. The observed salinity was always at the lower range of the simulated salinity, but the model never deviated from the observed salin-

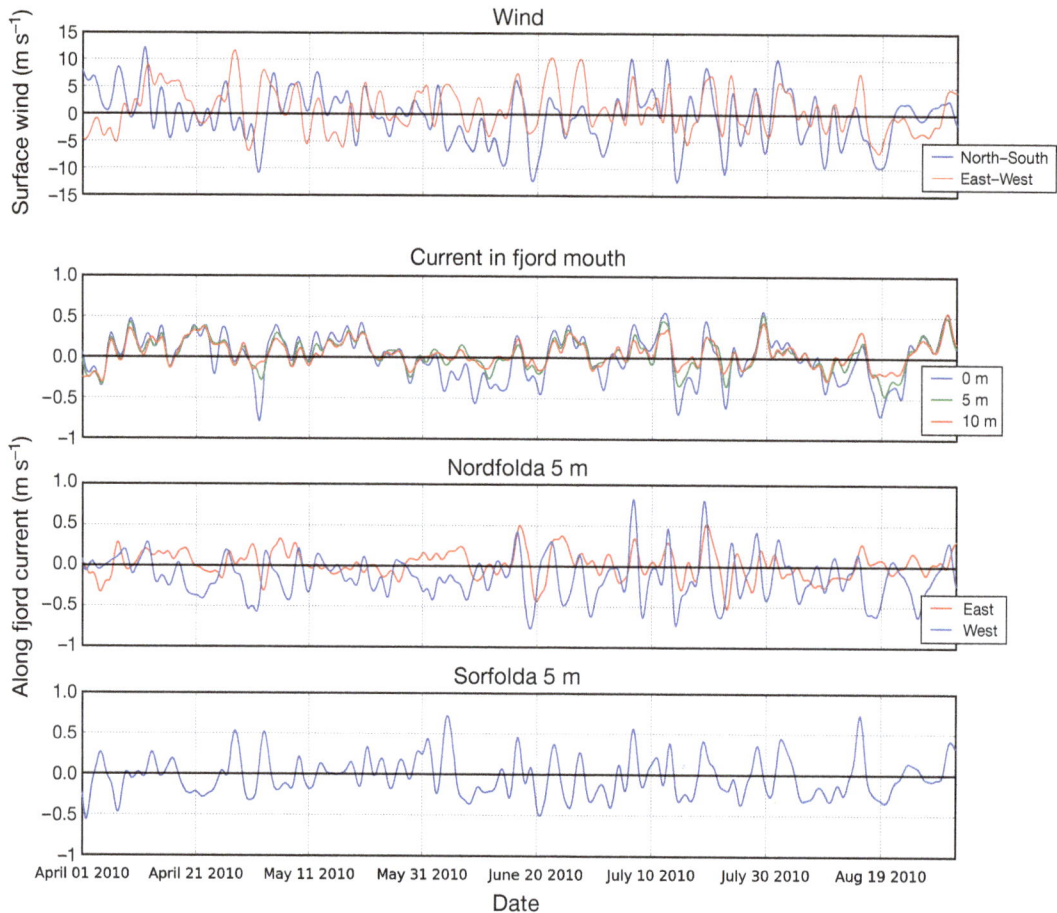

Fig. 3. Simulated surface wind at Vestfjorden outside the Folda fjord system and along fjord current in the middle of the mouth of Folda (0, 5 and 10 m depth, compass direction 60°), Nordfolda (east and west side, 5 m, compass direction 45°) and Sørfolda (5 m, compass direction 135°). Positive currents are directed into the fjord

ity by more than 1 unit. The hourly model results revealed relatively large variations in temperature and salinity over the course of a day. These fluctuations were the result of tidal movements of the water masses and were largest close to the surface. The temperature fluctuations were most obvious close to the fjord mouth and increased with the fjord temperature during spring. In June, tidal temperature fluctuations were typically around 2°C at the surface close to the fjord mouth due to the tidal movement of water masses with a horizontal temperature gradient between the fjord- and coastal water. At the inner parts of the fjord where the salinity gradient is strongest, the fluctuation in salinity within a day was often up to 4 salinity units. Even though the model underestimated the vertical temperature and salinity gradient it was able to reproduce the currents close to that of the observations (Fig. 4). The measured current was directed into the fjord 41 % of the time with a mean velocity of 0.09 m s^{-1}, and the mean velocity directed out of the fjord was 0.08 m s^{-1}. The fjord model pre-

dicted currents at the corresponding position to be directed into the fjord 51 % of the time with a magnitude of 0.11 m s^{-1}. The modelled mean velocity out of the fjord was 0.07 m s^{-1}. Over the full simulation period, the model slightly overestimated the currents into the fjord compared to the field observations.

Sensitivity of the salmon lice dispersion model

In the model, the vertical distribution of salmon lice (and hence horizontal dispersion since the horizontal current varied with depth) was affected by the parameterization of lice behaviour (Fig. 5). The largest distribution difference occurred by changing the lice's swimming velocity. When the simulated lice were given a vertical swimming capacity of 10^{-4} m s^{-1} (i.e. slow swimmers), they were not able to aggregate as close to the surface as they were by increasing the swimming velocity. A difference in swimming velocity between 10^{-3} m s^{-1} (dynamic mixing) and 5 ×

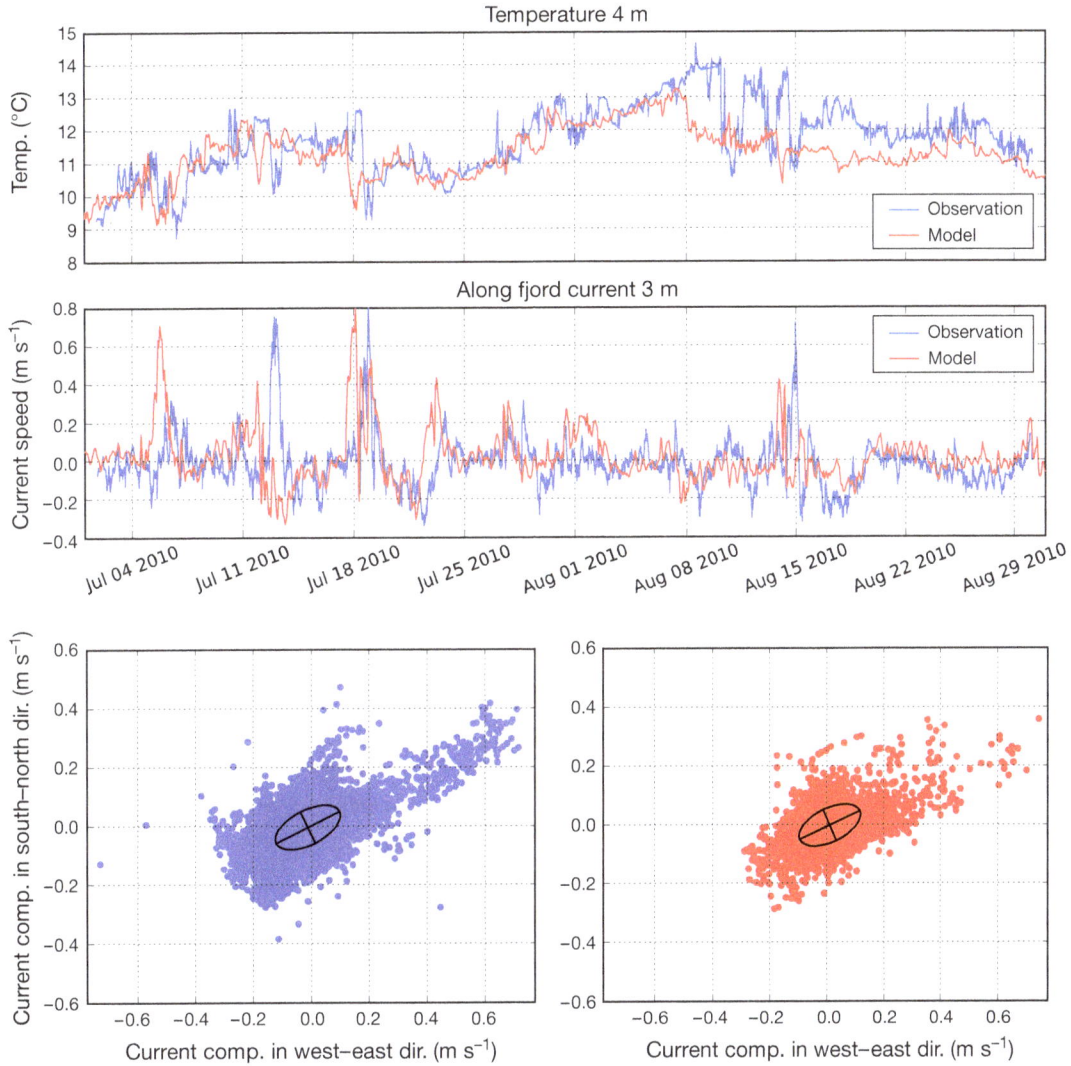

Fig. 4. Observed and simulated temperatures (top panel) and current in Nordfolda (green cross in Fig. 1). Along fjord current with positive current heading into the fjord (middle panel) and scatterplots of the velocity, with ellipse marking the standard deviation along and across the main current direction (lower panels). Observed data in blue, simulated data at the corresponding location in red

10^{-3} m s^{-1} (fast swimmers) altered the vertical distribution only to a small degree. As the simulated lice were reflected at the surface, a further increase in swimming velocity did not increase the number of salmon lice at the surface. By decreasing the light sensitivity of the lice to 0.1 µmol photons s^{-1} m^{-2} (i.e. less light sensitive), the vertical distribution was altered only to a minor extent. The irradiance level at the position of the salmon lice exceeded 0.1 µmol photons s^{-1} m^{-2} most of the day, triggering them to swim upwards. Simulations run using light levels similar to those in southern Norway at 60.0° N (60° N-light) with shorter days but similar light intensity at mid-day did not give results that differed from the simulation with light levels calculated from the actual position (i.e. dynamic mixing).

The simulated turbulence in the fjord was greater in Nordfolda than in Sørfolda. However, overall the water was less turbulent inside the fjord system than in the earlier assumptions in which the vertical mixing was set by random walk and limited by a constant diffusion coefficient of 10^{-3} m^2 s^{-1} (Asplin et al. 2014, Johnsen et al. 2014). On the other hand, this constant value was often exceeded at the coast outside the fjord. Depending on the driving forces, the turbulence level varied from close to zero to 0.1 m^2 s^{-1}. The vertical lice distribution fluctuated more in time by modifying the model from the assumption of constant turbulent mixing in the vertical (random walk), in which the salmon lice swam up during the day and dispersed in the vertical during the night, to using the instant vertical mixing parameter as simulated by

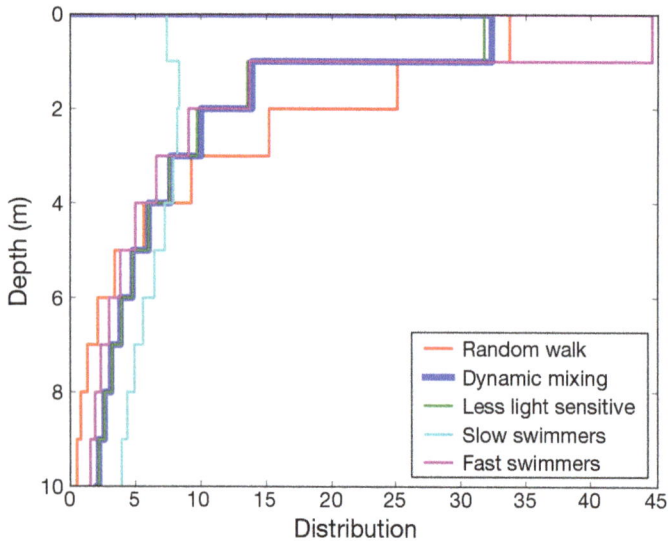

Fig. 5. Aggregated vertical distribution of salmon lice *Lepeoph-theirus salmonis* copepodids during simulations from all farms with parameters as defined in Table 1

the current model (dynamic mixing). When the salmon lice experienced greater displacement from vertical mixing than they were able to overcome with their swimming capacity, the they sank down to avoid the high turbulence. Thus the mean distribution of salmon lice was deeper for the dynamic mixing simulation compared to the random walk simulation (Fig. 6).

The total horizontal area affected by the dispersed salmon lice from all fish farms in the fjord ranged between 3780 and 4190 km^2, depending on the simulated vertical behaviour of the lice (mortality included). In total, 95% of infectious lice were found within 25% of the total influenced area. The largest influence area occurred with the random walk simulation, while the slow swimmers simulation displayed the smallest area.

The retention of salmon lice within the fjord system was defined as the fraction of infectious salmon lice residing inside the fjord mouth (indicated by the line in Fig. 1). The time-averaged retention between the simulations was 82–86%. The random walk simulation had the greatest retention while the slow swimmers simulation had the smallest. Retention fluctuated over time with the exchange of water into and out of the fjord. Salmon lice retention varied over time between 60 to >95% for the dynamic mixing simulation. As the simulated salmon lice sank when turbulent mixing exceeded their vertical swimming capacity, the lice were distributed deeper during periods with increased mixing. Nordfolda was more exposed to vertical mixing than Sørfolda, and hence the vertical distribution was deeper there. Turbulence was strongest during inflow episodes to the fjord system, and therefore the salmon lice were distributed deeper during inflow than outflow, and were unable to increase their retention within the fjord by sinking down at times with large vertical mixing.

The time-averaged aggregation of infectious copepodids along land was 28–29% for all of the simulations, but large fluctuations in time also occurred here. The dynamic mixing simulation results showed more salmon lice residing mid-fjord in Nordfolda than the other simulation results. In the following

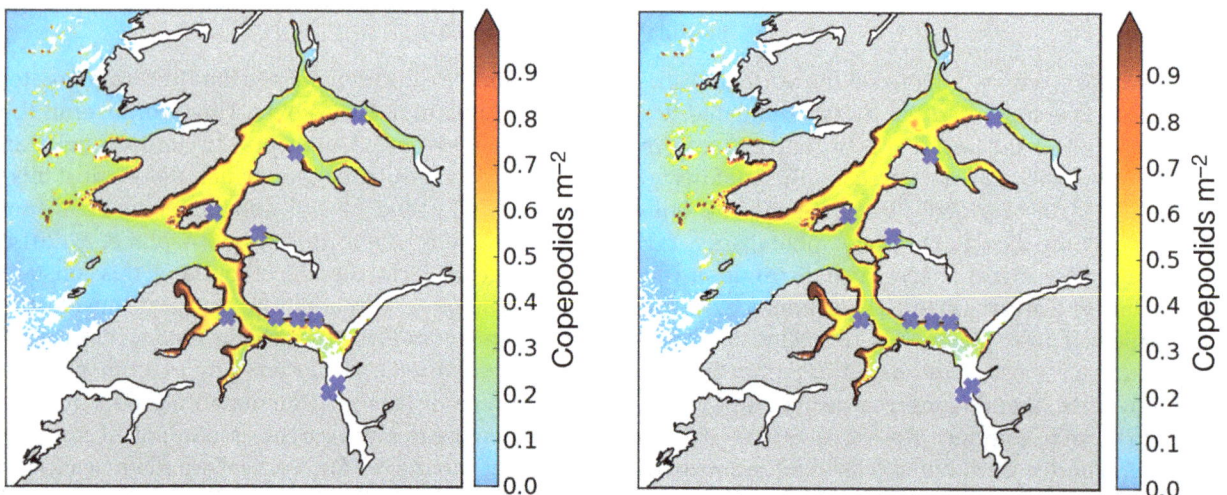

Fig. 6. Normalized horizontal distribution of salmon lice *Lepeophtheirus salmonis* copepodids from simulation with no mortality (right) and mortality of 0.17% d^{-1} (left) during nauplii and copepodid stages. All sources release the same number of particles

section, the presented model results are from the dynamic mixing parameterization.

By including a constant mortality of 0.17 ind. d^{-1}, 35% of the salmon lice reached the copepodid stage. Survival to copepodid stage is dependent on temperature, but was the same as reported for Hardangerfjord in May 2009 (Johnsen et al. 2014). Mortality during the planktonic stages (both nauplii and copepodid) decreased the density field of infectious copepodids to 14% of the simulation with no mortality (1 April to 31 August 2010). However, by normalizing the density fields it became apparent that the distributions were qualitatively almost identical (Fig. 6).

Salmon lice distribution and validation

The salmon lice simulation was conducted using the dynamic mixing parameterization, including mortality, for the time period 1 April to 31 August 2010, releasing realistic numbers of lice from all active farms in the region. During the summer, the model predicted lice to be present in the whole fjord system. The distribution of lice fluctuated widely both in time and space, but as a general feature the largest densities of infectious lice occurred inside the fjord, with an aggregation along land (Fig. 7). The land aggregation was most pronounced in Sørfolda. The number of salmon lice copepdids residing within the same area as the captured sea trout was also highly variable in time (Fig. 8). The

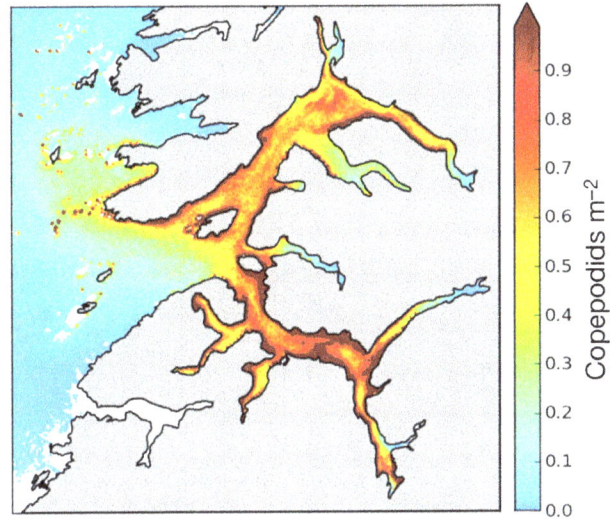

Fig. 7. Mean density of salmon lice *Lepeophtheirus salmonis* released from fish farms from 1 April to 31 August 2010. The released number of salmon lice was based on the reported number of lice on all farms in the fjord system

periods assumed to correspond to the infestation of the wild fish are marked with grey in Fig. 8. The observed number of lice on wild fish was greatest in Norfolda during Week 29. There was a good agreement between the modelled salmon lice infestation and the number of lice observed on the wild fish, with the exception of Week 26, where the model overestimated the number of lice in Nordfolda (Fig. 9).

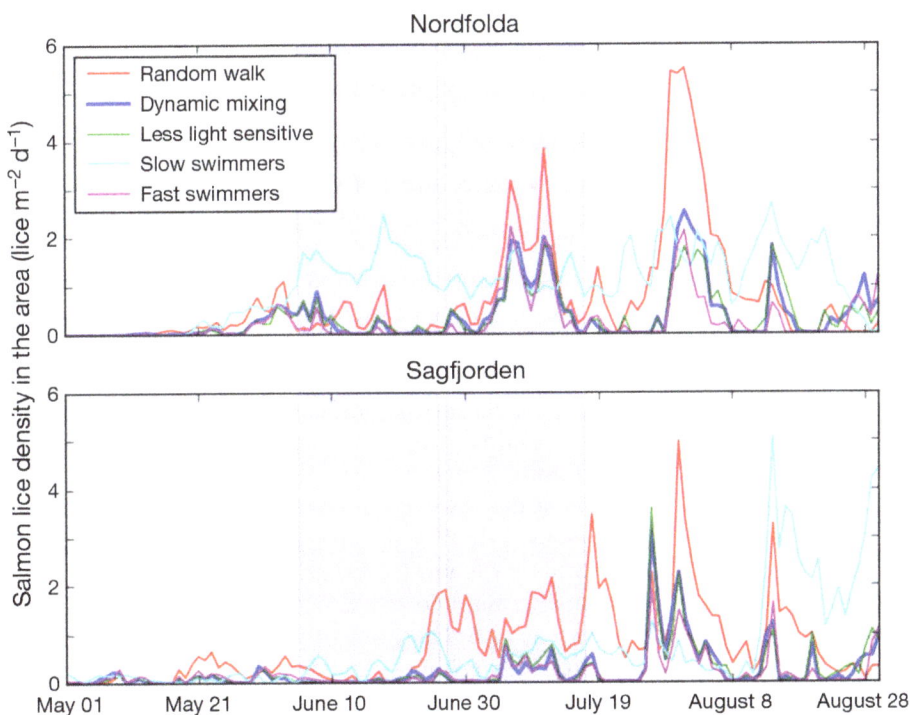

Fig. 8. Simulated density of salmon lice *Lepeophtheirus salmonis* copepodids residing within the blue area in Fig. 1. Shaded area: assumed infection time for the captured wild fish

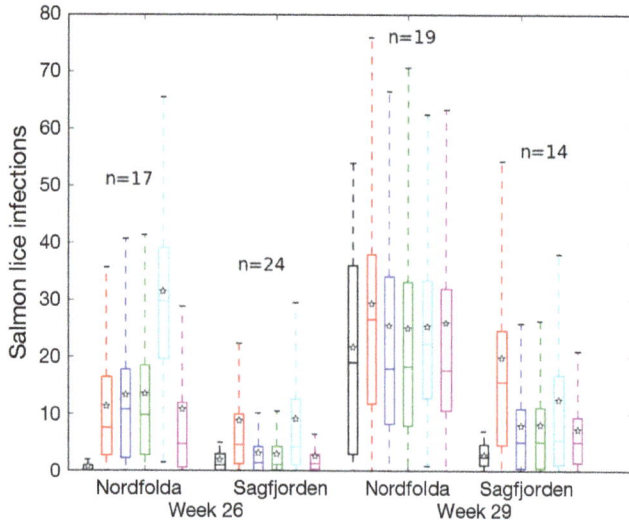

Fig. 9. Observed (black box) and simulated (coloured boxes, colour codes as in Fig. 8) salmon lice *Lepeophtheirus salmonis* infections on wild sea trout *Salmo trutta* (n = no. of fish). Note that the boxes show the 25th and 75th percentiles for the variation in infestation for the observations, and variation between the horizontal grids for the model; whiskers show extremes

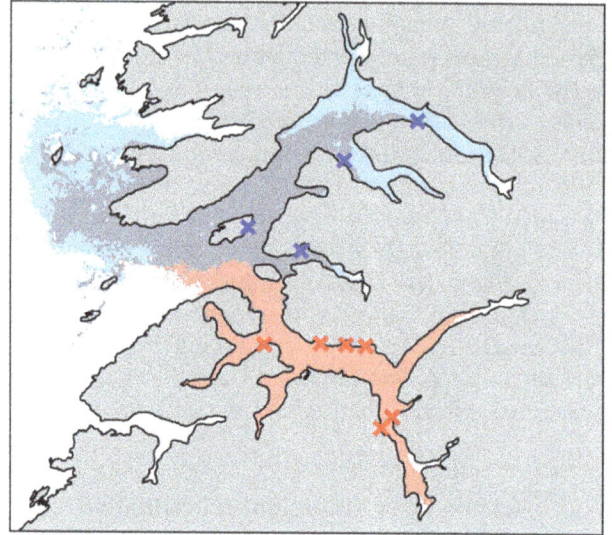

Fig. 10. Area influenced by 85 % of the largest densities of infectious salmon lice *Lepeophtheirus salmonis* released by farms (crosses) in Nordfolda (blue) and Sørfolda (red, grey if overlapping with blue). Area calculation is based on abundance date re-run by the dynamic mixing simulation, with mortality included and assuming all farms released the same amount of salmon lice

Potential dispersion of lice

The simulated abundance of salmon lice in Folda was greatly affected by the production cycles on the different farms, in which those farms containing many large fish released more lice into the fjords than the ones having smaller (or less) fish. In order to evaluate the potential dispersion between farms on a general basis, simulations were re-run with the condition that all farms released the same amount of salmon lice. In the following analysis, the dynamic mixing simulation was used and mortality was included. By simulating the release of the same amount of salmon lice it became clear that the farms in Nordfolda were more exposed to lice originating from farms in Sørfolda than the other way around. The reason for this difference was that the currents transported more water from Sørfolda into Nordfolda than vice versa (Fig. 10).

By the time the salmon lice reached the infectious copepodid stage, they had been transported away from their release positions. In total, 20 and 24 % of the lice released from Farms C and E (see Fig. 1) were found <10 km from their initial release position. For the remaining farms, the number never exceeded 6 %. In general, salmon lice released close to the head of the fjord were transported furthest away, whereas those released closer to the mouth created an infection pressure that covered a larger area (Fig. 11). This is a result of the fjords being narrower

closer to the head. The area in which salmon lice can disperse is more restricted close to the fjord head than in the more open parts of the fjord. Currents generated by tides, wind and density differences are directed along the fjord in the inner region, efficiently transporting the lice away from their source.

Connectivity between the farms

In order to evaluate connectivity between the salmon farms in terms of export and import of salmon lice, all copepodids entering a 3 × 3 model grid (600 × 600 m) at each farm were counted. The fraction of all released lice that reached a farm during the copepodid stage was assumed to represent a potential infection. The number of potential infections was calculated between all farms and is presented in a connectivity matrix (Fig. 12). The sum of the columns in the matrix (given in Table 2) can be read as the lice exposure the farm experienced, and likewise the sum of the rows of the matrix indicates the potential infectivity from that farm (Table 2). Based on this, Farm G was most exposed to salmon lice from the other farms, and the lice from Farm H were most likely to find a host fish at a farm location within the fjord. The colour of the boxes along the diagonal marks the farms' potential internal infection, which

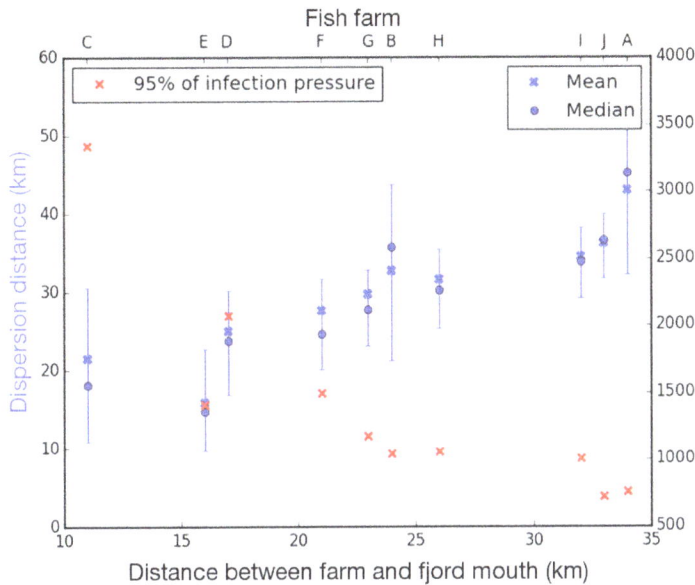

Fig. 11. Blue markers show mean and median dispersion distances of infectious salmon lice *Lepeophtheirus salmonis*. Blue error bars: 25th and 75th percentiles. Red crosses: the area in which 95% of the infection pressure occurs

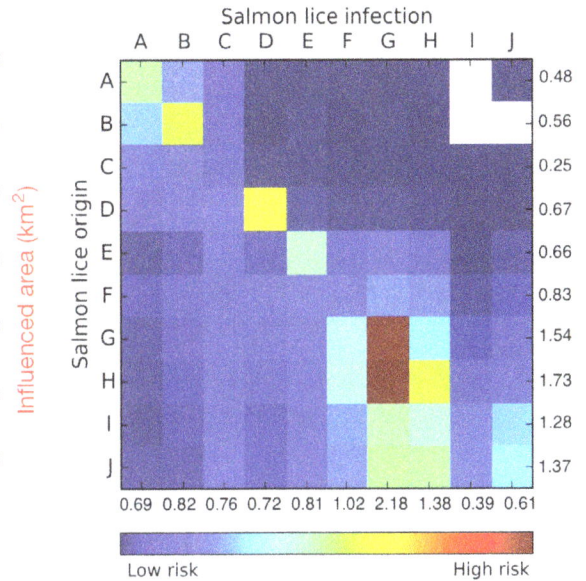

Fig. 12. Connectivity matrix of simulated salmon lice *Lepeophtheirus salmonis* infestation between fish farms in the Folda fjord system between 1 April and 31 August 2010

was 7 to 57% of the total potential infection (on average 32%). Farm D was the only site that exhibited a larger internal than external infection (Table 2). The farms in Sørfolda (Farms E–J) displayed stronger connectivity than those in Nordfolda.

DISCUSSION

Fjord circulation

The fjord model was able to recreate the circulation in the fjords realistically, based on comparisons with field observations (Fig. 4). This has also been shown earlier for the Hardanger fjord (Johnsen et al. 2014). Currents are formed by contributions from wind, tides, freshwater runoff and internal waves (horizontal pressure) of remote origin. For transportation of salmon lice over longer distances, the relatively long-lasting current components are most efficient and the currents associated with internal waves are important (Asplin et al. 1999, 2014). Internal waves propagate with land on the right side on the way into the fjords, and with the upper layer current both into and out of the fjord depending on whether the internal wave is at a depressed or elevated phase (Asplin et al. 1999). In Sørfolda, the waves cover the full width of the fjord, while in the wider Nordfolda they generate

stronger currents along land, thus acting as a retention eddy. As a result, salmon lice in Sørfolda aggregated along land, while in the wider Nordfolda high densities of lice were also found in the centre of the fjord. The inflow episodes were externally generated by northerly winds outside the fjord or by density differences between the fjord and the coastal water masses, demonstrating the importance of realistic boundary conditions (both surface wind and the open boundary) when simulating fjord circulations. We did not find any systematic differences in circulation dynamics in this northern Norwegian fjord compared to fjords in southern Norway.

Sensitivity analysis of salmon lice dispersion

By simulating the dispersion of salmon lice we can estimate lice densities over larger areas. If the connection between lice abundance in the water masses and lice infestations on fish is known, we can also predict the infestation pressure on wild fish. Although the model is already useful, the entire model system (including the behaviour of the planktonic lice) can still be improved. To evaluate which of the parameters had the strongest influence on the results, a sensitivity analysis evaluating different lice behaviours was conducted. Salmon lice are known to

Table 2. Simulated potential infection, potential export and ratio internal:external salmon lice *Lepeophtheirus salmonis* infection for the Folda fjord system, 1 April to 31 August 2010. A–J: fish farms; see Fig. 1 for locations

	A	B	C	D	E	F	G	H	I	J
Potential infections (10^{-3} of all simulated lice)	0.69	0.82	0.76	0.72	0.81	1.02	2.18	1.38	0.39	0.61
Potential export (10^{-3} of all simulated lice)	0.48	0.56	0.25	0.67	0.66	0.83	1.54	1.73	1.28	1.37
Internal infection (%)	42	42	7	57	35	12	30	26	35	35

be positively phototactic and can sense light at a very low level (Bron et al. 1993, Heuch et al. 1995, Gravil 1996, Flamarique et al. 2000). To the human eye, the light difference between the northern and southern Norwegian summer is obvious. However, for salmon lice that sense light at much lower levels, the water surface appears as bright during the night as during the day in both northern and southern Norway. The critical level at which the lice were triggered to swim towards the surface did not alter the model results. A further decrease in the light sensing level during the simulation resulted in a deeper distribution of the lice, but the highest densities of lice were always found at the surface. Different behaviours have been observed in laboratory experiments, in which salmon lice copepodids were reported to be more active than the nauplii stages (Gravil 1996), especially the late nauplii stage prior to moulting (Johannessen 1977, Pike et al. 1993). Variations in vertical distributions have also been reported in the literature. Penston et al. (2008) found more salmon lice nauplii at 5 m depth than at the surface, and á Norði et al. (2015) found only copepodids in a near-surface sampling. The same result was found by Gravil (1996) during the first nauplii stage, however for the second nauplii stage the situation was reversed. In our model, the only result showing fewer lice at the surface than deeper down was the simulation parameterized with slow swimming lice. This indicates that reduced swimming during the nauplii stage, not being less phototactic, might cause the deeper distribution of salmon lice nauplii found in nature. Salmon lice are known to be negatively geotactic (Gravil 1996), and therefore dispersion to unrealistic depths in simulations can be avoided by including such a response in the model.

With an exception of the slow swimming simulation, the vertical distribution of salmon lice copepodids in the model appears similar to the distribution observed in nature (based on pumping and net trawling; Gravil 1996, Hevrøy et al. 2003, Penston et al. 2004) and to the vertical infection gradient found by Hevrøy et al. (2003). Since the simulated level of tur-

bulence generally decreased with depth and was lower within the fjord compared to the constant value used in earlier model simulations, the dynamic mixing simulation provides the most realistic results. This simulation gave the most dynamic vertical distribution of lice, indicating that a formulation of turbulence and vertical mixing should be included as precisely as possible. The simplification of turbulence in the vertical random walk did not alter the retention of salmon lice within the fjord or the aggregation along land dramatically for the whole study area, but the dynamic mixing simulation showed greater mid-fjord densities in Nordfolda.

In the dynamic mixing simulation, the lice were assumed to sink down and out of the highly turbulent water. It is known that other zooplankton that are approximately the same size as salmon lice sink down during windy conditions (Ellertsen et al. 1984, Lagadeuc et al. 1997, Incze et al. 2001). Zooplankton benefit from avoiding very turbulent water since the risk of encountering a predator increases with turbulence (Visser et al. 2008). However, salmon lice differ from other zooplankton as the increased encounter rate can increase the likelihood of meeting a host. Hence a level of turbulence that is unfavourable for other species might be favourable for salmon lice. It has been suggested that the salmon lice might seek turbulent water associated with high current velocities to be transported with the tidal currents (Costello 2006). In our study area, the largest turbulent episodes were associated with internal waves propagating into the fjords, but if the lice actively seek out these currents they would be transported into the fjord, where the sea trout mainly reside. Conversely, high levels of turbulence decrease the contact time between lice and fish, potentially making it difficult for the lice to detect and attach to the host. Salmon lice infestation success has been found to relate to the current velocities in a dome-shaped curve (Samsing et al. 2015). About 30 % of the simulated salmon lice were found close to shore (i.e. the area in which sea trout reside and feed) during their infectious stage. Further research into the vertical behaviour of salmon lice and their infectivity under turbulent conditions or high currents would be of interest.

The mortality of the planktonic lice is naturally high, but exact mortality rates are unknown and hence impossible to simulate. By including constant mortality, the number of infectious copepodids in the fjord was reduced to 14% compared to the simulation with no mortality. However the geographical distribution was qualitatively unchanged. If dispersion were calculated using a unidirectional current field, the age of the salmon lice would increase with distance from the release position. As the mortality rate was held constant in time and space, the number of salmon lice would be reduced by age, and hence the density of lice would be lower further away from the release point. This is not the case due to the fluctuating nature of fjord circulation, where water transport and exchange occurs in pulses, dispersing into the fjords. Mortality is likely not constant in time and space due to the changing abundance of predators. This was not investigated here, but these effects might be important, especially on longer scales both in time (i.e. seasonal fluctuations) and distance (i.e. along the Norwegian coastline).

Salmon lice dispersion in Folda

Since the vertical distribution of salmon lice and simulated current are close to the ones found in earlier studies and similar to observations, it is reasonable to believe that the modelled distribution of salmon lice accurately depicts their actual distribution in the fjord. This is also supported by the fit between simulated and observed number of lice infestations on wild fish, despite the uncertainty of where the captured sea trout became infested (Fig. 9).

The assumed residence areas of the fish (Fig. 1) were based on the local rivers and earlier field studies on post-smolts tagged with acoustic transmitters (Finstad et al. 2005, Middlemas et al. 2009). Within fjords, post-smolts appear to prefer to remain close to shore rather than exploiting open waters within the mid-fjord areas (Thorstad et al. 2004, 2007). The mean distance to shore for post-smolts immediately following their entry into the marine environment was reported to be 125 m (Thorstad et al. 2004). In a study in northern Norway, results indicated that sea trout remained primarily within the inner parts of the fjord during the summer months (Rikardsen et al. 2007, Jensen et al. 2014).

The water volume to which a fish was exposed (which was used to calculate infestations from the modelled density field of salmon lice in the infection area) was calculated based on variables (such as swimming speed of the fish, infection time, infestation radii of the lice and infestation success) selected from the literature, but will probably vary with environmental conditions. For instance, the swimming speed of a salmonid fish depends on temperature as well as the size of the fish (Brett 1965, Brett & Glass 1973, Webb et al. 1984). Similarly, salmon lice infestation success reportedly depends on temperature, salinity and swimming speed of the fish (Tucker et al. 2000, Bricknell et al. 2006, Samsing et al. 2015). The salmon lice counted at the farms were only reported once a month. Hatching time for the eggs in the model was set as a function of temperature, but the number of eggs produced per egg string was held constant. We believe that more frequent counting of the lice on the fish farms and possibly a variable number of eggs produced through the season would increase the quality of the model. Beginning in 2013, lice counting at fish farms on a weekly basis became mandatory. Salmon lice are known to produce fewer eggs on caged fish than on wild, and even fewer (~150 eggs per egg string) if the caged fish are medically treated (Jackson & Minchin 1992). The lice also produce more eggs per egg string during winter than summer (Tully 1989, Gravil 1996, Heuch et al. 2000); however, 150 eggs per string is considered to be a reasonable estimate.

Despite limitations in terms of both data input and observations of salmon lice infestations on fish, the model appeared to capture the observed infestation pressure (Fig. 9). The dynamic mixing, less light sensitive and fast swimmers simulations fit the observations better than the random walk and slow swimmers simulations. Theoretically, they should be more realistic as their dispersion includes dynamic turbulence. However, it must be kept in mind that comprehensive model validation was not possible due to the limited wild fish dataset, as only 14–24 wild fish were collected. Since the model is generic (i.e. without any tuning of the currents or hydrography to measurements), the model should be equally suitable for the entire Norwegian coast on a similar scale.

Open water observations indicate that salmon lice infestations occur in pulses (Costelloe et al. 1995, Penston et al. 2004). Our model supports these findings, and explains these pulses as a result of moving water masses. The large concentrations along the shore indicated by the model results are supported by the findings of McKibben & Hay (2004) and Penston et al. (2008), who reported the largest amounts of infectious copepodids along shorelines. As this is the area where wild salmon migrate to the sea, and where sea trout reside, it is the area where salmon lice naturally have the greatest chance of success (Pemberton 1976, Thorstad et al. 2004, 2007).

The wild fish in the Folda fjord system are influenced by the salmon lice infestations that occur at the farms in the area. On average, migrating salmon smolts were estimated to have a population reduction of 0, 28 and 31% in 2010, 2011 and 2012 caused by infestations of salmon lice (Taranger et al. 2014b). The corresponding numbers for sea trout were on average 28, 48 and 49% (with large geographical differences). In order to reduce the lice-induced mortality of the wild fish (especially for sea trout and charr), coordinated management of the fish farms in the area should be considered.

CONCLUSIONS

Numerical models can provide valuable and quantitative information of both spatial and temporal salmon lice dispersion, connectivity between fish farms and infestation levels of wild fish. Hence, with this model we were able to define the dispersal of salmon lice and the regional influence from specific farm locations. The models also make it possible to test different area management scenarios (i.e. production zones, disease management areas or fire breaks) in order to optimize sizes of management areas and the distance between these. In general, the lice released from farms in the Folda fjord system are efficiently transported away from their source and disperse over a large area. The median transport distance of the lice is on the same scale as that found in the Hardangerfjord, Norway (Asplin et al. 2011) and in the Broughton Arcipelago of British Columbia, Canada (Brooks 2005) but about 5 times greater than that found in Loch Torridon and Shieldag, and Loch Linnhe, Scotland (Gillibrandt & Willis 2007, Amundrud & Murray 2009, Salama et al. 2015). In Folda, the dispersion distance increased with distance from the fjord mouth, but the influence area decreased because the fjord area is smallest here. This illustrates the challenges involved in presenting dispersion results for management purposes. The smallest influence area might seem desirable, but it also represents the most efficient transport of lice away from the source.

Even if the fjord experiences pulses of inflowing water, the general circulation pattern transports more lice from Sørfolda to Nordfolda than vice versa. Thus

delousing treatments would be most effective if they were coordinated and began in the southern part of the fjord system.

The dispersion of salmon lice over large areas leads to a rapid dilution in the water masses. Less than 1% of the simulated salmon lice came close to a salmon farm location during their infectious stage. In comparison, the simulated connectivity between farms was 10 times larger in Loch Fyne, Scotland, although the relationship of internal/external exposure was about the same (Adams et al. 2015). Differences in connectivity between farms were found, with Farm G being 3 times more exposed to lice than Farm J, while Farm H exported almost 7 times more lice to the other farms than Farm C. Despite less than 6% of the infectious salmon lice eventually residing <10 km from their release positions, 6 out of 10 farms had greater internal exposure than exposure from any other single farm. However, external exposure must be counted as the sum of exposure from all other farms; when calculating the ratio between internal exposure and the sum of exposure from all other farms only Farm D had larger internal than external exposure. Adams et al. (2015) recognised that decreasing the infestations at the farm sites with the largest influx of copepodids most efficiently reduced the total amount of lice in an area. By removing/relocating or minimizing the lice burden at the fish farms marked by red in Fig. 12, the fish farms could reduce the need for delousing treatments and the infection pressure on wild fish in the fjord. In general, the simulations indicate that mid-fjord locations are less exposed to salmon lice than those closer to the shore, suggesting that these locations would be more suitable for successful aquaculture activities.

Acknowledgements. The field data used in this research was collected by master students at the University of Tromsø in collaboration with the Institute of Marine Research. Special thanks to Øivind Østensen and Pål Arne Bjørn for assisting with this work.

LITERATURE CITED

á Norði G, Simonsen K, Danielsen E, Eliasen K and others (2015) Abundance and distribution of planktonic *Lepeophtheirus salmonis* and *Caligus elongatus* in a fish farming region in the Faroe Islands. Aquacult Environ Interact 7:15–27

ä Adams TP, Black KD, MacIntyre C, MacIntyre I, Dean R (2012) Connectivity modelling and network analysis of sea lice infection in Loch Fyne, west coast of Scotland. Aquacult Environ Interact 3:51–63

ä Adams T, Proud R, Black K (2015) Connected networks of

sea lice populations: dynamics and implications for control. Aquacult Environ Interact 6:273–284

Albretsen J, Sperrevik AK, Staalstrøm A, Sandvik AD, Vikebø F, Asplin L (2011) NorKyst-800 report no. 1: user manual and technical descriptions. IMR Res Rep Ser Fisken og havet 2/2011. Institute of Marine Research, Bergen

ä Amundrud TL, Murray AG (2009) Modelling sea lice dispersion under varying environmental forcing in a Scottish sea loch. J Fish Dis 32:27–44

ä Asplin L, Salvanes AGV, Kristoffersen JB (1999) Nonlocal wind driven fjord–coast advection and its potential effect on plankton and fish recruitment. Fish Oceanogr 8: 255–263

Asplin L, Boxaspen K, Sandvik AD (2004) Modelled distribution of salmon lice in a Norwegian fjord. ICES CM 2004/P:11. International Council for the Exploration of the Sea, Copenhagen

Asplin L, Boxaspen KK, Sandvik AD (2011) Modeling the distribution and abundance of planktonic larval stages of Lepeophtheirus salmonis in Norway. In: Jones S, Beamish R (eds) Salmon lice: an integrated approach to understanding parasite abundance and distribution. Wiley-Blackwell, Oxford, p 31–50

ä Asplin L, Johnsen IA, Sandvik AD, Albretsen J, Sundfjord V, Aure J, Boxaspen KK (2014) Dispersion of salmon lice in the Hardangerfjord. Mar Biol Res 10:216–225

ä Bjorn PA, Finstad B, Kristoffersen R (2001) Salmon lice infection of wild sea trout and Arctic char in marine and freshwaters: the effects of salmon farms. Aquacult Res 32: 947–962

ä Brett JR (1965) The relation of size to rate of oxygen consumption and sustained swimming speed of sockeye salmon (Oncorhynchus nerka). J Fish Res Board Can 22: 1491–1501

ä Brett JR, Glass NR (1973) Metabolic rates and critical swimming speeds of sockeye salmon (Oncorhynchus nerka) in relation to size and temperature. J Fish Res Board Can 30:379–387

ä Bricknell IR, Dalesman SJ, O'Shea B, Pert CC, Mordue Luntz AJ (2006) Effect of environmental salinity on sea lice Lepeophtheirus salmonis settlement success. Dis Aquat Org 71:201–212

Bron JE, Sommerville C, Rae GH (1993) Aspects of the behaviour of copepodid larvae of the salmon louse Lepeophtheirus salmonis (Krøyer, 1837). In: Boxshall GA, Defaye D (eds) Pathogens of wild and farmed fish: sea lice. Ellis Horwood, Chichester, p 125–142

ä Brooks KM (2005) The effects of water temperature, salinity, and currents on the survival and distribution of the infective copepodid stage of sea lice (Lepeophtheirus salmonis) originating on Atlantic salmon farms in the Broughton Archipelago of British Columbia, Canada. Rev Fish Sci 13:177–204

ä Burridge L, Weis JS, Cabello F, Pizarro J, Bostick K (2010) Chemical use in salmon aquaculture: a review of current practices and possible environmental effects. Aquaculture 306:7–23

ä Costello MJ (2006) Ecology of sea lice parasitic on farmed and wild fish. Trends Parasitol 22:475–483

ä Costelloe J, Costelloe M, Roche N (1995) Variation in sea lice infestation on Atlantic salmon smolts in Killary Harbour, west coast of Ireland. Aquacult Int 3:379–393

ä Dawson LHJ, Pike AW, Houlihan DF, McVicar AH (1998) Effects of salmon lice Lepeophtheirus salmonis on sea

trout Salmo trutta at different times after seawater transfer. Dis Aquat Org 33:179–186

Directorate of Fisheries (2013) Forskrift om bekjempelse av lakselus i akvakulturanlegg. Industry and Fisheries Ministry, Bergen

Ellertsen B, Fossum P, Solemdal P, Sundby S, Tilseth S (1984) A case study on the distribution of cod larvae and availability of prey organisms in relation to physical processes in Lofoten. Fløedevigen Rapportserie 1:453–477

ä Finstad B, Bjorn PA, Grimnes A, Hvidsten NA (2000) Laboratory and field investigations of salmon lice [Lepeophtheirus salmonis (Kroyer)] infestation on Atlantic salmon (Salmo salar L.) post-smolts. Aquacult Res 31:795–803

ä Finstad B, Økland F, Thorstad EB, Bjørn PA, McKinley RS (2005) Migration of hatchery reared Atlantic salmon and wild anadromous brown trout post smolts in a Norwegian fjord system. J Fish Biol 66:86–96

Flamarique IN, Browman HI, Belanger M, Boxaspen K (2000) Ontogenetic changes in visual sensitivity of the parasitic salmon louse Lepeophtheirus salmonis. J Exp Biol 203:1649–1657

ä Gillibrandt PA, Willis KJ (2007) Dispersal of sea louse larvae from salmon farms: modelling the influence of environmental conditions and larval behaviour. Aquat Biol 1: 63–75

Gravil HR (1996) Studies on the biology and ecology of the free swimming larval stages of Lepeophtheirus salmonis (Kroyer, 1838) and Caligus elongatus Nordmann, 1832 (Copepoda: Caligidae). PhD thesis, University of Stirling

▶ Heuch PA, Karlsen E (1997) Detection of infrasonic water oscillations by copepodids of Lepeophtheirus salmonis (Copepoda Caligida). J Plankton Res 19:735–747

ä Heuch PA, Mo TA (2001) A model of salmon louse production in Norway: effects of increasing salmon production and public management measures. Dis Aquat Org 45: 145–152

ä Heuch PA, Parsons A, Boxaspen K (1995) Diel vertical migration: a possible host-finding mechanism in salmon louse (Lepeophtheirus salmonis) copepodids? Can J Fish Aquat Sci 52:681–689

ä Heuch PA, Nordhagen JR, Schram TA (2000) Egg production in the salmon louse [Lepeophtheirus salmonis (Krøyer)] in relation to origin and water temperature. Aquacult Res 31:805–814

▶ Heuch PA, Doall MH, Yen J (2007) Water flow around a fish mimic attracts a parasitic and deters a planktonic copepod. J Plankton Res 29(Suppl 1):i3–i16

Heuch PA, Bjorn PA, Finstad B, Asplin L, Holst JC (2009) Salmon lice infection of farmed and wild salmonids in Norway: an overview. Integr Comp Biol 49:E74

ä Heuch PA, Gettinby G, Revie CW (2011) Counting sea lice on Atlantic salmon farms - empirical and theoretical observations. Aquaculture 320:149–153

ä Hevrøy EM, Boxaspen K, Oppedal F, Taranger GL, Holm JC (2003) The effect of artificial light treatment and depth on the infestation of the sea louse Lepeophtheirus salmonis on Atlantic salmon (Salmo salar L.) culture. Aquaculture 220:1–14

ä Incze LS, Hebert D, Wolff N, Oakey N, Dye D (2001) Changes in copepod distributions associated with increased turbulence from wind stress. Mar Ecol Prog Ser 213:229–240

ä Jackson D, Minchin D (1992) Aspects of the reproductive output of two caligid copepod species parasitic on cultivated salmon. Invertebr Reprod Dev 22:87–90

Jackson D, Minchin D (1993) Lice infestation of farmed salmon in Ireland. In: Boxshall GA, Defaye D (eds) Pathogens of wild and farmed fish: sea lice. Ellis Horwood, Chichester, p 188–201

➤ Jensen JLA, Rikardsen AH, Thorstad EB, Suhr AH, Davidsen JG, Primicerio R (2014) Water temperatures influence the marine area use of *Salvelinus alpinus* and *Salmo trutta*. J Fish Biol 84:1640–1653

Johannessen A (1977) Early stages of *Lepeophtheirus salmonis* (Copepoda, Caligidae). Sarsia 63:169–176

🪰 Johnsen IA, Fiksen Ø, Sandvik AD, Asplin L (2014) Vertical salmon lice behaviour as a response to environmental conditions and its influence on regional dispersion in a fjord system. Aquacult Environ Interact 5:127–141

➤ Johnson SC, Albright LJ (1991) The developmental stages of *Lepeophtheirus salmonis* (Krøyer, 1837) (Copepoda: Caligidae). Can J Zool 69:929–950

ä Lagadeuc Y, Bouté M, Dodson JJ (1997) Effect of vertical mixing on the vertical distribution of copepods in coastal waters. J Plankton Res 19:1183–1204

➤ McHenery JG, Saward D, Seaton DD (1991) Lethal and sublethal effects of the salmon delousing agent dichlorvos on the larvae of the lobster (*Homarus gammarus* L.) and herring (*Clupea harengus* L.). Aquaculture 98:331–347

ä McKibben MA, Hay DW (2004) Distributions of planktonic sea lice larvae *Lepeophtheirus salmonis* in the inter-tidal zone in Loch Torridon, Western Scotland in relation to salmon farm production cycles. Aquacult Res 35:742–750

ä Middlemas SJ, Stewart DC, Mackay S, Armstrong JD (2009) Habitat use and dispersal of post smolt sea trout *Salmo trutta* in a Scottish sea loch system. J Fish Biol 74:639–651

Ministry of Trade, Industry and Fisheries (2015) Forutsigbar og miljømessig bærekraftig vekst i norsk lakse- og ørretoppdrett. Meld. St. 16 (2014–2015), Ministry of Trade, Industry and Fisheries, Oslo

ä Mitchelson-Jacob G, Sundby S (2001) Eddies of Vestfjorden, Norway. Cont Shelf Res 21:1901–1918

ä Murray AG, Gillibrand PA (2006) Modelling salmon lice dispersal in Loch Torridon, Scotland. Mar Pollut Bull 53: 128–135

Murray AG, Amundrud TL, Penston MJ, Pert CC, Middlemas SJ (2011) Abundance and distribution of larval sea lice in Scottish coastal waters. In: Jones S, Beamish R (eds) Salmon lice: an integrated approach to understanding parasite abundance and distribution. Wiley-Blackwell, Oxford, p 51–81

ä Myksvoll MS, Sundby S, Ådlandsvik B, Vikebø FB (2011) Retention of coastal cod eggs in a fjord caused by interactions between egg buoyancy and circulation pattern. Mar Coast Fish 3:279–294

ä Pemberton R (1976) Sea trout in North Argyll sea lochs, population, distribution and movements. J Fish Biol 9: 157–179

ä Penston MJ, McKibben MA, Hay DW, Gillibrand PA (2004) Observations on open water densities of sea lice larvae in Loch Shieldaig, western Scotland. Aquacult Res 35: 793–805

ä Penston MJ, Millar CP, Zuur A, Davies IM (2008) Spatial and temporal distribution of *Lepeophtheirus salmonis* (Kroyer) larvae in a sea loch containing Atlantic salmon, *Salmo salar* L., farms on the north-west coast of Scotland. J Fish Dis 31:361–371

➤ Pike AW (1989) Sea lice—major pathogens of farmed Atlantic salmon. Parasitol Today 5:291–297

ä Pike AW, Wadsworth SL (1999) Sealice on salmonids: their biology and control. Adv Parasitol 44:233–337

Pike AW, Mackenzie K, Rowand A (1993) Ultrastructure of the frontal filament in chalimus larvae of *Caligus elongatus* and *Lepeophtheirus salmonis* from Atlantic salmon, *Salmo salar*. In: Boxshall GA, Defaye D (eds) Pathogens of wild and farmed fish: sea lice. Ellis Horwood, Chichester, p 99–113

Rikardsen AH, Diserud OH, Elliott J, Dempson JB, Sturlaugsson J, Jensen AJ (2007) The marine temperature and depth preferences of Arctic charr (*Salvelinus alpinus*) and sea trout (*Salmo trutta*), as recorded by data storage tags. Fish Oceanogr 16:436–447

ä Salama NK, Collins CM, Fraser JG, Dunn J, Pert CC, Murray AG, Rabe B (2013) Development and assessment of a biophysical dispersal model for sea lice. J Fish Dis 36: 323–337

🪰 Salama NKG, Murray AG, Rabe B (2015) Simulated environmental transport distances of *Lepeophtheirus salmonis* in Loch Linnhe, Scotland, for informing aquaculture area management structures. J Fish Dis, doi:10.1111/jfd.12375

➤ Samsing F, Solstorm D, Oppedal F, Solstorm F, Dempster T (2015) Gone with the flow: current velocities mediate parasitic infestation of an aquatic host. Int J Parasitol 45: 559–565

Schram TA (1993) Supplementary descriptions of the developmental stages of *Lepeophtheirus salmonis* (Krøyer, 1837) (Copepoda: Caligidae). In: Boxshall GA, Defaye D (eds) Pathogens of wild and farmed fish: sea lice. Ellis Horwood, Chichester, p 30–47

➤ Serra-Llinares RM, Bjørn PA, Finstad B, Nilsen R, Harbitz A, Berg M, Asplin L (2014) Salmon lice infection on wild salmonids in marine protected areas: an evaluation of the Norwegian 'National Salmon Fjords'. Aquacult Environ Interact 5:1–16

Skartveit A, Olseth JA (1988) Varighetstabeller for timevis belysning mot 5 flater på 16 norske stasjoner. Meteorological Report Series 7, University of Bergen

🪰 Stien A, Bjørn PA, Heuch PA, Elston DA (2005) Population dynamics of salmon lice *Lepeophtheirus salmonis* on Atlantic salmon and sea trout. Mar Ecol Prog Ser 290: 263–275

➤ Stigebrandt A, Aure J (1989) Vertical mixing in basin waters of fjords. J Phys Oceanogr 19:917–926

🪰 Stormoen M, Skjerve E, Aunsmo A (2013) Modelling salmon lice, *Lepeophtheirus salmonis*, reproduction on farmed Atlantic salmon, *Salmo salar* L. J Fish Dis 36:25–33

Stucchi DJ, Guo M, Foreman MGG, Czajko P, Galbraith M, Mackas DL, Gillibrand PA (2011) Modeling sea lice production and concentrations in the Broughton Archipelago, British Columbia. In: Jones S, Beamish R (eds) Salmon lice: an integrated approach to understanding parasite abundance and distribution. Wiley-Blackwell, Oxford, p 117–150

Svedberg HA (2011) The abundance, spatial distribution and dispersion of salmon lice (*Lepeophtheirus salmonis*, Krøyer) in Folda fjord over two years. MSc thesis, University of Tromsø

🪰 Taranger GL, Karlsen Ø, Bannister RJ, Glover KA and others (2014a) Risk assessment of the environmental impact of Norwegian Atlantic salmon farming. ICES J Mar Sci 72: 997–1021

Taranger GL, Svåsand T, Kvamme BO, Kristiansen T, Boxaspen K (2014b) Risikovurdering norsk fiskeoppdrett 2013. Institute of Marine Research, Bergen

Thorstad EB, Økland F, Finstad B, Sivertsgård R, Bjørn PA,

McKinley RS (2004) Migration speeds and orientation of Atlantic salmon and sea trout post-smolts in a Norwegian fjord system. Environ Biol Fish 71:305–311

► Thorstad EB, Økland F, Finstad B, Sivertsgård R, Plantalech N, Bjørn PA, McKinley RS (2007) Fjord migration and survival of wild and hatchery-reared Atlantic salmon and wild brown trout post-smolts. Hydrobiologia 582:99–107

Thorstad EB, Todd CD, Bjørn PA, Gargan PG and others (2014) Effects of salmon lice on sea trout: a literature review. NINA Report 1044, Norwegian Institute for Nature Research, Trondheim

► Tucker CS, Sommerville C, Wootten R (2000) The effect of temperature and salinity on the settlement and survival of copepodids of Lepeophtheirus salmonis (Krøyer, 1837) on Atlantic salmon, Salmo salar L. J Fish Dis 23:309–320

► Tully O (1989) The succession of generations and growth of the caligid copepods Caligus elongatus and Lepeophtheirus salmonis parasitising farmed Atlantic salmon smolts (Salmo salar L.). J Mar Biol Assoc UK 69:279–287

► Visser AW (1997) Using random walk models to simulate the vertical distribution of particles in a turbulent water column. Mar Ecol Prog Ser 158:275–281

► Visser AW, Mariani P, Pigolotti S (2008) Swimming in turbulence: zooplankton fitness in terms of foraging efficiency and predation risk. J Plankton Res 31:121–133

Webb PW, Kostecki PT, Stevens ED (1984) The effect of size and swimming speed on locomotor kinematics of rainbow trout. J Exp Biol 109:77–95

4

Evaluating genetic traceability methods for captive-bred marine fish and their applications in fisheries management and wildlife forensics

Jonas Bylemans[1,2,*], Gregory E. Maes[1,3], Eveline Diopere[1], Alessia Cariani[4], Helen Senn[5], Martin I. Taylor[6], Sarah Helyar[7], Luca Bargelloni[8], Alessio Bonaldo[9], Gary Carvalho[10], Ilaria Guarniero[9], Hans Komen[11], Jann Th. Martinsohn[12], Einar E. Nielsen[13], Fausto Tinti[4], Filip A. M. Volckaert[1], Rob Ogden[14]

[1]Laboratory of Biodiversity and Evolutionary Genomics, University of Leuven, Ch. Deberiotstraat 32, 3000 Leuven, Belgium

[2]Institute for Applied Ecology, University of Canberra, Canberra, ACT 2612, Australia

[3]Centre for Sustainable Tropical Fisheries and Aquaculture, Comparative Genomics Centre, College of Marine and Environmental Sciences, Faculty of Science and Engineering, James Cook University, Townsville, 4811 QLD, Australia

[4]Department of Biological, Geological and Environmental Sciences, University of Bologna, Ravenna 48123, Italy

[5]WildGenes Laboratory, Royal Zoological Society of Scotland, Edinburgh EH12 6TS, UK

[6]School of Biological Sciences, University of East Anglia, Norwich NR4 7TJ, UK

[7]Institute for Global Food Security, Queen's University Belfast, Belfast BT9 5BN, UK

[8]Department of Public Health, Comparative Pathology, and Veterinary Hygiene, University of Padova, Viale dell'Università 16, 35020 Legnaro, Italy

[9]Department of Veterinary Medical Sciences DIMEVET, University of Bologna, 40064 Bologna, Italy

[10]Molecular Ecology and Fisheries Genetics Laboratory, School of Biological Sciences, Environment Centre Wales, Bangor University, Bangor, Gwynedd LL57 2UW, UK

[11]Animal Breeding and Genomics Centre, Wageningen University, PO Box 338, 6700AH Wageningen, The Netherlands

[12]JRC.G.4 – Maritime Affairs, Institute for the Protection and Security of the Citizen (IPSC), Joint Research Centre (JRC), European Commission, Via Enrico Fermi 2749, 21027 Ispra VA, Italy

[13]Section for Marine Living Resources, National Institute of Aquatic Resources, Technical University of Denmark, Vejlsøvej 39, 8600 Silkeborg, Denmark

[14]TRACE Wildlife Forensics Network, Royal Zoological Society of Scotland, Edinburgh EH12 6TS, UK

ABSTRACT: Growing demands for marine fish products is leading to increased pressure on already depleted wild populations and a rise in aquaculture production. Consequently, more captive-bred fish are released into the wild through accidental escape or deliberate releases. The increased mixing of captive-bred and wild fish may affect the ecological and/or genetic integrity of wild fish populations. Unambiguous identification tools for captive-bred fish will be highly valuable to manage risks (fisheries management) and tracing of escapees and seafood products (wildlife forensics). Using single nucleotide polymorphism (SNP) data from captive-bred and wild populations of Atlantic cod *Gadus morhua* L. and sole *Solea solea* L., we explored the efficiency of population and parentage assignment techniques for the identification and tracing of captive-bred fish. Simulated and empirical data were used to correct for stochastic genetic effects. Overall, parentage assignment performed well when a large effective population size characterized the broodstock and escapees originated from early generations of captive breeding. Consequently, parentage assignments are particularly useful from a fisheries management perspective to monitor the effects of deliberate releases of captive-bred fish on wild populations. Population assignment proved to be more efficient after several generations of captive breeding, which makes it a useful method in forensic applications for well-established aquaculture species. We suggest the implementation of a case-by-case strategy when choosing the best method.

KEY WORDS: Aquaculture · Conservation genetics · Escapees · Fisheries management · Wildlife forensics

---

*Corresponding author: Jonas.Bylemans@canberra.edu.au

INTRODUCTION

Aquaculture is one of the fastest growing food-producing sectors and will remain so in the foreseeable future due to a growing human demand for animal protein and lipids (Braithwaite & Salvanes 2010) and the limits that have been reached for wild-capture fisheries production (FAO Fisheries and Aquaculture Department 2014). This has led to various challenges related to the aquaculture industry, including organic, chemical and pharmaceutical pollution (Seymour & Bergheim 1991), infectious diseases (Murray & Peeler 2005), feed supply (Naylor et al. 2000, 2009, Natale et al. 2013) and escapees (Kitada et al. 2009, Glover 2010, Glover et al. 2011, Noble et al. 2014).

Accidental escapees (Bekkevold et al. 2006, Glover et al. 2013, Noble et al. 2014) or deliberate releases (Bell et al. 2008, Kitada et al. 2009) of captive-bred marine fish may impact the environment, and the ecological and genetic integrity of wild fish populations (Braithwaite & Salvanes 2010, Laikre et al. 2010). First, a decrease in genetic diversity, and consequently a lower evolutionary potential, has been observed in wild marine fish populations which have been invaded by captive-bred conspecifics (Hindar et al. 1991, Weir & Grant 2005, Glover et al. 2013). Given that recent studies have indicated surprisingly fine-scale local genetic adaptation in marine fish (André et al. 2011, Nielsen et al. 2012, Vandamme et al. 2014), the introgression of captive-bred fish can be detrimental to the long-term survival of wild fish populations. Second, introgression might disrupt adaptive gene complexes, which reduces the fitness of hybrids and in turn may compromise the persistence of locally adapted populations (McGinnity et al. 2003, Danancher & Garcia-Vazquez 2011, Lamaze et al. 2013). Managing and mitigating risks and assessing the impacts of released/escaped captive-bred fish on local wild populations are thus of utmost importance to ensure the long-term sustainability of aquaculture and fisheries industries. Third, aquaculture companies might have legal obligations to report escapees and failure to comply with these regulations might result in fines (Glover 2010). As such, the ability to trace back escapees to the farm of origin constitutes a highly valuable asset in delivering evidence for legal action (Glover et al. 2008, Glover 2010). Finally, an increase in international trade and consumer awareness in recent decades has highlighted the need for accurate labelling of seafood products. Mislabelling to increase profits has been extensively documented in the seafood industry (Jacquet & Pauly 2008, Hanner et al. 2011, Mariani et al. 2014). Given that market prices of wild-caught marine fish species are generally higher than aquaculture sourced fish, fraudulent labelling captive-bred fish as 'wild-caught' may increase income for the perpetrator (Cline 2012, Warner et al. 2013). Hence, genetic identification methods for farmed and wild marine fish species would be extremely valuable in aquaculture and fisheries management and wildlife forensics.

For a large variety of commercially reared species, escapees and deliberate releases have been reported (Liao et al. 2003, Bell et al. 2008, Jensen et al. 2010, Danancher & Garcia-Vazquez 2011). However, due to their long breeding history and the availability of genetic tools, research on tracing and quantifying escapees has focused mainly on salmonids (Glover 2010, Glover et al. 2013) and only recently on sea bass and sea bream (Arechavala-Lopez et al. 2013, Somarakis et al. 2013, Brown et al. 2015). Extending standardized traceability methods to other commercially exploited marine fish species will thus advance research into the effects of escapees and restocking programmes.

The lack of a long breeding history in most cultured marine fish species complicates the genetic discrimination between wild and captive-bred marine fish, especially when the identification of the hatchery of origin is required. The recent domestication history of many marine fish results in similar allele frequencies in captive-bred and wild populations, which lowers the discrimination power of genetic markers (Duarte et al. 2007). Likewise, the absence of long-term selective breeding programmes reduces the likelihood of finding species-specific 'domestication' markers (Karlsson et al. 2011, Gjedrem et al. 2012). Stochastic and selective breeding processes in aquaculture and recent developments in genetic traceability tools can however facilitate discrimination between captive-bred and wild fish. The common use of a relatively small broodstock and the unwanted high variance in reproductive success within the hatchery will result in increased genetic differentiation between captive-bred and wild populations and a lower genetic diversity within the captive-bred population (Porta et al. 2006a,b, 2007). Within the marine environment, provided that a solid genetic baseline is available, wild fish can be individually assigned to their region and/or population of origin with high precision using gene-associated single nucleotide polymorphism (SNP) markers (Nielsen et al. 2012). Genetic background information is increasingly available for commercially important fish species (Nielsen et al. 2009, Abadía-Cardoso et al. 2013, Clemento et al. 2014), which makes the use of simulation studies possible to assess the discrimination power of existing genetic markers for wild and captive-bred fish. Finally, while the rate of genetic drift at neutral markers depends on the effective population size (N_e) of the broodstock, SNP markers associated with important aquaculture traits (such as growth and disease resistance) are subjected to directional selection which will increase the degree of differentiation between wild and captive-bred populations (Glover et al. 2010). Such markers may introgress at different rates compared to selectively neutral markers (Lamaze et al. 2012, Hohenlohe et al. 2013), thus providing crucial insights into both the fitness and molecular consequences of escapees and restocking programmes.

Multiple approaches are available for identifying and discriminating between captive-bred and wild marine fish (Manel et al. 2005). The 2 main methods used to date are individual assignment (IA) and parentage-based tagging (PBT) (Manel et al. 2005, Jones et al. 2010). Most commonly used, IA methods rely on allele frequency differences between populations to assign an individual to its most likely source (Ogden 2008, Glover 2010, Nielsen et al. 2012). However, in order to achieve highly robust assignments, IA requires some level of genetic differentiation between populations and extensive genetic reference data (Manel et al. 2005, Nielsen et al. 2012). In contrast, PBT utilizes the genetic variation within the complete data to determine the most likely parental pair for a particular genotype and can achieve high assignment success even when genetic differentiation among populations is insufficient for IA (Jones & Ardren 2003, Steele et al. 2013).

Our study focuses on Atlantic cod *Gadus morhua* L., 1758 and sole *Solea solea* L., 1958, 2 commercially important fish of the Northeast Atlantic Ocean for which extensive genetic resources are available (Nielsen et al. 2012). Both species have a widespread distribution across the Northeast Atlantic Ocean, and their high commercial value has resulted in an increased interest in captive-breeding programmes, restocking, stock enhancement and sea ranching (Howell 1997, Kjesbu et al. 2006, Björnsson 2011). More specifically, declines in wild-caught Atlantic cod and advances in captive breeding and feed formulation have led to an increase in global aquaculture production, reaching 22 000 tons in 2010 (Rosenlund & Skretting 2006, Thurstan et al. 2010, FAO Fisheries and Aquaculture Department 2015a). Although cod aquaculture has recently decreased

due to large catches on the northern fishing grounds (FAO Fisheries and Aquaculture Department 2015a), the use of traditional cage farming in cod aquaculture and the substantial interest in stock enhancement and sea ranching programmes continues to represent a significant risk for interactions between wild and hatchery-reared cod (Bekkevold et al. 2006, Jørstad et al. 2008, Björnsson 2011). Similarly, recent advances and changing economic perspective have increased the interest in sole aquaculture, with production peaking at 125 tons in 2010 but decreasing in recent years (Howell 1997, Imsland et al. 2003, FAO Fisheries and Aquaculture Department 2015b). Although intensive land-based recirculation systems are currently preferred in flatfish aquaculture, there is considerable interest to reduce production costs through less intensive systems (e.g. cage farming, stock enhancement and sea ranching) (Brown 2002, Kitada & Kishino 2006, Sparrevohn & Støttrup 2007). Hence, for both focal species, there is a considerable risk of introgression between captive-bred individuals and local wild populations.

In this study, we aimed to evaluate the utility of IA and PBT approaches to discriminate between captive-bred marine fish and natural fish populations. To achieve this, we used a combination of simulated and empirical SNP datasets to perform a series of assignment experiments in each species, across a range of potential scenarios. The level of genetic differentiation between captive-bred and wild marine fish will vary due to: (1) the number of captive-bred generations (F_n) prior to escape or release, (2) the number of broodstock and the strength of reproductive variance between broodstock individuals, which both influence N_e, and (3) genetic (and geographical) differences between the hatchery population and locally occurring wild populations with which the escapees will intermingle. We investigated each of these potential variables to evaluate their relative impact on assignment power under IA and PBT approaches. In addition, the effect of (in)complete reference samples was also assessed given that the availability and representative nature of reference samples will also affect traceability outcomes.

From the outset, we anticipated that increasing F_n, decreasing N_e and a distinct genetic origin of the broodstock will all favour IA, given that IA relies on the realized level of genetic differentiation between populations to make robust assignments. On the other hand, the performance of PBT will be negatively impacted by those parameters that reduce the genetic diversity within the captive-bred population (i.e. high F_n and low N_e) due to the difficulty of excluding candidate parents from real parents. Therefore, in addition to evaluating the relative performance of the 2 approaches, we were interested in examining possible thresholds of F_n and N_e across which the optimal approach for determining fish origin actually changes.

MATERIALS AND METHODS

Sampling

Wild samples of 10 Atlantic cod and 14 sole populations have been previously collected from European waters and genotyped (Nielsen et al. 2012). An Atlantic cod broodstock (A_{cod-BS}) (n = 92) sourced from the ICES region 27.V.b2 – Faroe Bank was sampled from the Fiskaaling aquaculture research station (Faroe Islands). Atlantic-sourced (ICES 27.IV.c – Southern North Sea) sole were sampled from a Dutch experimental breeding farm, SOLEA in IJmuiden, and consisted of 2 full-sib families with 4 broodstock individuals ($A_{sole-BS}$) (n = 4) and their first-generation offspring ($F_n = 1$) ($A_{sole-F1}$) (n = 92) (Blonk et al. 2009). Captive-bred sole samples originating from the Mediterranean Sea (FAO 37.2.1 – North Adriatic) were obtained from a pilot farm of the Laboratory of Aquaculture, Department of Veterinary Medical Sciences of the University of Bologna, Italy, and included samples from a broodstock ($M_{sole-BS}$) (n = 26) and first-generation offspring ($F_n = 1$) ($M_{sole-F1}$) (n = 96), obtained from 4 batch spawnings ($M_{sole-F1-B1}$, $M_{sole-F1-B2}$, $M_{sole-F1-B3}$ and $M_{sole-F1-B4}$). More details on all populations used in this study are found in Supplement 1 (www.int-res.com/articles/suppl/q008 p131_supp.pdf).

Genotypic data

Gene-associated SNP markers were available for: the wild populations of Atlantic cod (1258 SNPs), the wild populations of sole (427 SNPs), A_{cod-BS} (427 SNPs), $A_{sole-BS}$ and $A_{sole-F1}$ (423 SNPs) (Table 1) (Nielsen et al. 2012, Diopere et al. 2014). Additional genotyping of the $M_{sole-BS}$ and $M_{sole-F1}$ samples was conducted using VeraCode™ technology on the BeadExpress platform (Illumina), following the manufacturer's instructions. Of the 427 available SNPs, the 192 most informative SNPs, showing high genetic discrimination values (F_{ST} values) between the Mediterranean populations, were genotyped (Nielsen et al. 2012). Quality assessment and genotype calling

was performed using GenomeStudio v.2009.2 software (Illumina). Three individuals from the $A_{sole-BS}$, initially genotyped with the wild populations using the SAM assay (GoldenGate, GG) on the iScan platform (Illumina) and with the highest GG call rate for the selected panel (Diopere et al. 2014), were included as cross-platform genotyping controls to ensure comparability between the archived and newly generated data.

Marker selection

In order to obtain marker panels with sufficient assignment power and to ensure that they are easily transferrable between laboratories, a subset of 96 highly informative SNPs were selected based on the practical limitations of common genotyping platforms (Supplement 2 at www.int-res.com/articles/suppl/q008p131_supp.pdf). Given that cod data were only used in IA analyses (see 'Tracing escapees' below), and the ability of markers to distinguish between populations provides a good indication of their power in IA analyses, the available SNPs for cod were ranked based on the pairwise F_{ST} values calculated among the wild cod populations using FSTAT v.2.9.3 (Goudet 1995). The Atlantic and Mediterranean sole data (wild and aquaculture) were used in both IA and PBT. To maximize the traceability power of selected sole SNPs for IA, markers were first ranked based on the pairwise F_{ST} values obtained from comparisons between the wild Atlantic and Mediterranean populations respectively. PBT analyses, on the other hand, require markers with a high genetic variability within a population to make robust assignments. Consequently, a second ranking of markers was based on their polymorphic information content (PIC) calculated with Cervus v.3.0 (Marshall et al. 1998) using the com-

Table 1. Available and newly generated single nucleotide polymorphism (SNP) genotypic datasets for wild and captive-bred populations of *Solea solea* and *Gadus morhua*. Atl-Aqua = aquaculture population sourced from the Atlantic Ocean, Med-Aqua = aquaculture population sourced from the Mediterranean Sea

Species	Origin	SNP genotypic data	
		Available	Source
Sole	Wild	427	Nielsen et al. (2012)
	Atl-Aqua	423	Diopere et al. (2014)
	Med-Aqua	181	Current study
Cod	Wild	1258	Nielsen et al. (2012)
	Atl-Aqua	427	Nielsen et al. (2012)

bined data from the respective broodstocks ($A_{sole-BS}$ and $M_{sole-BS}$) and their genetically similar wild populations (GER and ADR1 respectively). The top 96 ranking SNPs were used in all further analyses and further reduced genotypic datasets were used in the assignment analyses to determine the assignment power of the selected loci (Table 2).

This selection procedure for highly informative markers is unlikely to suffer from high-grading bias (Anderson 2010a, Waples 2010) for 3 reasons. First, assignment power was estimated from a different (holdout) set of samples to those used for SNP selection. Second, outlier SNPs were defined using initially high sampling sizes ($n \approx 40$) from various geographical locations, which reduces the effects of random sampling errors (Nielsen et al. 2012). Third, the use of 2 rigorous outlier detection methods and annotation information provides confidence that the high F_{ST} values of the selected markers are more likely to result from diversifying selection (i.e. real differentiation) rather than being at the extremes of a neutral marker F_{ST} distribution (Waples 2010, Nielsen et al. 2012).

Simulations of hatchery data

To formally evaluate the individual and combined impacts of F_n, N_e and the availability of reference data on the traceability efficiency of IA and PBT analyses, various breeding scenarios were simulated for both species. Data were simulated using the previously selected 96 SNPs of the A_{cod-BS} and $M_{sole-BS}$ (including 7 individuals that died before reproduc-

tion) genotypes for cod and sole respectively (Fig. 1). An initial parental broodstock (P_{1-SIM}) and 4 offspring generations (F_{1-SIM}, F_{2-SIM}, F_{3-SIM} and F_{4-SIM}) were simulated under the assumption of perfect Hardy-Weinberg (HW) equilibrium. Different simulation series were performed using various N_e values (N_e = 5, 10, 20 or 50) to simulate drift due to reproductive variance. HYBRIDLAB v.1.0 (Nielsen et al. 2006) was used to simulate offspring genotypes used in the IA analyses. For the PBT analyses, Nookie v.1.0 (Anderson 2014) was used to simulate offspring genotypes because it generates individual genotypes 'bred' from specific parental pairs which are required for parentage assignment in simulated generations, rather than simply simulating individuals from a pool of population allele frequencies. Comparisons of genetic diversity showed that datasets generated through both programs were comparable (Supplement 3 at www.int-res.com/articles/suppl/q008p131_supp.pdf).

Comparative data analyses

A detailed comparison of the traceability results based on both the simulated and empirical datasets is important to determine the optimal traceability approach for a specific scenario. To be able to compare simulated and empirical results, population genetic parameters (F_{ST} values, observed heterozygosity H_{obs} and expected heterozygosity H_{exp}) associated with each dataset have to be understood as they will strongly influence the traceability power of the datasets. With the most comprehensive empirical data being available for the Mediterranean captive-bred sole, population genetic parameters were calculated for the broodstock ($M_{sole-BS}$ and P_{1-SIM} [N_e = 5, 10, 20, 50]) and first-generation offspring ($M_{sole-F1}$ and F_{1-SIM} [N_e = 5, 10, 20, 50]). The Northern Adriatic population (ADR1), as the original source of $M_{sole-BS}$, was included in the analysis as a reference. Genetic diversity (H_{obs} and H_{exp}) was calculated for each independent dataset ($M_{sole-BS}$, $M_{sole-F1-B1}$, $M_{sole-F1-B2}$, $M_{sole-F1-B3}$, $M_{sole-F1-B4}$, P_{1-SIM} [N_e = 5, 10, 20, 50], F_{1-SIM} [N_e = 5, 10, 20, 50] and ADR1) using

Table 2. *Solea solea* and *Gadus morhua*. Datasets used in population (IA = individual assignment) and parentage (PBT = parentage-based tagging) analysis to test for effects of sampling regimes and traceability scenarios. Sampling Regime 1: reference data of the aquaculture population is available for the parental generation. Sampling Regime 2: reference data of the aquaculture population is limited to the founding broodstock. Scenario A: aquaculture broodstock originated from a genetically distinct population than the local wild populations. Scenario B: aquaculture broodstock originated from a local, genetically similar wild population. The number of single nucleotide polymorphisms (SNPs) used in each analysis is indicated between parentheses. SD = simulated data, EAD = empirical Atlantic data, EMD = empirical Mediterranean data, na = not applicable

Traceability method	Species	Sampling Regime		Scenario	
		1	2	A	B
IA	Sole	SD (96, 52)	SD (96)	na	EAD (96, 30, 1)
	Cod	SD (96)	SD (96)	na	na
PBT	Sole	SD (96, 48)	na	EAD (50, 30, 21)	EAD (50, 35, 30)
				EMD (40, 35, 30)	EMD (40, 35, 30)
	Cod	na	na	na	na

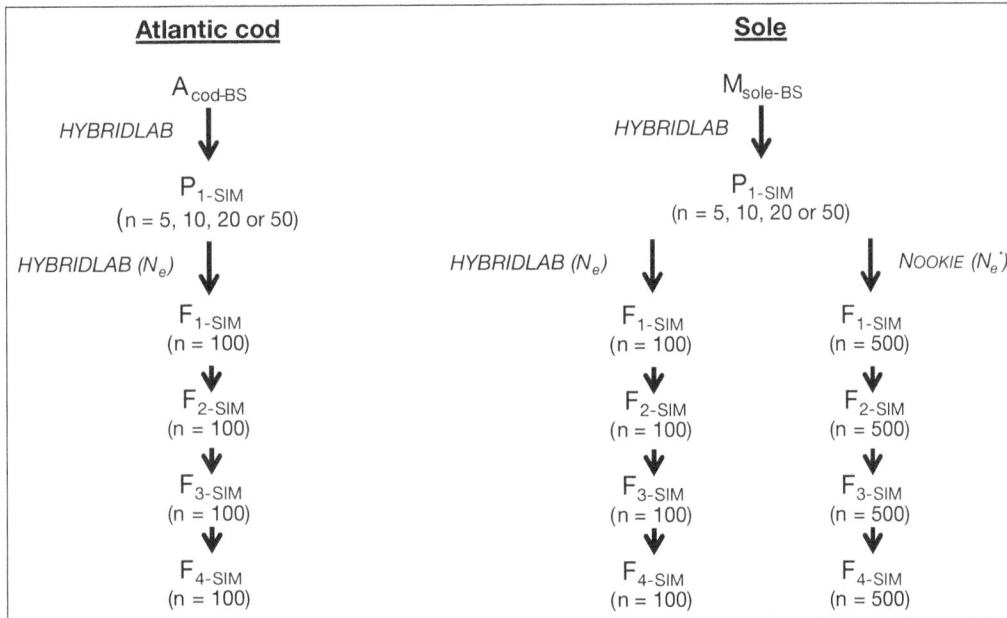

Fig. 1. *Gadus morhua* and *Solea solea*. Simulations used to generate captive-bred offspring genotypes. HYBRIDLAB v.1.0 (Nielsen et al. 2006) was used to generate the initial broodstock (P_{1-SIM}). Offspring genotypes used in individual assignment (IA) and parentage-based tagging (PBT) analyses were simulated with HYBRIDLAB and Nookie v.1.0 (Anderson 2014), respectively. A_{cod-BS} = Atlantic cod broodstock, $M_{sole-BS}$ = Mediterranean sole broodstock, F_{1-SIM}, F_{2-SIM}, F_{3-SIM}, F_{4-SIM} = 4 offspring generations, n = number of individuals, N_e (effective population size) = 5, 10, 20 or 50 and N_e^* = 4 or 50

Genetix v.4.05 (Belkhir et al. 2004). The realized levels of genetic differentiation between the simulated and empirical datasets was evaluated by calculating pairwise F_{ST} values and performing a discriminant analysis of principal components (DAPC) with the adegenet package in R v.3.0.2 (Jombart 2008, R Development Core Team 2010).

In addition to the comparative data analyses, the results also allow us to determine the N_e of the $M_{sole-BS}$. Using the H_{obs} and H_{exp} values obtained for $M_{sole-F1-B1}$, $M_{sole-F1-B2}$, $M_{sole-F1-B3}$ and $M_{sole-F1-B4}$, the N_e can be calculated for $M_{sole-BS}$ during each batch spawning event using the equation from Luikart & Cornuet (1999):

$$N_e = H_{exp}/[2(H_{obs} - H_{exp})] \qquad (1)$$

Tracing escapees

Assignment efficiency is strongly influenced by the realized levels of genetic differentiation between the captive-bred and wild populations (IA) and the amount of genetic variability within the captive-bred population (PBT). By using the simulated datasets which are characterized by differences in the N_e and F_n, 2 parameters that significantly affect genetic differentiation and genetic variability, the effects of

these changes could be evaluated. In addition, the origin of the captive-bred population will also influence the traceability outcomes. To assess the effects of genetic dissimilarities between captive-bred and wild populations, 2 traceability scenarios were used: A, the broodstock originated from a genetically distinct population than the local wild populations; and B, the broodstock originated from a local, genetically similar wild population (Table 2).

From a forensic perspective, the ability to assign captive-bred fish back to their origin will be influenced by the nature and availability of reference samples, which may be challenging in well-established aquaculture species (Glover et al. 2009). For the purpose of our study, 2 simplified sampling regimes were used to evaluate the effect of missing data from previous captive-bred generations: Sampling Regime 1, in which data from the parental generation, which produced the escapees, is available, and escapees can thus be assigned to their parental generation or to the wild populations; Sampling Regime 2, in which data is restricted to the founding broodstock (often the case in operational hatcheries) and escapees can only be assigned to the founding broodstock or the wild populations. The lack of multiple captive-bred generations in the empirical data restricted the analyses of the empirical data to Sampling Regime 1. Furthermore, PBT relies on the identification of parent–offspring relationships and will thus only be valuable under the assumptions of Sampling Regime 1.

IA and PBT analyses were performed using both simulated and empirical datasets (see Table 2). For the analyses of the simulated data, escapees were assumed to be flagged (i.e. genotypes of escapees are known) to obtain a baseline traceability efficiency, while for the analysis of the empirical data, escapees were mixed within a single wild population to create a more realistic scenario. IA analyses used the simulated datasets of both species and the empirical Atlantic sole data. PBT analyses were performed using sole data only, as Atlantic cod family data was unavailable.

IA analysis

IA analyses were performed with GeneClass2 v.2.0 using the 'assign/exclude population as origin of individuals' option (Piry et al. 2004). The threshold value was set to $p = 0.05$ and only individuals assigned to a population with rank 1 were considered. The probability of an individual being assigned to all possible reference populations was calculated using the Monte Carlo re-sampling method (Paetkau et al. 2004).

Using the simulated data of both species, assignment efficiency was evaluated under both sampling regimes. Input data for assignments consisted of 100 F_{n-SIM} genotypes (escapees) which could be assigned to either wild populations or their captive-bred population (i.e. their parental generation $F_{(n-1)-SIM}$ or their founding broodstock P_{1-SIM} for Sampling Regime 1 or 2 respectively).

For the analyses of the empirical data, 20 individuals from $A_{sole-F1}$ (10 from each full-sib family) representing the escapees were randomly selected and mixed with a genetically similar wild population (Scenario B) originating from the Belgian coast (BEL). Genotypes contained within this mixed population and the neighbouring wild populations (STO, GER, NOR, ENG, IS and GAS; see Supplement 1) could then be assigned to the remaining $A_{sole-F1}$ individuals.

PBT analysis

The parent–offspring relationships within the empirical aquaculture samples were obtained from previous studies (Blonk et al. 2009) and additional parentage testing (Supplement 4 at www.int-res.com/articles/suppl/q008p131_supp.pdf). Only the SNP genotypes of individuals for which reliable parent-offspring relationships could be obtained were used in further analyses to ensure that the effectiveness of PBT could be formally evaluated. PBT analyses were performed with the software SNPPIT v.1.0 (Anderson 2010b), using only genotypic information (i.e. sex, age, year of sampling, etc. were considered unknown) and a genotyping error rate of 0.5% per allele.

Using the simulated data and the wild populations of sole, the effect of N_e and F_n on the assignment success was evaluated under the assumption of Sampling Regime 1. Input files consisted of a list of putative parents (all wild populations and $F_{(n-1)-SIM}$) and offspring to be assigned (F_{n-SIM}) (i.e. the escapees).

Empirical analyses were performed with the Atlantic and Mediterranean sole data to determine the influence of the origin of the broodstock (Scenario A or B) on the traceability efficiency. Under Scenario A, the input file of putative parents contained the genotypes of $A_{sole-BS}$ or $M_{sole-BS}$ mixed with their respective source population (i.e. GER and ADR1 respectively). The offspring to be assigned contained a mixed population of 20 randomly selected $A_{sole-F1}$ or $M_{sole-F1}$ individuals added to genetically different wild populations (i.e. IS and THY respectively) and the remaining wild populations. In the case of Scenario B, assignment input was similar with the exception that the 20 randomly selected $A_{sole-F1}$ or $M_{sole-F1}$ individuals were mixed with a genetically similar wild population (i.e. BEL and ADR2 respectively).

RESULTS

Sampling and genotyping

Following complementary genotyping of the sole samples ($M_{sole-F1}$) with 192 SNPs, 181 SNPs passed the initial quality assessment. Of these, a panel of 96 highly informative SNP markers was selected and used in the analyses. An overview of all 96 selected SNPs used in the traceability analyses can be found in Supplement 2. Based on the re-genotyping of the 3 $A_{sole-BS}$ individuals at 181 loci, a genotyping discordance rate of 1.2% was obtained. Hence, in all further analyses, a genotyping error of 1% was used as an approximation.

Comparative data analyses

The comparative analyses of overall genetic diversity (H_{obs} and H_{exp}) showed no strong deviation between H_{obs} and H_{exp} in the P_{1-SIM} (N_e = 5, 10, 20, 50) and F_{1-SIM} (N_e = 5, 10, 20, 50) data (Fig. 2). However, in the $M_{sole-F1}$ data, a heterozygote excess was observed ($H_{obs} > H_{exp}$), suggesting that within $M_{sole-BS}$, a low number of individuals contributed to the next generation. Based on this heterozygote excess, the N_e is estimated to be 2.16, 2.10, 1.58 and 1.67 for $M_{sole-F1-B1}$, $M_{sole-F1-B2}$, $M_{sole-F1-B3}$ and $M_{sole-F1-B4}$ respectively.

Pairwise F_{ST} values and the DAPC show that both $M_{sole-BS}$ and ADR1 have a similar genetic composition (Fig. 3; Supplement 5 at www.int-res.com/articles/suppl/q008p131_supp.pdf). However, strong genetic differentiation is observed between $M_{sole-F1}$ and their population of origin ($M_{sole-BS}$ and ADR1). A comparison of the simulated data (P_{1-SIM}) and the wild populations ($M_{sole-BS}$ and ADR1) shows an increase in genetic differentiation when a strong bottleneck was applied (from N_e = 50 to N_e = 5), and the same pattern can be observed in the derived F_{1-SIM} samples. Furthermore, the F_{ST} values are generally higher between the $M_{sole-F1}$ batches ($M_{sole-F1-B1}$, $M_{sole-F1-B2}$, $M_{sole-F1-B3}$, $M_{sole-F1-B4}$) than between the F_{1-SIM} data (Fig. 3), and the same pattern can be observed with the DAPC (i.e. F_{1-SIM} clusters are positioned closer together than $M_{sole-F1}$ clusters). One exception is the low genetic differentiation between $M_{sole-F1-B3}$ and $M_{sole-F1-B4}$, which is due to the same parents having produced these batches (Supplement 4). The results

suggest that the simulated data provides a good baseline (broodstock under HW equilibrium) for the validation of the traceability methods under real-life scenarios.

Tracing escapees

IA analysis

The success rate of correctly assigning escapees to the previous aquaculture generation (Sampling Regime 1) ranged from 73 to 100% across all simulated datasets (Fig. 4). Results clearly indicate that the assignment success increased with increasing genetic drift (smaller N_e) and increasing generational distance from the original broodstock generation (higher F_n). Under the assumptions of Sampling Regime 2, the assignment success increased with increasing genetic drift, but no change in assignment success was observed with increasing generational distance from the broodstock (Fig. 4). However, in sole, an increasing F_n resulted in a decrease of assignment performance when a large effective population size (N_e = 50) was employed.

The population assignment analyses based on the empirical data of the Atlantic farmed sole and their neighbouring wild populations revealed that 81% of escapees were correctly assigned using 1 SNP (average assignment score: 40), while a 100% assignment was achieved with only 30 SNPs (average assignment score: 100).

PBT analysis

PBT analyses (SNPPIT) using the simulated data of sole showed that a panel of 48 SNP loci was sufficient to obtain an assignment success of ≥99%. Assignment success decreased (i.e. increasing number of non-excluded parent-offspring trios) with an increasing number of breeding generations (F_n), especially when N_e was small (Table 3).

The PBT results based on the empirical sole data show that under Scenario A, a dataset of 30 and 40 highly polymorphic SNPs was sufficient to trace back the Atlantic and Mediterranean aquaculture escapees, respectively (Table 4). Under the assumptions of Scenario B, a total of 35 highly poly-

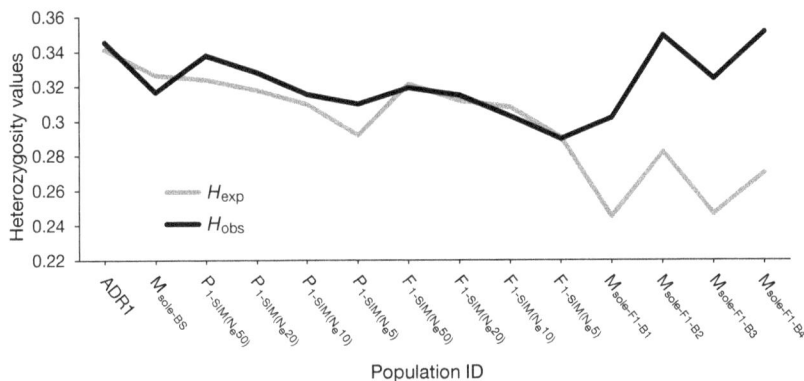

Fig. 2. Observed (H_{obs}) and expected (H_{exp}) heterozygosity values in population samples of Mediterranean sole *Solea solea*. ADR1 = natural population from Adriatic Sea; $M_{sole-BS}$ = broodstock composed of wild fish from Adriatic Sea; $P_{1-SIM(N_eX) (X = 5-50)}$ = simulated parental populations at varying degrees of effective population size N_e; $F_{1-SIM(N_eX) (X = 5-50)}$ = simulated first offspring populations at varying degrees of effective population size N_e; $M_{sole-F1-BX (X = 1-4)}$ = 4 actual offspring batches from aquaculture

	M_{soleBS}	$M_{soleF1-B1}$	$M_{soleF1-B2}$	$M_{soleF1-B3}$	$M_{soleF1-B4}$	$P_{1-SIM}(N_e50)$	$P_{1-SIM}(N_e20)$	$P_{1-SIM}(N_e10)$	$P_{1-SIM}(N_e5)$	$F_{1-SIM}(N_e50)$	$F_{1-SIM}(N_e20)$	$F_{1-SIM}(N_e10)$	$F_{1-SIM}(N_e5)$	ADR1
M_{soleBS}	0.000													
$M_{soleF1-B1}$	0.078	0.000												
$M_{soleF1-B2}$	0.062	0.145	0.000											
$M_{soleF1-B3}$	0.057	0.161	0.080	0.000										
$M_{soleF1-B4}$	0.043	0.135	0.053	0.010	0.000									
$P_{1-SIM}(N_e50)$	0.002	0.068	0.061	0.050	0.040	0.000								
$P_{1-SIM}(N_e20)$	0.006	0.092	0.076	0.058	0.047	0.003	0.000							
$P_{1-SIM}(N_e10)$	0.009	0.085	0.076	0.061	0.048	0.005	0.007	0.000						
$P_{1-SIM}(N_e5)$	0.012	0.072	0.070	0.053	0.044	0.006	0.011	0.010	0.000					
$F_{1-SIM}(N_e50)$	0.002	0.048	0.046	0.035	0.029	0.002	0.003	0.003	0.003	0.000				
$F_{1-SIM}(N_e20)$	0.005	0.059	0.049	0.035	0.029	0.005	0.001	0.003	0.003	0.007	0.000			
$F_{1-SIM}(N_e10)$	0.007	0.058	0.056	0.043	0.036	0.008	0.004	0.000	0.002	0.008	0.008	0.000		
$F_{1-SIM}(N_e5)$	0.017	0.070	0.074	0.053	0.047	0.020	0.012	0.005	0.000	0.022	0.022	0.016	0.000	
ADR1	0.010	0.070	0.058	0.062	0.050	0.011	0.015	0.015	0.013	0.011	0.016	0.021	0.030	0.000

Fig. 3. Pairwise F_{ST} values matrix among the simulated and empirical datasets of Mediterranean aquaculture sole *Solea solea*. □: F_{ST} = 0; □: 0 < F_{ST} < 0.01; □: 0.01 ≤ F_{ST} < 0.05; ▨: 0.05 ≤ F_{ST} < 0.1; ▩: 0.1 ≤ F_{ST}. See Fig. 2 for abbreviations

morphic SNPs were sufficient for the identification of all aquaculture escapees for both broodstocks (Table 4).

DISCUSSION

Our study shows that a panel of highly informative, gene-associated SNP markers can discriminate between wild and captive-bred marine fish, even without extensive domestication of the species of interest. Furthermore, the results show that IA and

PBT analyses can both be valuable tools for wildlife forensics and fisheries management, depending on the genetic history of the relevant captive populations.

Potential of SNP markers for traceability

Biallelic SNP markers are generally considered less informative than microsatellite markers. However, SNPs are highly abundant and evenly distributed throughout the genome (Morin et al. 2004).

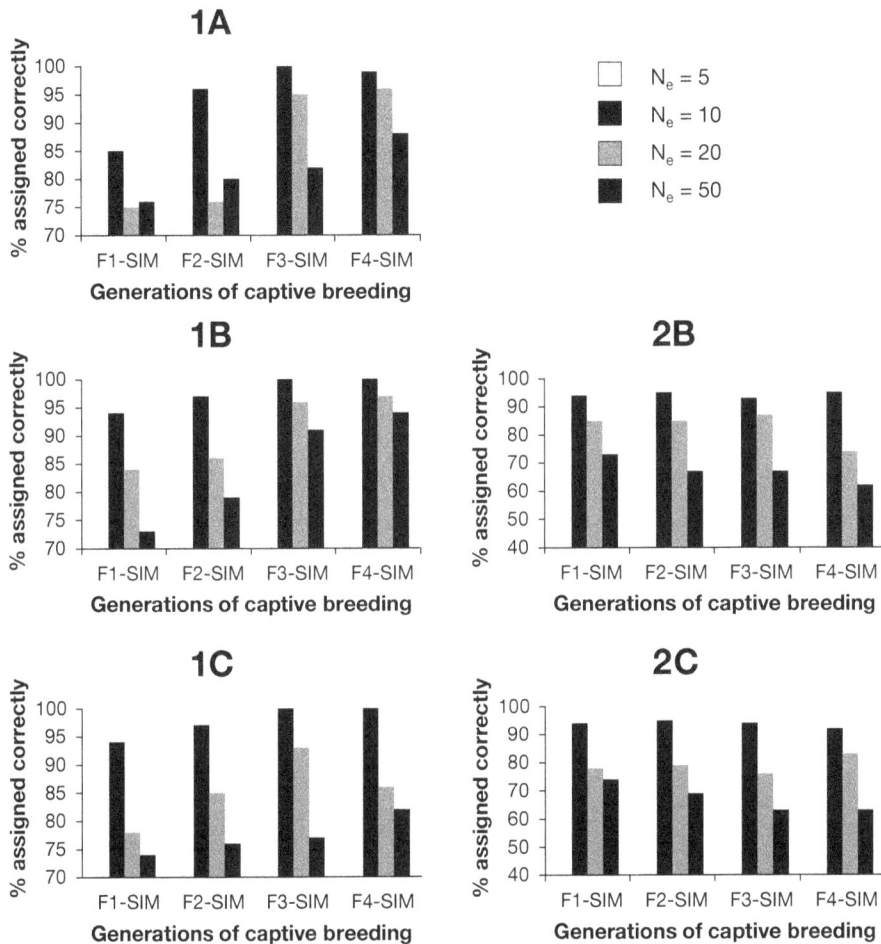

Fig. 4. *Gadus morhua* and *Solea solea*. Percentage of correctly assigned individuals in individual assignment (IA) analysis using simulated datasets. 1A, 1B, 1C: Sampling Regime 1 (reference data of aquaculture population is available for parental generation); 2B, 2C: Sampling Regime 2 (reference data of aquaculture population is limited to the founding broodstock); with results based on (A) sole data using 52 single nucleotide polymorphisms (SNPs), (B) sole data using 96 SNPs, (C) cod data using 96 SNPs. F_{1-SIM}, F_{2-SIM}, F_{3-SIM}, F_{4-SIM} = 4 offspring generations, N_e = effective population size

Hence, low polymorphism levels can be compensated through the development of a large number of gene-associated SNPs which can detect even small population genetic differences (Nielsen et al. 2012). Additionally, SNP genotyping can be highly automated and does not require extensive calibrations for marker exchange (Hauser & Seeb 2008). These characteristics make SNP markers ideal for the development of universally applicable genetic traceability tools, which inherently rely on the availability of robust reference data (Helyar et al. 2011, Nielsen et al. 2012). In the case of tracing captive-bred marine fish, SNPs can be used to detect subtle genetic differences between wild and captive-bred populations, even after just a few generations of captive breeding. Consequently, there is ample opportunity to use SNP-based tracing in fisheries management and wildlife forensics. From a management perspective, SNPs can be employed to monitor the effects of accidental/deliberate releases of captive-bred fish on wild populations. SNP-based tracing will also have forensic applications, as it will be a useful tool in the fight against mismanagement practices in aquaculture and the mislabelling of seafood products, since universal markers for the identification of captive-bred individuals can be developed (Karlsson et al. 2011).

Applications of IA and PBT analyses

Our results demonstrate that IA and PBT perform optimally under different scenarios. The performance of IA analyses improves with increased genetic differentiation between the aquaculture and wild populations as a result of increased generational breeding (high F_n) and/or a low N_e in the broodstock. PBT analyses, on the other hand, perform better when a high N_e characterizes the broodstock and/or generational breeding is low. This is as expected, given that candidate parents are less likely to be excluded from being the real parents due to loss of genetic diversity (low N_e and/or high F_n). As a result of the performance differences, the suitability of IA

Table 3. *Solea solea*. Parentage-based tagging (PBT) analysis using software package SNPPIT to identify escapees based on simulated sole data. $F_{n\text{-SIM}}$ = number of captive-bred generations that were simulated; $F_{1\text{-SIM}}$, $F_{2\text{-SIM}}$, $F_{3\text{-SIM}}$, $F_{4\text{-SIM}}$ = 4 offspring generations that were simulated; N_e = effective population size; na = not applicable

Loci	N_e	$F_{n\text{-SIM}}$	% assigned to correct population	Proportion assignments with p > 0.05	Number of non-excluded parentage trios ($\times 10^3$)	Number of non-excluded trios from the wrong population ($\times 10^3$)
96	50	$F_{1\text{-SIM}}$	100	0.00	0.69	0.05
		$F_{2\text{-SIM}}$	100	0.04	4.98	0.07
		$F_{3\text{-SIM}}$	100	0.04	5.95	0.09
		$F_{4\text{-SIM}}$	100	0.04	7.70	0.13
	4	$F_{1\text{-SIM}}$	100	0.00	0.54	0.04
		$F_{2\text{-SIM}}$	100	0.34	471.06	0.18
		$F_{3\text{-SIM}}$	100	0.28	400.90	1.84
		$F_{4\text{-SIM}}$	na	na	na	na
48	50	$F_{1\text{-SIM}}$	99	0.01	6.65	1.94
		$F_{2\text{-SIM}}$	100	0.11	81.82	5.18
		$F_{3\text{-SIM}}$	99	0.10	122.12	8.41
		$F_{4\text{-SIM}}$	100	0.25	240.94	10.69
	4	$F_{1\text{-SIM}}$	100	0.00	1.72	1.14
		$F_{2\text{-SIM}}$	99	0.11	81.83	5.18
		$F_{3\text{-SIM}}$	na	na	na	na
		$F_{4\text{-SIM}}$	na	na	na	na

and PBT analyses is strongly dependent on the ultimate goal of genetic tracing studies. Hence, our results are important for wildlife forensics and fisheries management to determine the optimal assignment strategy.

A common goal of fisheries management is the preservation or restoration of commercially important fish populations to levels which will produce a long-term maximum sustainable yield (MSY) (FAO Fisheries and Aquaculture Department 2008). Since the number of overexploited marine fish populations con-

tinues to increase, stock enhancement and sea ranching programmes have become popular management actions (Bell et al. 2008, FAO Fisheries and Aquaculture Department 2012). Consequently, the release of first-generation captive-bred juvenile fish which are genetically similar to the local wild populations has increased (Bell et al. 2008). Given that PBT analyses have a high identification efficiency for first-generation escapees and can detect hybridization between wild and captive-reared conspecifics, they can be used to jointly evaluate the levels of introgression (enforcement action) and the efficiency of restocking, stock enhancement and sea ranching programmes (management action).

Robust, forensically validated and universally applicable traceability tools can also be used in wildlife forensics to support legal actions against mismanagement of aquaculture facilities, which increases the chance of escapees, or the mislabelling of seafood products for financial profits (Ogden 2008, Glover 2010, Hanner et al. 2011). Our results indicate that both IA and PBT are potentially valuable provided that the aquaculture history of the species of interest is taken into account. IA analyses are a powerful tool for species with a long aquaculture history since cap-

Table 4. *Solea solea*. Parentage-based tagging (PBT) approach using software package SNPPIT for identifying escapees based on the empirical sole aquaculture data. Scenario A: broodstock originated from a genetically different population than the local wild populations, Scenario B: broodstock originated from a local wild population. SNP = single nucleotide polymorphism

Broodstock origin	Scenario	Number of SNPs	Escapees		Natural individuals	
			% assigned to both parents	% significantly assigned	% assigned to at least 1 parent	% significantly assigned
Atlantic Ocean	A	50	100	100	6	0
	A	30	100	100	21	0
	A	21	85	35	26	2
	B	50	100	100	6	0
	B	35	100	100	15	0
	B	30	100	95	19	0
Mediterranean Sea	A	40	100	100	52	0
	A	35	100	95	51	0
	A	30	95	80	74	0.7
	B	40	100	100	84	0
	B	35	100	100	81	0
	B	30	95	80	90	0

tive breeding has resulted in a strong genetic differentiation between captive-bred and wild fish populations (Bekkevold et al. 2006, Karlsson et al. 2011). However, most marine fish species have only recently been bred in captivity and thus forensic tools need to be able to differentiate between genetically similar captive-bred species and wild conspecifics. Our findings suggest that PBT can be used for these recently domesticated fish species, since assignment success was high after only a single generation of captive breeding. This is in line with expectations, since PBT was originally developed to identify the source of salmon released into rivers and is thus capable of differentiating between genetically similar hatchery populations (Anderson & Garza 2006). Genetic assignment methods have already been successfully applied in a forensic context (Wong & Hanner 2008, Glover 2010). However, real-life situations often complicate genetic tracing studies (Glover et al. 2009). As such, the presence of multiple (genetically similar) putative source farms and the lack of extensive genetic reference data will reduce the assignment efficiency of both IA and PBT. Although the latter is less problematic for IA analysis, PBT unequivocally requires genotypic information from all parental individuals that have contributed to the subsequent generation. The increased use of genetic broodstock management and selective breeding programmes might partially resolve this but the feasibility of using PBT in a forensic context remains controversial (Blonk et al. 2010, Vandeputte et al. 2011).

Validation of traceability approaches

Validating traceability methods requires a detailed comparison between expected (simulations) and observed (empirical) results. The assignment success rates of the analyses based on the $F_{1\text{-SIM}}$ and the empirical data reveal that overall, a higher success rate is obtained in the empirical analyses. The fact that relatively more SNP makers are needed for unambiguous assignments in the simulated data can be explained by a high reproductive skew in real aquaculture production ($A_{\text{sole-BS}}$ and $M_{\text{sole-BS}}$), which is difficult to simulate with currently available software packages. From the N_e values estimated based on the observed heterozygote excess in the $M_{\text{sole-F1}}$, we conclude that on average, 2 parental individuals contributed to each offspring batch, and these finding are supported by the results from the additional parentage testing (Supplement 4). Furthermore, comparing the genetic differentiation between the em-

pirical and simulated data (DAPC) suggests that within the $M_{\text{sole-BS}}$, an N_e of between 5 and 10 is the most likely, which is supported by the N_e estimates found in the $A_{\text{sole-BS}}$ by Blonk et al. (2009).

Other evidence supporting the methodology employed here arises from the comparison of current results with earlier studies. Vandeputte et al. (2011) recorded a decrease in the assignment power when comparing theoretical, simulated and empirical parentage assignments using microsatellite data. However, our study has clearly indicated that large-scale SNP genotyping (i.e. genome scan) combined with a selection procedure for highly informative gene-associated markers (high F_{ST} values and PIC) can increase the assignment power in empirical studies. This is consistent with the findings of previous studies which recorded similarly high assignment efficiencies with only a small number of markers (Nielsen et al. 2012). Hence, the methodology presented here will be valuable for future traceability studies where sufficient genetic background information is available for the species of interest. With low-cost high-throughput genotyping-by-sequencing methods now available to be implemented in breeding programmes (Davey et al. 2011), the cost of developing a large battery of markers should not impede applications to fisheries management and wildlife forensics.

CONCLUSIONS

This study has evaluated the relative power of parentage-based tagging (PBT) and individual assignment (IA) for identifying the population of origin of marine aquaculture fish under a range of scenarios, highlighting the benefits and disadvantages of each. PBT potentially offers the strongest line of traceability evidence, as the identification of a specific parental pair with high confidence is likely to be more powerful than a combined population assignment and exclusion approach under IA, particularly where aquaculture and wild populations have not diverged significantly. The results presented here have shown that PBT analyses will be particularly valuable in fisheries management to evaluate the genetic effects and the impact of accidental and/or deliberately released captive-bred fish. However, current aquaculture practices restrict the practical application of PBT due to the requirement for complete broodstock sampling; consequently, in most marine fish aquaculture scenarios, IA analyses are considered to be of more practical use for future traceability applications. Ultimately, the availability of genetic background information and the aim of the study will determine whether IA or PBT will be the method of choice.

Acknowledgements. We thank Fiskaaling aquaculture research station (Faroe Islands) for providing samples of farmed cod. Additionally, we thank R. J. W. Blonk for providing samples and parentage information of the hatchery-reared Atlantic sole. Research was funded by the European Commission Joint Research Centre (JRC) through the AQUAGEN project (https://aquagen.jrc.ec.europa.eu; tender No. IPSC_2010_04_11_NC) and benefited from the EU FP7 project FishPopTrace (KBBE-2007-212399) and Aqua-Trace (KBBE-311920). E.D. and G.E.M. were funded by the Research Foundation-Flanders (FWO-Vlaanderen) during the course of this research.

LITERATURE CITED

ä Abadía-Cardoso A, Anderson EC, Pearse DE, Garza JC (2013) Large-scale parentage analysis reveals reproductive patterns and heritability of spawn timing in a hatchery population of steelhead (*Oncorhynchus mykiss*). Mol Ecol 22:4733–4746

ä Anderson EC (2010a) Assessing the power of informative subsets of loci for population assignment: standard methods are upwardly biased. Mol Ecol Resour 10:701–710

Anderson EC (2010b) Computational algorithms and user-friendly software for parentage-based tagging of Pacific salmonids. Final report to Pacific Salmon Commission's Chinook Technical Committee. NOAA Southwest Fisheries Science Center, Santa Cruz, CA

Anderson EC (2014) Simple simulation of Mendelian inheritance of unlinked markers from specified matings and populations. Source code for Nookie software. http://github.com/eriqande/nookie

ä Anderson EC, Garza JC (2006) The power of single-nucleotide polymorphisms for large-scale parentage inference. Genetics 172:2567–2582

ä André C, Larsson LC, Laikre L, Bekkevold D and others (2011) Detecting population structure in a high gene-flow species, Atlantic herring (*Clupea harengus*): direct, simultaneous evaluation of neutral vs putatively selected loci. Heredity 106:270–280

Arechavala-Lopez P, Fernandez-Jover D, Black KD, Ladoukakis E, Bayle-Sempere JT, Sanchez-Jerez P, Dempster T (2013) Differentiating the wild or farmed origin of Mediterranean fish: a review of tools for sea bream and sea bass. Rev Aquacult 5:137–157

ä Bekkevold D, Hansen MM, Nielsen EE (2006) Genetic impact of gadoid culture on wild fish populations: predictions, lessons from salmonids, and possibilities for minimizing adverse effects. ICES J Mar Sci 63:198–208

Belkhir K, Borsa P, Chikhi L, Raufaste N, Bonhomme F (2004) GENETIX 4.05, logiciel sous Windows TM pour la génétique des populations. Lab génome, Popul Interact CNRS Umr 5000, Université de Montpellier II, Montpellier

Bell JD, Leber KM, Blankenship HL, Loneragan NR, Masuda R (2008) A new era for restocking, stock enhancement and sea ranching of coastal fisheries resources. Rev Fish Sci 16:1–9

ä Björnsson B (2011) Ranching of wild cod in 'herds' formed with anthropogenic feeding. Aquaculture 312:43–51

ä Blonk RJW, Komen J, Kamstra A, Crooijmans RPMA, van Arendonk JAM (2009) Levels of inbreeding in group mating captive broodstock populations of Common sole, (*Solea solea*), inferred from parental relatedness and contribution. Aquaculture 289:26–31

ä Blonk RJW, Komen H, Kamstra A, van Arendonk JAM (2010) Estimating breeding values with molecular relatedness and reconstructed pedigrees in natural mating populations of common sole, *Solea solea*. Genetics 184:213–219

Braithwaite VA, Salvanes AGV (2010) Aquaculture and restocking: implications for conservation and welfare. Anim Welf 19:139–149

ä Brown N (2002) Flatfish farming systems in the Atlantic region. Rev Fish Sci 10:403–419

ä Brown C, Miltiadou D, Tsigenopoulos C (2015) Prevalence and survival of escaped European seabass *Dicentrarchus labrax* in Cyprus identified using genetic markers. Aquacult Environ Interact 7:49–59

ä Clemento AJ, Crandall ED, Garza JC, Anderson EC (2014) Evaluation of a single nucleotide polymorphism baseline for genetic stock identification of Chinook salmon (*Oncorhynchus tshawytscha*) in the California Current large marine ecosystem. Fish Bull 112:112–130

ä Cline E (2012) Marketplace substitution of Atlantic salmon for Pacific salmon in Washington State detected by DNA barcoding. Food Res Int 45:388–393

ä Danancher D, Garcia-Vazquez E (2011) Genetic population structure in flatfishes and potential impact of aquaculture and stock enhancement on wild populations in Europe. Rev Fish Biol Fish 21:441–462

ä Davey JW, Hohenlohe PA, Etter PD, Boone JQ, Catchen JM, Blaxter ML (2011) Genome-wide genetic marker discovery and genotyping using next-generation sequencing. Nat Rev Genet 12:499–510

▶ Diopere E, Maes GE, Komen H, Volckaert FAM, Groenen MAM (2014) A genetic linkage map of sole (*Solea solea*): a tool for evolutionary and comparative analyses of exploited (flat)fishes. PLoS ONE 9:e115040

ä Duarte CM, Marbá N, Holmer M (2007) Rapid domestication of marine species. Science 316:382–383

FAO Fisheries and Aquaculture Department (2008) FAO technical guidelines for responsible fisheries. FAO, Rome

FAO Fisheries and Aquaculture Department (2012) The state of world fisheries and aquaculture. FAO, Rome

FAO Fisheries and Aquaculture Department (2014) The state of world fisheries and aquaculture—opportunities and challenges. FAO, Rome

FAO Fisheries and Aquaculture Department (2015a) Species fact sheets: *Gadus morhua* (Linnaeus, 1758). FAO, Rome

FAO Fisheries and Aquaculture Department (2015b) Species fact sheets: *Solea solea* (Quensel, 1806). FAO, Rome

ä Gjedrem T, Robinson N, Rye M (2012) The importance of selective breeding in aquaculture to meet future demands for animal protein: a review. Aquaculture 350–353:117–129

▶ Glover KA (2010) Forensic identification of fish farm escapees: the Norwegian experience. Aquacult Environ Interact 1:1–10

ä Glover KA, Skilbrei OT, Skaala Ø (2008) Genetic assignment identifies farm of origin for Atlantic salmon *Salmo salar* escapees in a Norwegian fjord. ICES J Mar Sci 65:

912–920

ä Glover KA, Hansen MM, Skaala Ø (2009) Identifying the source of farmed escaped Atlantic salmon (*Salmo salar*): Bayesian clustering analysis increases accuracy of assignment. Aquaculture 290:37–46

ä Glover KA, Hansen MM, Lien S, Als TD, Høyheim B, Skaala Ø (2010) A comparison of SNP and STR loci for delineating population structure and performing individual genetic assignment. BMC Genet 11:2 doi:10.1186/1471-2156-11-2

► Glover KA, Dahle G, Jorstad KE (2011) Genetic identification of farmed and wild Atlantic cod, *Gadus morhua*, in coastal Norway. ICES J Mar Sci 68:901–910

► Glover KA, Pertoldi C, Besnier F, Wennevik V, Kent M, Skaala Ø (2013) Atlantic salmon populations invaded by farmed escapees: quantifying genetic introgression with a Bayesian approach and SNPs. BMC Genet 14:74 doi: 10.1186/1471-2156-14-74

Goudet J (1995) FSTAT (version 1.2): a computer program to calculate F-statistics. J Hered 86:485–486

► Hanner R, Becker S, Ivanova NV, Steinke D (2011) FISH-BOL and seafood identification: geographically dispersed case studies reveal systemic market substitution across Canada. Mitochondrial DNA 22:106–122

ä Hauser L, Seeb JE (2008) Advances in molecular technology and their impact on fisheries genetics. Fish Fish 9: 473–486

ä Helyar SJ, Hemmer-Hansen J, Bekkevold D, Taylor MI and others (2011) Application of SNPs for population genetics of nonmodel organisms: new opportunities and challenges. Mol Ecol Resour 11:123–136

ä Hindar K, Ryman N, Utter F (1991) Genetic effects of cultured fish on natural fish populations. Can J Fish Aquat Sci 48:945–957

ä Hohenlohe PA, Day MD, Amish SJ, Miller MR and others (2013) Genomic patterns of introgression in rainbow and westslope cutthroat trout illuminated by overlapping paired-end RAD sequencing. Mol Ecol 22:3002–3013

ä Howell BR (1997) A re-appraisal of the potential of the sole, *Solea solea* (L.), for commercial cultivation. Aquaculture 155:355–365

ä Imsland AK, Foss A, Conceicão LEC, Dinis MT and others (2003) A review of the culture potential of *Solea solea* and *S. senegalensis*. Rev Fish Biol Fish 13:379–407

► Jacquet JL, Pauly D (2008) Trade secrets: renaming and mislabeling of seafood. Mar Policy 32:309–318

ä Jensen Ø, Dempster T, Thorstad EB, Uglem I, Fredheim A (2010) Escapes of fishes from Norwegian sea-cage aquaculture: causes, consequences and prevention. Aquacult Environ Interact 1:71–83

ä Jombart T (2008) adegenet: a R package for the multivariate analysis of genetic markers. Bioinformatics 24:1403–1405

► Jones AG, Ardren WR (2003) Methods of parentage analysis in natural populations. Mol Ecol 12:2511–2523

ä Jones AG, Small CM, Paczolt KA, Ratterman NL (2010) A practical guide to methods of parentage analysis. Mol Ecol Resour 10:6–30

► Jørstad KE, van der Meeren T, Paulsen OI, Thomsen T, Thorsen A, Svåsand T (2008) 'Escapes' of eggs from farmed cod spawning in net pens: recruitment to wild stocks. Rev Fish Sci 16:285–295

ä Karlsson S, Moen T, Lien S, Glover KA, Hindar K (2011) Generic genetic differences between farmed and wild Atlantic salmon identified from a 7K SNP-chip. Mol Ecol Resour 11:247–253

► Kitada S, Kishino H (2006) Lessons learned from Japanese marine finfish stock enhancement programmes. Fish Res 80:101–112

► Kitada S, Shishidou H, Sugaya T, Kitakado T, Hamasaki K, Kishino H (2009) Genetic effects of long-term stock enhancement programs. Aquaculture 290:69–79

► Kjesbu OS, Taranger GL, Trippel EA (2006) Gadoid mariculture: development and future challenges. ICES J Mar Sci 63:187–191

► Laikre L, Schwartz MK, Waples RS, Ryman N (2010) Compromising genetic diversity in the wild: unmonitored large-scale release of plants and animals. Trends Ecol Evol 25:520–529

⚓ Lamaze FC, Sauvage C, Marie A, Garant D, Bernatchez L (2012) Dynamics of introgressive hybridization assessed by SNP population genomics of coding genes in stocked brook charr (*Salvelinus fontinalis*). Mol Ecol 21:2877–2895

⚓ Lamaze FC, Garant D, Bernatchez L (2013) Stocking impacts the expression of candidate genes and physiological condition in introgressed brook charr (*Salvelinus fontinalis*) populations. Evol Appl 6:393–407

⚓ Liao IC, Su MS, Leaño EM (2003) Status of research in stock enhancement and sea ranching. Rev Fish Biol Fish 13: 151–163

⚓ Luikart G, Cornuet JM (1999) Estimating the effective number of breeders from heterozygote excess in progeny. Genetics 151:1211–1216

⚓ Manel S, Gaggiotti OE, Waples RS (2005) Assignment methods: matching biological questions with appropriate techniques. Trends Ecol Evol 20:136–142

⚓ Mariani S, Ellis J, O'Reilly A, Bréchon AL, Sacchi C, Miller DD (2014) Mass media influence and the regulation of illegal practices in the seafood market. Conserv Lett 7: 478–483

⚓ Marshall TC, Slate J, Kruuk LE, Pemberton JM (1998) Statistical confidence for likelihood-based paternity. Mol Ecol 7:639–655

⚓ McGinnity P, Prodohl P, Ferguson K, Hynes R and others (2003) Fitness reduction and potential extinction of wild populations of Atlantic salmon, *Salmo salar*, as a result of interactions with escaped farm salmon. Proc R Soc Lond B Biol Sci 270:2443–2450

⚓ Morin PA, Luikart G, Wayne RK (2004) SNPs in ecology, evolution and conservation. Trends Ecol Evol 19:208–216

⚓ Murray AG, Peeler EJ (2005) A framework for understanding the potential for emerging diseases in aquaculture. Prev Vet Med 67:223–235

⚓ Natale F, Hofherr J, Fiore G, Virtanen J (2013) Interactions between aquaculture and fisheries. Mar Policy 38: 205–213

⚓ Naylor RL, Goldburg RJ, Primavera JH, Kautsky N and others (2000) Effect of aquaculture on world fish supplies. Nature 405:1017–1024

⚓ Naylor RL, Hardy RW, Bureau DP, Chiu A and others (2009) Feeding aquaculture in an era of finite resources. Proc Natl Acad Sci USA 106:15103–15110

⚓ Nielsen EE, Bach LA, Kotlicki P (2006) Hybridlab (version 1.0): a program for generating simulated hybrids from population samples. Mol Ecol Notes 6:971–973

⚓ Nielsen EE, Hemmer-Hansen J, Larsen PF, Bekkevold D (2009) Population genomics of marine fishes: identifying adaptive variation in space and time. Mol Ecol 18: 3128–3150

⚓ Nielsen EE, Cariani A, Aoidh E, Maes GE and others (2012) Gene-associated markers provide tools for tackling illegal fishing and false eco-certification. Nat Commun 3: 851–856

► Noble TH, Smith-Keune C, Jerry DR (2014) Genetic investigation of the large-scale escape of a tropical fish, barramundi *Lates calcarifer*, from a sea-cage facility in northern Australia. Aquacult Environ Interact 5:173–183

► Ogden R (2008) Fisheries forensics: the use of DNA tools for improving compliance, traceability and enforcement in the fishing industry. Fish Fish 9:462–472

► Paetkau D, Slade R, Burden M, Estoup A (2004) Genetic assignment methods for the direct, real-time estimation of migration rate: a simulation-based exploration of accuracy and power. Mol Ecol 13:55–65

► Piry S, Alapetite A, Cornuet JM, Paetkau D, Baudouin L, Estoup A (2004) GENECLASS2: a software for genetic assignment and first-generation migrant detection. J Hered 95:536–539

► Porta J, Porta JM, Martínez-Rodríguez G, Alvarez MC (2006a) Genetic structure and genetic relatedness of a hatchery stock of Senegal sole (*Solea senegalensis*) inferred by microsatellites. Aquaculture 251:46–55

► Porta J, Porta JM, Martínez-Rodríguez G, Alvarez MC (2006b) Development of a microsatellite multiplex PCR for Senegalese sole (*Solea senegalensis*) and its application to broodstock management. Aquaculture 256:159–166

► Porta J, Porta JM, Martínez-Rodríguez G, Alvarez MC (2007) Substantial loss of genetic variation in a single generation of Senegalese sole (*Solea senegalensis*) culture: implications in the domestication process. J Fish Biol 71:223–234

R Development Core Team (2010) R : a language and environment for statistical computing. R Foundation for Statistical Computing, Vienna

► Rosenlund G, Skretting M (2006) Worldwide status and perspective on gadoid culture. ICES J Mar Sci 63:194–197

► Seymour EA, Bergheim A (1991) Towards a reduction of pollution from intensive aquaculture with reference to the farming of salmonids in Norway. Aquacult Eng 10:73–88

► Somarakis S, Pavlidis M, Saapoglou C, Tsigenopoulos C, Dempster T (2013) Evidence for 'escape through spawning' in large gilthead sea bream *Sparus aurata* reared in commercial sea-cages. Aquacult Environ Interact 3: 135–152

► Sparrevohn CR, Støttrup JG (2007) Post-release survival and feeding in reared turbot. J Sea Res 57:151–161

► Steele CA, Anderson EC, Ackerman MW, Hess MA, Campbell NR, Narum SR, Campbell MR (2013) A validation of parentage-based tagging using hatchery steelhead in the Snake River basin. Can J Fish Aquat Sci 70: 1046–1054

► Thurstan RH, Brockington S, Roberts CM (2010) The effects of 118 years of industrial fishing on UK bottom trawl fisheries. Nat Commun 1:15, doi:10.1038/ncomms1013

► Vandamme SG, Maes GE, Raeymaekers JAM, Cottenie K and others (2014) Regional environmental pressure influences population differentiation in turbot (*Scophthalmus maximus*). Mol Ecol 23:618–636

► Vandeputte M, Rossignol MN, Pincent C (2011) From theory to practice: empirical evaluation of the assignment power of marker sets for pedigree analysis in fish breeding. Aquaculture 314:80–86

► Waples RS (2010) High-grading bias: subtle problems with assessing power of selected subsets of loci for population assignment. Mol Ecol 19:2599–2601

Warner K, Timme W, Lowell B, Hirshfield M (2013) Oceana study reveals seafood fraud nationwide. Oceana. http://oceana.org/en/news-media/publications/reports/oceana-study-reveals-seafood-fraud-nationwide

► Weir LK, Grant JWA (2005) Effects of aquaculture on wild fish populations : a synthesis of data. Environ Rev 13: 145–168

► Wong EHK, Hanner RH (2008) DNA barcoding detects market substitution in North American seafood. Food Res Int 41:828–837

5

Going beyond the search for solutions: understanding trade-offs in European integrated multi-trophic aquaculture development

Adam D. Hughes*, Kenneth D. Black

Scottish Association for Marine Sciences, Dunbeg, Oan, Argyll PA37 1QA, UK

ABSTRACT: There has been significant interest in the development of integrated multi-trophic aquaculture (IMTA) in Europe. Much of this interest has come from academia and regulators, and while elements within the European aquaculture industry have expressed an interest, to date, the adoption of the concept has been limited. Part of the attraction for regulators and academics is the ecological/economic win/win that is associated with eco-innovation solutions. However, if we are to understand why there has been limited uptake of IMTA in Europe, perhaps it is necessary to look at the issue in terms of trade-offs for the individual farmer or company. Using this viewpoint, we investigate the balance of trade-offs for the individual farmer or company to diversify from a traditional fin-fish production business into an IMTA system. In doing so, we reveal that the balance of trade-offs is currently not sufficiently positive to motivate the large-scale uptake of IMTA in Europe, and we contrast this against the situation in Asia where the balance of trade-offs gives better support for the adoption and practice of IMTA. By better understanding the trade-offs for the individual, it is possible to better understand the conditions that will promote the development of IMTA in Europe.

KEY WORDS: Cage culture · Eco-intensification · Integrated multi-trophic aquaculture · IMTA · Social licence · Extractive aquaculture

INTRODUCTION

Integrated multi-trophic aquaculture (IMTA) is both conceptually a simple idea and also highly appealing to regulators: the waste products from one food production process (in this case, fin-fish production) is acquired and assimilated by other organisms and converted into valuable products. This process both eliminates waste and increases the productivity of the food production system (Troell et al. 2003, Neori et al. 2004, Chopin et al. 2006). This win/win situation has its roots deeply buried in the eco-efficiency philosophy that aims to simultaneously in-

crease both the economic and environmental performances of an industry or business (Ehrenfeld 2005). Alternately, IMTA can be thought of in terms of eco-intensification, where the productivity per unit input is increased (Amano & Ebihara 2005). In Europe, the model for fed fin-fish aquaculture has been very linear, in line with a fast replacement economy where the inputs to the industry lead to consumption of natural resources with high energy and water consumption, with externalised wastes. This is in contrast to the principles of IMTA, which aim to create an industry-based spiral or loop system (now termed the circular economy) that minimises energy

*Corresponding author: adam.hughes@sams.ac.uk

flows, losses, and environmental deterioration, without restricting economic growth or social progress (Boulding 1966, Stahel 1982). The win/win that IMTA represents has been cited a number of times as a solution to some of the problems that are facing the European fin-fish aquaculture industry, such as ecological damage, economic stability, and dependence on commercial feed (Klinger & Naylor 2012, Chopin et al. 2013, Granada et al. 2015).

Despite a strong tradition in Asia (Chan 1993) and the fact that the IMTA concept in various guises has been in the scientific literature for at least 40 yr, since the early 1970s (Ryther et al. 1972, 1975, Ahn et al. 1998, Buschmann et al. 2001), there is almost no commercial uptake of IMTA in Europe. This is against an increasing academic interest in Europe in the concept of IMTA (OECD 2010). Given the significant uptake of the concept of the circular economy within Europe as a whole (World Economic Forum 2014) and the eco-efficiency potential of IMTA technology, this lack of uptake is, on the face of it, hard to understand. Several studies have elucidated possible reasons why there has not been a transition from academic promise to commercial reality (Troell et al. 2003, 2009) and have included reasons such as the performance of the extractive organisms or the economic performance of the systems. However, most previous studies have identified gaps in the scientific knowledge, but it will not be scientists who implement IMTA at a commercial scale, but rather companies and individuals within those countries. Therefore, this paper attempts to better understand the commercial motivation for the adoption of IMTA. The question is even more pertinent given the fact that aquaculture production in Europe is stagnating, with growth over the last decade only around 1 % per annum (Anon 2009, 2015). This is in stark contrast to the picture in Asia, where aquaculture is the fastest-growing food production sector (FAO 2012) and IMTA is commonplace.

Here, we argue that there is a need to move past this win/win conceptual framework and its view that IMTA is a solution: this framework is flawed or at best is unhelpful. If we move beyond a solution-based mind-set, we may explain the contrasting implementation of IMTA in Europe and Asia. Instead of considering IMTA as a 'solution' for European aquaculture, it is perhaps better to quantify the trade-offs involved in its adoption. Sowell (1995) argued that in social systems, there are no solutions, there are just trade-offs between different conditions, situations or states. Thus, instead of thinking 'What will remove particular negative features in an existing situation to create a

solution?', it is more useful to frame the question as 'What must be sacrificed to achieve this particular improvement?' (Sowell 1995). Instead of thinking of IMTA as a solution and wondering why there is no industry adoption of the technology, we consider the trade-offs between the benefits and costs of adopting IMTA at the level of an existing fin-fish farmer or company. Applying this analysis to regions where IMTA is more common, we can try to understand what needs to shift in that balance of trade-offs to foster the adoption of IMTA in Europe. For the sake of this thought experiment, we will assume the scenario of an existing European (including Norway) fin-fish farmer wishing to develop a simple system of IMTA consisting of fish, mussels, and seaweed in a temperate open-water system.

EXAMPLES OF TRADE-OFFS INVOLVED IN THE ADOPTION OF IMTA BY A EUROPEAN FIN-FISH PRODUCER

Increased productivity

One of the benefits cited for IMTA is an increase in productivity, but what does that really mean at the level of the farmer or the company? Already, fin-fish producers in Europe and especially salmon producers in the Atlantic have efficient, fully industrialised operations (Vassdal & Holst 2011, Asche et al. 2013a), and there is no mechanism for IMTA to increase the productivity of their fish operation *per se*. However, considering productivity per unit input and measuring other outputs as well as fish, there are real opportunities to increase productivity. This increase in productivity happens firstly because there is an increase in production from the lower trophic species that are grown alongside the fin-fish and secondly because there is good evidence to suggest that these lower trophic species are able to utilise the nutrient from the fin-fish and are more productive when grown alongside fed aquaculture (Lander et al. 2012, Sanderson et al. 2012, Irisarri et al. 2015), although this experience has not been universal (Cheshuk et al. 2003), indicating that the integration may require some tuning. Using a mass balance approach, the production of 1 tonne (t) of salmon releases approximately 50 kg of nitrogen into the environment (Wang et al. 2012), which could support the growth of 10 t of seaweed or 5 t of mussels over the course of the production cycle of the salmon (Holdt & Edwards 2014). In the case of seaweed, this equates to a 1000 % increase in biomass (wet weight relative to the fish production) produced and an increase in protein production by 166 % (based on the N content of fish being 3 %; Wang et al. 2014). As such, there is a clear increase in productivity of the whole system per unit feed (if the farmer could convert 100 % of emissions to product). However, to make this meaningful to the farmer, there needs to be a proven case that these products can be sold at above their production costs plus an acceptable margin.

There is a scenario in which IMTA may play a role in increasing feed efficiency of the main fin-fish species. The provision of feed is the single largest contributor to resource use and emissions from open-cage salmon production (Grosholz et al. 2015). If instead of framing productivity in terms of per unit feed, we frame it in terms of per unit of fish meal or fish oil from wild fish stocks (biotic depletion), then the recycling of nitrogen lost to the environment back into marine proteins and lipids, and the subsequent reincorporation into fish feed, offers opportunities to increase the productivity in a way that may be meaningful to the farmer. The incorporation of seaweed into fin-fish aquafeeds has been shown to be possible at an experimental level (Wahbeh 1997, Yildirim et al. 2009, Marinho et al. 2013). As previously stated, an IMTA system with 100 % efficiency in capturing nitrogen would allow for significant protein production using seaweed or mussels, but in addition, there is the potential for significant production of marine lipids. From the 10 t (w/w) of seaweed produced for every tonne of fish, it would be possible to produce 164 kg of protein and 9 kg of marine lipids (based on the production of *Alaria esculenta*; Mæhre et al. 2014), and in theory, these components could be recycled back into fish feed. This has the potential to significantly reduce the environmental impact of the industry by further reducing reliance on marine proteins and lipids from wild-harvest fisheries. However, the reality is much more complex and logistical; legal (in Europe) and economic constraints make this unfeasible in the foreseeable future. Furthermore, it is unlikely that this recycling would be of benefit to the individual farmer in the short term.

Reduced environmental impact

In the eco-efficiency win/win, the second win is reduced environmental impact. This is achieved through the ability of the extractive organisms to make use of waste products of the fin-fish production as nutrient and energy. As such, these waste streams are assimilated into the tissues of the extractive organisms and are removed from the environment. In our scenario, there are 2 waste product streams of interest: dissolved nutrients and particulate organic matter (POM: fish faeces and uneaten pellets). The dissolved waste stream consists mainly of ammonia (Sanderson et al. 2008), which can be detected close to the fish cages but can quickly attenuate (Merceron et al. 2002). In the case of the POM, the physical extent of the plume of suspended particles is difficult to detect (Cranford et al. 2013) and may not extend much beyond a few hundred metres from the farm (Brager et al. 2015). Any direct ecological benefit with direct trophic transfer of nutrient needs to take place within this limited zone around the fish farm. Furthermore, the bioremediation potential within this zone has been shown to be limited for both mussels and for seaweed (Broch et al. 2013, Cranford et al. 2013). In addition, both mussel cultivation and seaweed cultivation have their own environmental impacts on the benthos (Eklöf et al. 2006, Wilding 2012, Ren et al. 2014). Therefore, IMTA is likely to increase the total benthic impact of any one farm, if that farm now incorporates mussel and seaweed production (Troell & Norberg 1998), but the benthic impact per unit of production (salmon plus mussels or seaweed) would be significantly reduced. However, if the benthic footprint of the fin-fish and the extractive organisms (the mussel or seaweeds) overlap, this would then in fact increase the environmental impact in this zone locally, through the additive effect of the deposition from the fin-fish and the deposition from the extractive organisms. Because fish farms are regulated regarding their benthic impact, this possible increase possesses a significant risk to the fin-fish producer.

Increased space requirement

Most fin-fish aquaculture in Europe is intensive (FAO 2012), while the extractive species usually used in IMTA are extensive cultures, with much lower levels of production per unit area. For example, cage culture of fish produces between 1125 to 1750 t ha^{-1}, mussels 76 t ha^{-1} and for aquatic plants 1 t ha^{-1} (Bostock et al. 2010), although other estimates (Hughes et al. 2012) give higher values for kelps. Because the availability of sites is cited as a limiting factor for the development of European aquaculture (IUCN 2009), the decision to use the available space for the production of anything other than the primary fin-fish product would seem a paradoxical decision for a fish farmer. This is further compounded by the value of those respective crops. Using FAO data, in Europe, the value of bivalves and aquatic plants per tonne is approximately 45% and 11%, respectively, of the value of fin-fish (FAO 2012). This is of course the value of the product and not the profit realised by the farmer. The availability of sites for aquaculture development is not a simple factor of available space but is entirely mediated through the local regulatory system, which designates the availability of sites. The availability of fin-fish sites may be partially or entirely decoupled from the availability of sites for extractive species or from IMTA sites, and therefore, the fin-fish farmer is not making a simple decision of fin-fish versus extractive species. However, given that effort must be expended both in gaining licences for fin-fish and for extractive species, the value per unit area of fin-fish would suggest that effort is better spent obtaining additional fin-fish licences where available. It has been estimated that the ratio of wet weight of kelp biomass to salmon required to sequester the nitrogen output of the salmon ranges between 6.7 and 12.9 depending on the kelp species: when converted into a space equivalent, this ranges between 0.1 and 0.13 ha per tonne of fish (Reid et al. 2013). In terms of sequestering 10% (as a nominal value) of the nitrogen from a 1000 t salmon farm, this would require approximately 10 to 13 ha of seaweed cultivation. These values are roughly in line with lower estimates for the space required to sequester 10% of the dissolved nitrogen from Danish fish farms (Holdt & Edwards 2014). Using these values and Bostock et al.'s (2010) estimates of space required for salmon production, 1 ha of salmon production would require between 17 and 23 ha of seaweed to sequester 10% of the nitrogen output. However, in their study, Holdt & Edwards (2014) argue that mussel cultivation is approximately 220% more efficient than seaweed cultivation per unit area (Holdt & Edwards 2014).

When considering these values in relation to the spatial configuration of an IMTA system, thought needs to be given to the large amount of sea room that modern fin-fish farming requires around the cages. With well boats now up to 75 m in length, there is the need for significant amounts of sea room around cage groups. This need combined with the extensive amount of space required for extractive organisms will mean that the majority of extractive organisms in an IMTA system will be outside the zone where the outputs from the fin-fish cages can be measured or any direct trophic linkage can be assumed. There is, however, a case in which this additional requirement for space that IMTA represents could be viewed as a benefit to the fish-farmer. There has been a general

move from smaller to larger farms as the aquaculture industry has developed in Europe, and this shift has left a number of smaller farm sites vacant. As licenced sites have become scarcer, there has been increasing pressure from the regulators to bring these sites back into production. Because these smaller sites are no longer cost-effective for large producers, there is a risk that the sites will be reassigned to smaller producers. This reassignment is a significant risk to larger producers in terms of disease control and biosecurity, if those sites are in the same water body as their larger sites. One option for the large producer is to use these sites for non-fin-fish production of extractive species and for the sites to act as a 'fire break' between fin-fish sites in terms of bio-security. This use opens up another possibility that IMTA can be considered not just at a farm scale but also in terms of a water-body scale, where the direct trophic linkage between the fin-fish and the extractive organisms is unproven (or unprovable) but instead a mass balance approach is taken. In this approach, the amount of nitrogen that enters the system through the fin-fish cultivation is balanced against the nitrogen removed from the system by the mussels and seaweed, irrespective of actual distance, as long as they are within the same water body (Reid et al. 2013).

One important consideration for space and IMTA is the development of benthic IMTA, because this IMTA would probably sit within the footprint of the existing farm, and as such, the space requirements would be small (Robinson et al. 2011). Initial modelling studies show that benthic IMTA could well be an appropriate technology to improve productivity and to reduce benthic enrichment (Cubillo et al. 2016). However, the authors are unaware of any such technology commercially available in Europe at the moment.

Increased social licence

Currently, there is no legislative or regulatory requirement for a fin-fish producer to implement IMTA in Europe, despite legislation in Demark to reduce the environmental emission of fish farms, which is prompting IMTA development (Holdt & Edwards 2014). However, to operate effectively within a community and to expand, an industry requires a social licence to operate, going beyond what is just required for strict compliance with the regulation or law (Gunningham et al. 2004). One of the barriers to the development of aquaculture in Europe is limited access to new sites. The availability of new sites will ultimately be determined by how the fish-farming industry is thought of in the society in which it operates and how fish farming reflects the values of the society in which it operates (Hamouda et al. 2005), and it can be argued that aquaculture increasingly requires a social licence to operate (Leith et al. 2014). Negative public perceptions of aquaculture are based around the industrialisation of the ocean (Mazur & Curtis 2008) and the emissions associated with it (Katranidis et al. 2003) and the fact that the public prioritize the reduction of environmental damage associated with aquaculture (Whitmarsh & Wattage 2006). Public acceptance of aquaculture is a function of the perceived value in terms of economic benefit weighed against the negative perceptions, such as environmental degradation (Whitmarsh & Palmieri 2009).

The conceptually simple idea of IMTA with the win/win of reduced environmental damage and increased economic benefit offers the opportunity to shift this perceived balance toward increased benefit and reduced pollution and so increase the social licence of IMTA-related aquaculture. This effect has been shown in a number of studies where, after explanation of the principles of IMTA, there is an increase in positive social perceptions (Ridler et al. 2007, Barrington et al. 2010). This pathway from better environmental performance to increased social licence to increased availability of aquaculture licences can already be seen in Norway, where the Norwegian government has created 45 'green aquaculture' licences in 2013 (Nikitina 2015). These licences are subject to strict environmental criteria on sea lice, escape risk, and other controls of environmental impacts.

Increased complexity

Much of the fin-fish aquaculture in Europe is highly industrialised and optimised, and profit margins on the fish have historically been somewhat volatile (Andersen et al. 2008, Asche et al. 2013b, Iotti & Bonazzi 2015). Adding more species to a site will add new layers of complexity to the system. If the fin-fish operation (the core business) is to remain profitable, it is crucial that these new complexities do not reduce the efficiency of the fin-fish production. Some of these complexities are logistical in terms of the additional infrastructure that is required, such as mussel and seaweed longlines. As previously discussed, any impingement of the IMTA infrastructure on the requirement of large boats or ships to access the fin-fish cages may reduce the efficiency of the

fin-fish operation. There will also be an increase in the complexity of the biosecurity of the site when dealing with organisms with different production cycles. Disease is a major constraint on the industry, costing the industry as a whole approximately $US6 billion annually (Brummett et al. 2014), so any new production system must not increase the risk of disease. There is a lack of clear evidence about the role extractive species may play as a reservoir for infectious agents or the role they may play in eliminating or reducing the risk of disease. In the case of *Vibrio anguillarum* (vibriosis), blue mussels *Mytilus edulis* were shown to accumulate the vibrio in their digestive glands (Pietrak et al. 2010). Mussel pseudofaeces contained concentrated and infectious *V. anguillarum*. Juvenile cod exposed to infected faecal material suffered 60 to 80% mortality. This result indicated that in the co-culture of mussels and fin-fish, mussels may act as a reservoir of infections for *V. angillarium* (Pietrak et al. 2010). However, in the case of infectious salmon anaemia virus (ISAV), blue mussels were shown not to accumulate the virus and may deactivate the virus (Skår & Mortensen 2007, Molloy et al. 2014). There has been significant interest in the ability of bivalves to ingest sea lice larvae (Molloy et al. 2011, Webb et al. 2013), and if the efficacy of this as a lice-control method could be proved, it would be a big driver for the adoption of IMTA by salmon farmers. Another important component when considering increased complexity is the human resource capacity within any one farm or company. Fin-fish farming, shellfish farming, and seaweed farming are all skilled professions, and while there is a degree of complementarity in the skill sets among the 3, there are also large differences in required skills. A fin-fish farmer may need to 'buy in' expertise from outside, creating further complexity and cost.

Increased profitability

Full and comprehensive economic analyses of integrated aquaculture are difficult to find for European or western context. The existing studies are based on models, and simulations suggest that IMTA can increase the profitability for the individual operator when the market conditions are right and can provide a measure of resilience during periods of unfavourable conditions (Whitmarsh et al. 2006, Ridler et al. 2007). However, both these studies are based on hypothetical farms. A study from Sanggou Bay, China, based on real data and not models, showed that there was a significant increase in profitability in an IMTA operation based on scallop and seaweed production, compared to monocultures of those species alone (Shi et al. 2013). The other route to increased profitability for a fish farmer is if s/he can sell his/her main products for more as a result of being produced through IMTA. In a public perceptions survey of integrated aquaculture (Barrington et al. 2010), restaurateurs stated that they would be willing to pay up to 10 % more for environmentally friendly seafood. Another survey showed that 38 % of New York seafood consumers would be prepared to pay 10 % extra for IMTA produced mussels if they carried appropriate labelling (Shuve et al. 2009). However, it is impossible to predict if these results would be borne out in a real marketplace.

HOW DO THESE TRADE-OFFS COMPARE WITH THE SITUATION IN ASIA?

At a very gross level, a first-order calculation shows that in Asia, the balance of fin-fish aquaculture to extractive aquaculture (where molluscs and aquatic plants categories from the FAO database are considered extractive organisms) is approximately 1:1, whereas in Europe, it is 3.5:1 (FAO 2012). This difference coupled with a higher relative value of extractive organisms (molluscs and seaweed in Asia compared to fin-fish; FAO 2012) means that the economic case for choosing a production system that boosts the growth of extractive organisms is much stronger. As such, there are fundamental differences in how these

trade-offs impact the industry in Asia. While there is no increase in productivity for fin-fish under IMTA, there is evidence to support the benefit to shellfish and seaweeds, which are a larger proportion of the Asian industry and have a higher value to this industry. Also, in Asia, the main area of environmental concern for the aquaculture industry and regulators is the impact associated with aquaculture and the water column as opposed to the benthic impacts (Hu et al. 2010, Keesing et al. 2011), and the link between IMTA (seaweed and mussels) and reduced environmental impacts is much clearer for the water-column impacts than for benthic impacts. In terms of the negative trade-offs, the increased space requirement is less of an issue to an industry more biased toward extensive production, and the increase in complexity of an IMTA operation is more manageable with a less-mechanised and more labour-intensive industry that is characteristic of Asian aquaculture. From this initial characterisation, it would appear that the trade-offs for IMTA are more positive for Asia compared to Europe (Table 1). For the European industry to embrace IMTA, there needs to be development of economically and technically viable benthic IMTA which will ameliorate the seabed impact of fin-fish culture. There also needs to be a better financial case made for the adoption of IMTA based on empirical evidence. This, combined with an increased social licence for companies practicing IMTA that translates into a greater licenced area and biomass, would significantly increase the development of IMTA in Europe.

Table 1. The relative balance of trade-offs associated with IMTA in Europe and Asia

	Europe	Asia
Positive trade-off (for individual company)		
Increased productivity	No benefit to core business (fin-fish) Relatively lower value of extractive species	Benefit to core business (shellfish and seaweed) Relatively higher value of extractive species
Better environmental performance relative to industry's concerns	Not proven for most environmental impacts, likely to increase main environmental constraint (benthic footprint)	Evidence for improved water quality (main environmental constraint)
Increased social licence	Evidence to support	No evidence
Increased profitability	Not proven	Evidence to support
Negative trade-offs		
Increased space requirement	Availability of space is a major constraint to the industry development, with limited opportunity to increase production per site due to regulation	Availability of space is a major constraint to the industry development. Limited space is driving an increase in productivity
Increased complexity	Highly industrialised industry less able to deal with increased complexity	Less industrialised production, with higher levels of human labour and therefore greater flexibility

CONCLUSION

If IMTA is to be adopted by an individual farmer, the trade-offs will have to provide a net benefit or, to paraphrase Sowell (Sowell 1995), 'What must the finfish sacrifice in order to achieve the benefits of IMTA?' There are plenty of examples of where other industries have seen this balance of trade-offs as positive and have developed new environmental standards that have increased productivity and reduced environmental damage (Porter & Van der Linde 1995, Florida 1996). The question then rises how the trade-offs are weighed to determine if there is a positive balance to the adoption of IMTA by society. It is important to note that these trade-offs and their weight is entirely scale- and context-dependent (McShane et al. 2011) and will vary according to national and international market conditions or regulations. Currently, there are few regulatory drivers in Europe to incentivise the adoption of IMTA at the level of an individual or a company. In fact, at the moment, the balance of trade-offs seems to be against the individual farmer adopting IMTA. For this to alter for the individual farmer, there needs to be a greater body of evidence of a financial benefit to the farmer, better systems to reduce the increase in complexity, and better support from policy and regulation to reinforce the increase in social licence associated with IMTA. If these trade-offs were to be considered at a national level or at an industry level, the outcomes might well be dramatically different. It is beyond the scope of this paper to make that weighing for any individual company, but it is possible to look at where the balance of evidence lies for both Europe and Asia and see some of the reasons why IMTA may be more prevalent in Asia and what needs to change in Europe before IMTA is more widely adopted.

Acknowledgements. The authors acknowledge the valuable comments supplied by the IDREEM consortium during the writing of this manuscript, in particular Marc Shorten, Angelica Mendoza and Mariachiara Chiantore. The research leading to these results was undertaken as part of the IDREEM project and received funding from the European Union's Seventh Framework Programme (FP7/2007-2013) under grant agreement number 308571.

LITERATURE CITED

ä Ahn O, Petrell RJ, Harrison PJ (1998) Ammonium and nitrate uptake by *Laminaria saccharina* and *Nereocystis luetkeana* originating from a salmon sea cage farm. J Appl Phycol 10:333–340

ä Amano K, Ebihara M (2005) Eco-intensity analysis as sustainability indicators related to energy and material flow. Manag Environ Qual 16:160–166

ä Andersen TB, Roll KH, Tveteras S (2008) The price responsiveness of salmon supply in the short and long run. Mar Resour Econ 23:425–437

ä Asche F, Guttormsen AG, Nielsen R (2013a) Future challenges for the maturing Norwegian salmon aquaculture industry: an analysis of total factor productivity change from 1996 to 2008. Aquaculture 396–399:43–50

ä Asche F, Roll KH, Sandvold HN, Sørvig A, Zhang D (2013b) Salmon aquaculture: larger companies and increased production. Aquacult Econ Manage 17:322–339

ä Barrington K, Ridler N, Chopin T, Robinson S, Robinson B (2010) Social aspects of the sustainability of integrated multi-trophic aquaculture. Aquacult Int 18:201–211

ä Bostock J, McAndrew B, Richards R, Jauncey K and others (2010) Aquaculture: global status and trends. Philos Trans R Soc Lond B 365:2897–2912

Boulding KE (1966) The economics of the coming spaceship earth. In: Jarrett H (ed) Environmental quality issues in a growing economy. Johns Hopkins University Press, Baltimore, MD, p 3–14

ä Brager LM, Cranford PJ, Grant J, Robinson SMC (2015) Spatial distribution of suspended particulate wastes at open-water Atlantic salmon and sablefish aquaculture farms in Canada. Aquacult Environ Interact 6:135–149

Breuer MEG (2015) European aquaculture. In: European Parliment (ed) Fact sheets on the European Union. European Parliament, Brussels

ä Broch OJ, Ellingsen IH, Forbord S, Wang X and others (2013) Modelling the cultivation and bioremediation potential of the kelp *Saccharina latissima* in close proximity to an exposed salmon farm in Norway. Aquacult Environ Interact 4:187–206

Brummett RE, Alvial A, Kibenge F, Forster J and others (2014) Reducing disease risk in aquaculture. Agriculture and Environmental Services Discussion Paper No. 9. World Bank Group, Washington, DC

Buschmann AH, Troell M, Kautsky N (2001) Integrated algal farming: a review. Cah Biol Mar 42:83–90

Chan GL (1993) Aquaculture, ecological engineering—lessons from China. Ambio 22:491–494

ä Cheshuk BW, Purser GJ, Quintana R (2003) Integrated open-water mussel (*Mytilus planulatus*) and Atlantic salmon (*Salmo salar*) culture in Tasmania, Australia. Aquaculture 218:357–378

Chopin T, Robinson S, Sawhney M, Bastarache S, Belyea E (2004) The AquaNet integrated multi-trophic aquaculture project: rationale of the project and development of kelp cultivation as the inorganic extractive component of the system. Bull Aquacult Asoc Can 104-3:11–18

ä Chopin T, MacDonald B, Robinson S, Cross S and others (2013) The Canadian integrated multi-trophic aquaculture network (CIMTAN)—a network for a new era of ecosystem responsible aquaculture. Fisheries (Bethesda, Md) 38:297–308

Commission of the European Communities (2009) Building a sustainable future for aquaculture: a new impetus for the strategy for the sustainable development of European aquaculture. Commission of the European Communities, Brussels

ä Cranford PJ, Reid GK, Robinson SMC (2013) Open water integrated multi-trophic aquaculture: constraints on the effectiveness of mussels as an organic extractive component. Aquacult Environ Interact 4:163–173

Cubillo AM, Ferreira JG, Robinson SMC, Pearce CM,

Corner RA, Johansen J (2016) Role of deposit feeders in integrated multi-trophic aquaculture—a model analysis. Aquaculture 453:54–66

ä Ehrenfeld JR (2005) Eco-efficiency: philosophy, theory, and tools. J Ind Ecol 9:6–8

ä Eklöf JS, Henriksson R, Kautsky N (2006) Effects of tropical open-water seaweed farming on seagrass ecosystem structure and function. Mar Ecol Prog Ser 325:73–84

FAO (Food and Agriculture Organization of the United Nations) (2012) The state of world fisheries and aquaculture—2012. FAO, Rome. www.fao.org/docrep/016/i2727e/i2727e00.htm

ä Florida R (1996) Lean and green: the move to environmentally conscious manufacturing. Calif Manage Rev 39:80

World Economic Forum (2014) Towards the circular economy: accelerating the scale-up across global supply chains. World Economic Forum with the Ellen MacArthur Foundation and McKinsey & Company, Geneva

Granada L, Sousa N, Lopes S, Lemos MFL (2015) Is integrated multitrophic aquaculture the solution to the sectors' major challenges?—a review. Rev Aquacult (in press), doi:10.1111/raq.12093

Grosholz E, Crafton RE, Fontana R, Pasari J, Williams S, Zabin C (2015) Aquaculture as a vector for marine invasions in California. Biol Invasions 17:1471–1484

ä Gunningham N, Kagan RA, Thornton D (2004) Social license and environmental protection: why businesses go beyond compliance. Law Soc Inq 29:307–341

ä Hamouda L, Hipel KW, Marc Kilgour D, Noakes DJ, Fang L, McDaniels T (2005) The salmon aquaculture conflict in British Columbia: a graph model analysis. Ocean Coast Manage 48:571–587

ä Holdt SL, Edwards MD (2014) Cost-effective IMTA: a comparison of the production efficiencies of mussels and seaweed. J Appl Phycol 26:933–945

ä Hu CM, Li DQ, Chen CS, Ge JZ and others (2010) On the recurrent *Ulva prolifera* blooms in the Yellow Sea and East China Sea. J Geophys Res C Oceans 115:C05017, doi:10.1029/2009JC005561

ä Hughes A, Kelly M, Black K, Stanley M (2012) Biogas from macroalgae: Is it time to revisit the idea? Biotechnol Biofuels 5:86

► Iotti M, Bonazzi G (2015) Profitability and financial sustainability analysis in Italian aquaculture firms by application of economic and financial margins. Am J Agric Biol Sci 10:18–34

ä Irisarri J, Fernández-Reiriz MJ, Labarta U, Cranford PJ, Robinson SMC (2015) Availability and utilization of waste fish feed by mussels *Mytilus edulis* in a commercial integrated multi-trophic aquaculture (IMTA) system: a multi-indicator assessment approach. Ecol Indic 48:673–686

IUCN (International Union for the Conservation of Nature) (2009) Aquaculture site selection and site management. In: IUCN (ed) Guide for the sustainable development of Mediterranean aquaculture 2. IUCN, Gland

ä Katranidis S, Nitsi E, Vakrou A (2003) Social acceptability of aquaculture development in coastal areas: the case of two Greek islands. Coast Manage 31:37–53

► Keesing JK, Liu DY, Fearns P, Garcia R (2011) Inter- and intra-annual patterns of *Ulva prolifera* green tides in the Yellow Sea during 2007-2009, their origin and relationship to the expansion of coastal seaweed aquaculture in China. Mar Pollut Bull 62:1169–1182

ä Klinger D, Naylor R (2012) Searching for solutions in aquaculture: charting a sustainable course. Annu Rev Environ

Resour 37:247

ä Lander TR, Robinson SMC, Macdonald BA, Martin JD (2012) Enhanced growth rates and condition index of blue mussels (*Mytilus edulis*) held at integrated multi-trophic aquaculture sites in the Bay of Fundy. J Shellfish Res 31:997–1007

ä Leith P, Ogier E, Haward M (2014) Science and social license: defining environmental sustainability of Atlantic salmon aquaculture in south-eastern Tasmania, Australia. Soc Epistemol 28:277–296

► Mæhre HK, Malde MK, Eilertsen KE, Elvevoll EO (2014) Characterization of protein, lipid and mineral contents in common Norwegian seaweeds and evaluation of their potential as food and feed. J Sci Food Agric 94:3281–3290

ä Marinho G, Nunes C, Sousa-Pinto I, Pereira R, Rema P, Valente LMP (2013) The IMTA-cultivated Chlorophyta *Ulva* spp. as a sustainable ingredient in Nile tilapia (*Oreochromis niloticus*) diets. J Appl Phycol 25:1359–1367

► Mazur NA, Curtis AL (2008) Understanding community perceptions of aquaculture: lessons from Australia. Aquacult Int 16:601–621

ä McShane TO, Hirsch PD, Trung TC, Songorwa AN and others (2011) Hard choices: making trade-offs between biodiversity conservation and human well-being. Biol Conserv 144:966–972

► Merceron M, Kempf M, Bentley D, Gaffet JD, Le Grand J, Lamort-Datin L (2002) Environmental impact of a salmonid farm on a well flushed marine site. I. Current and water quality. J Appl Ichthyol 18:40–50

ä Molloy SD, Pietrak MR, Bouchard DA, Bricknell I (2011) Ingestion of *Lepeophtheirus salmonis* by the blue mussel *Mytilus edulis*. Aquaculture 311:61–64

ä Molloy SD, Pietrak MR, Bouchard DA, Bricknell I (2014) The interaction of infectious salmon anaemia virus (ISAV) with the blue mussel, *Mytilus edulis*. Aquacult Res 45:509–518

ä Neori A, Chopin T, Troell M, Buschmann AH and others (2004) Integrated aquaculture: rationale, evolution and state of the art emphasizing seaweed biofiltration in modem mariculture. Aquaculture 231:361–391

Nikitina E (2015) The role of 'green' licences in defining environmental controls in Norwegian salmon aquaculture. MS thesis, The Arctic University of Norway, Tromsø

OECD (2010) Proceedings of the workshop on advancing the aquaculture agenda: policies to ensure a sustainable aquaculture sector. OECD Conference, Paris, 15–16 April 2010. OECD Publishing, Paris

Pietrak MR, Molloy SD, Bouchard DA, Singer JT, Bricknell I (2010) Interaction of a bacterial fish pathogen *Vibrio anguillarum* 02β with mussels *Mytilus edulis*. In: Proc Northeast Aquacult Conf Expo, 1–3 December 2010, Plymouth, MA

Porter ME, Van der Linde C (1995) Green and competitive: ending the stalemate. Harv Bus Rev 73:120–134

ä Reid GK, Chopin T, Robinson SMC, Azevedo P, Quinton M, Belyea E (2013) Weight ratios of the kelps, *Alaria esculenta* and *Saccharina latissima*, required to sequester dissolved inorganic nutrients and supply oxygen for Atlantic salmon, *Salmo salar*, in integrated multi-trophic aquaculture systems. Aquaculture 408–409:34–46

ä Ren LH, Zhang JH, Fang JG, Tang QS, Zhang ML, Du MR (2014) Impact of shellfish biodeposits and rotten seaweed on the sediments of Ailian Bay, China. Aquacult Int 22:811–819

ä Ridler N, Wowchuk M, Robinson B, Barrington K and others (2007) Integrated multi-trophic aquaculture (IMTA): a potential strategic choice for farmers. Aquacult Econ Manage 11:99–110

Robinson S, Martin J, Cooper J, Lander T, Reid G, Powell F, Griffin R (2011) The role of three dimensional habitats in the establishment of integrated multi-trophic aquaculture (IMTA) systems. Bull Aquacult Assoc Can 109:23–29

▶ Ryther JH, Tenore KR, Dunstan WM, Huguenin JE (1972) Controlled eutrophication: increasing food production from sea by recycling human wastes. Bioscience 22:144

▶ Ryther JH, Goldman JC, Gifford CE, Huguenin JE and others (1975) Physical models of integrated waste recycling–marine polyculture systems. Aquaculture 5:163–177

▶ Sanderson JC, Cromey CJ, Dring MJ, Kelly MS (2008) Distribution of nutrients for seaweed cultivation around salmon cages at farm sites in north-west Scotland. Aquaculture 278:60–68

▶ Sanderson JC, Dring MJ, Davidson K, Kelly MS (2012) Culture, yield and bioremediation potential of *Palmaria palmata* (Linnaeus) Weber & Mohr and *Saccharina latissima* (Linnaeus) C.E. Lane, C. Mayes, Druehl & G.W. Saunders adjacent to fish farm cages in northwest Scotland. Aquaculture 354–355:128–135

▶ Shi H, Zheng W, Zhang X, Zhu M, Ding D (2013) Ecological–economic assessment of monoculture and integrated multi-trophic aquaculture in Sanggou Bay of China. Aquaculture 410–411:172–178

Shuve H, Caines E, Ridler N, Chopin T and others (2009) Survey finds consumers support integrated multitrophic aquaculture: effective marketing concept key. Global Aquacult Advoc 2009:22–23

▶ Skår CK, Mortensen S (2007) Fate of infectious salmon anaemia virus (ISAV) in experimentally challenged blue mussels *Mytilus edulis*. Dis Aquat Org 74:1–6

Sowell T (1995) The vision of the anointed: self-congratulation as a basis for social policy. Basic Books, New York, NY

Stahel WR (1982) The product life factor. An inquiry into the nature of sustainable societies: the role of the private sector (Series: 1982 Mitchell Prize Papers). Houston Area Research Center, the Woodlands, TX

▶ Troell M, Norberg J (1998) Modelling output and retention of suspended solids in an integrated salmon–mussel culture. Ecol Modell 110:65–77

▶ Troell M, Halling C, Neori A, Chopin T, Buschmann AH, Kautsky N, Yarish C (2003) Integrated mariculture: asking the right questions. Aquaculture 226:69–90

Troell M, Joyce A, Chopin T, Neori A, Buschmann AH, Fang JG (2009) Ecological engineering in aquaculture—potential for integrated multi-trophic aquaculture (IMTA) in marine offshore systems. Aquaculture 297:1–9

▶ Vassdal T, Holst HMS (2011) Technical progress and regress in Norwegian salmon farming: a Malmquist index approach. Mar Resour Econ 26:329–341

▶ Wahbeh MI (1997) Amino acid and fatty acid profiles of four species of macroalgae from Aqaba and their suitability for use in fish diets. Aquaculture 159:101–109

▶ Wang X, Olsen LM, Reitan KI, Olsen Y (2012) Discharge of nutrient wastes from salmon farms: environmental effects, and potential for integrated multi-trophic aquaculture. Aquacult Environ Interact 2:267–283

▶ Wang X, Broch OJ, Forbord S, Handa A and others (2014) Assimilation of inorganic nutrients from salmon (*Salmo salar*) farming by the macroalgae (*Saccharina latissima*) in an exposed coastal environment: implications for integrated multi-trophic aquaculture. J Appl Phycol 26:1869–1878

▶ Webb JL, Vandenbor J, Pirie B, Robinson SMC, Cross SF, Jones SRM, Pearce CM (2013) Effects of temperature, diet, and bivalve size on the ingestion of sea lice (*Lepeophtheirus salmonis*) larvae by various filter-feeding shellfish. Aquaculture 406–407:9–17

▶ Whitmarsh D, Palmieri MG (2009) Social acceptability of marine aquaculture: the use of survey-based methods for eliciting public and stakeholder preferences. Mar Policy 33:452–457

▶ Whitmarsh D, Wattage P (2006) Public attitudes towards the environmental impact of salmon aquaculture in Scotland. Eur Environ 16:108–121

▶ Whitmarsh DJ, Cook EJ, Black KD (2006) Searching for sustainability in aquaculture: an investigation into the economic prospects for an integrated salmon-mussel production system. Mar Policy 30:293–298

▶ Wilding TA (2012) Changes in sedimentary redox associated with mussel (*Mytilus edulis* L.) farms on the west-coast of Scotland. PLoS One 7:e45159

Yildirim O, Ergun S, Yaman S, Turker A (2009) Effects of two seaweeds (*Ulva lactuca* and *Enteromorpha linza*) as a feed additive in diets on growth performance, feed utilization, and body composition of rainbow trout (*Oncorhynchus mykiss*). Kafkas Univ Vet Fak Derg 15:455–460

6

Sources and export of nutrients associated with integrated multi-trophic aquaculture in Sanggou Bay, China

Ruihuan Li[1,5], Sumei Liu[1,2,*], Jing Zhang[3], Zengjie Jiang[4], Jianguang Fang[4]

[1]Key Laboratory of Marine Chemistry Theory and Technology, MOE, Ocean University of China/Qingdao Collaborative Innovation Center of Marine Science and Technology, Qingdao 266100, PR China

[2]Laboratory for Marine Ecology and Environmental Science,
Qingdao National Laboratory for Marine Science and Technology, Qingdao, PR China

[3]State Key Laboratory of Estuarine and Coastal Research, East China Normal University, Shanghai 200062, PR China

[4]Carbon Sink Fisheries Laboratory, Key Laboratory of Sustainable Utilization of Marine Fisheries Resources,
Ministry of Agriculture, Yellow Sea Fisheries Research Institute, Chinese Academy of Fishery Sciences, 106 Nanjing Road,
Qingdao 266071, PR China

[5]*Present address:* State Key Laboratory of Tropical Oceanography, South China Sea Institute of Oceanology,
Chinese Academy of Sciences, 164 West Xingang Road, Guangzhou 510301, PR China

ABSTRACT: Field observations were made from 2012 to 2014 at an integrated multi-trophic aquaculture (IMTA) site in Sanggou Bay (SGB), China, to characterize the nutrients associated with aquaculture activities, and to assess the effects of aquaculture on nutrient cycles in the bay. Dissolved inorganic and organic nutrient levels were measured in rivers, groundwater, and SGB. Seasonal variations in nutrient concentrations were detected in the rivers, particularly enrichment of dissolved inorganic nitrogen (DIN) and silicate (DSi). Nutrient concentrations showed considerable seasonal variation, with higher and significantly different concentrations occurring in autumn than in the other seasons. The composition and distribution of nutrients were also affected by the species being cultured. Dissolved organic nitrogen and phosphorus (DON and DOP) accounted for 27 to 87% of total dissolved nitrogen and 34 to 81% of total dissolved phosphorus, respectively. Phosphorus may be a potentially limiting nutrient for phytoplankton growth in summer. Nutrient budgets were developed based on a simple steady-state box model. These showed that bivalve aquaculture was the major source of PO_4^{3-} (contributing 64% of total influx) and led to increased riverine fluxes of PO_4^{3-}. The results indicated that substantial quantities of nitrogen and DSi accumulated in sediments or were transformed into other forms (e.g. phytoplankton cell composition or particles). Large quantities of DIN and PO_4^{3-} were removed from the bay through harvesting of seaweeds and bivalves, which represented up to 64 and 81% of total outflux, respectively. The results show that aquaculture activities play the most important role in nutrient cycling in SGB.

KEY WORDS: Nutrients · IMTA · Budgets · Aquaculture activities · Sanggou Bay

INTRODUCTION

With an annual average increase of 8.7% over the past 40 yr, aquaculture is the fastest-growing food production sector in the world, and is overtaking capture fisheries as a source of food fish (Herbeck et al. 2013). The rapid growth of aquaculture has given rise to a wide variety of environmental problems, including ecosystem degradation and water pollution (Neori et al. 2004). One of the largest of impacts of aquaculture effluents to local ecosystems is imbalance created in nutrient dynamics and eutrophic

*Corresponding author: sumeiliu@ouc.edu.cn

conditions (Marinho-Soriano et al. 2009, Bouwman et al. 2011). In addition, excess nutrients cause stress in the cultivated organisms, with deleterious effects including smaller size, reduced production, and mass mortality (Newell 2004, Mao et al. 2006). Due to increasing concerns about the environmental impacts of aquaculture, a new method of aquaculture with a smaller ecological footprint has been developed. Integrated multi-trophic aquaculture (IMTA) has the potential to mitigate the environmental impacts of aquaculture (Buschmann et al. 2008).

IMTA is described as the cultivation of aquatic species from different trophic levels within a shared water system (Bostock et al. 2010). Such systems significantly increase the sustainability of aquaculture and recycle waste nutrients from high trophic-level species into production of lower trophic-level crops of commercial value (Troell et al. 2009). Seaweeds are used in IMTA systems for their nutrient-absorbing and sequestering properties. Nutrients excreted and egested by bivalves can be absorbed by macroalgae and recycled into valuable biomass (Newell 2004, Buschmann et al. 2008), and this amount of nutrient waste can be effectively removed from the ecosystem. In addition, a number of studies have confirmed that suspension-feeding bivalves can exert top-down control on phytoplankton (Newell & Koch 2004, Wall et al. 2008); larger nanoplankton will be removed in comparison with smaller (<3 μm diameter) picoplankton species, thereby reducing turbidity (Newell 2004). The resulting increased light penetration can potentially enhance the production of benthic plants (Newell & Koch 2004). If high levels of dissolved inorganic nitrogen (DIN) regenerated by bivalves are sufficient to allow the relatively slow-growing nanoplankton to grow fast enough to overcome grazer control, primary production can be stimulated through recycling of nitrogen (Smaal et al. 2001). Some marine IMTA systems have been commercially successful at industrial scales, especially in Asia (China) (Troell et al. 2009).

China is the largest aquaculture producer in the world, with a total production of 34.1 million tons, which accounts for 62% of total global production and 51% of the global value (Yang et al. 2005, FAO 2010, Yuan et al. 2010, Yu et al. 2012). The area devoted to aquaculture increased from 11.2×10^4 ha in 1977 to 218×10^4 ha in 2012 (The People's Republic of China Ministry of Agriculture Fisheries Bureau 2013). The rapid growth of aquaculture has led to eutrophication of coastal waters (Wu et al. 2014), and to the occurrence of aquatic diseases that have resulted in major economic losses (Fei 2004); for example, in 1998, more than 10 billion Chinese Yuan (approximately US$ 1.5 billion) were lost because of mariculture disease (Fei 2004). To improve the environmental sustainability of aquaculture and benefit the local economy, IMTA was developed in China. Sea-ranching and suspended aquaculture are the 2 main forms of IMTA in China, and the latter is used in Sanggou Bay.

Sanggou Bay (SGB) is located in northern China and has been used for aquaculture for over 30 yr (Zhang et al. 2009). It has been estimated that more than 300 t of inorganic nitrogen have been excreted into the bay by cultivated and fouling animals (Troell et al. 2009). Studies of core sediments also indicated that the total nitrogen (TN) content has increased in recent decades as a consequence of aquaculture activities (Song et al. 2012). Bivalves clear seston particles >3 μm in diameter from natural water and are not supplied with additional feed in the bay. The absolute and relative abundances of dinoflagellate cells in the bay are lower inside the scallop culture area than outside (Zhang et al. 2005), and the phytoplankton community has changed as a result; meanwhile, the reduction in phytoplankton biomass has a negative impact on bivalve growth (Duarte et al. 2003, Shi et al. 2011a). In addition, kelp can compete with phytoplankton for nutrients, and 80 000 t of dried kelp can be produced annually through uptake of inorganic nitrogen from the bay (Zhang et al. 2009). In pursuing high levels of productivity, SGB has been subject to a rapid growth in aquaculture, with long-line culture of kelp having expanded to areas more than 8 km away from the coast, where the water depth is between 20 and 30 m (Troell et al. 2009, Fu et al. 2013).

Much attention has been focused on the carrying capacity of shellfish and kelp mariculture (Bacher et al. 2003, Nunes et al. 2003, Shi et al. 2011a), ecology (Song et al. 2007, Hao et al. 2012), nutrient levels (Wang 2012, Zhang et al. 2012), and nutrient fluxes at the sediment–water interface (Jiang et al. 2007, Sun et al. 2010) in SGB, but the effects of aquaculture activities on nutrient cycling have not been well studied in the bay. The objective of this study was to determine the amounts and composition of dissolved nutrients in the bay and associated rivers and groundwater, to assess the sources and transportation of nutrients, to evaluate the impact of aquaculture activities on nutrient cycling, and to discriminate the importance of internal nutrient inputs vs. physical transport, based on the land–ocean interactions in the coastal zone (LOICZ) nutrient model (Gordon et al. 1996).

MATERIALS AND METHODS

Study area

SGB (Fig. 1) is a semi-enclosed water body of approximately 144 km^2 at the eastern end of Shandong Peninsula, and has an average depth of 7.5 m (Zhang et al. 2009). The bay is characterized by semidiurnal tides having an average tidal range of 2 m, and is connected to the Yellow Sea through an 11.5 km wide channel (Mao et al. 2006, Jiang et al. 2007). It is dominated by land–ocean climate, with water temperatures ranging from 2 to 26°C (Kuang et al. 1996). Approximately 73.3% of annual precipitation in the area (819.6 mm) occurs during the wet season, from June to September. The average river discharge into the bay is 1.7–2.3 × 10^8 m^3 yr^{-1}, and this carries an annual sediment load of 17.1 × 10^4 t.

More than 70% of the area of SGB is currently used for aquaculture (Zhang et al. 2009, 2010, Fu et al. 2013). It is one of the largest aquaculture production sites in China, and is extensively used for the culture

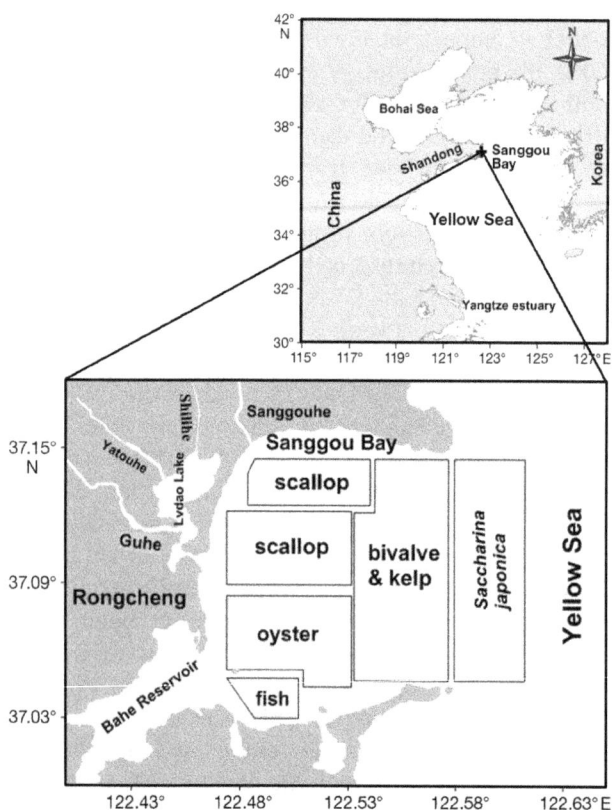

Fig. 1. Location of Sanggou Bay, China, and aquaculture activities, showing the regions of kelp (*Saccharina japonica*) monoculture; scallop, oyster, and fish monoculture; and multispecies aquaculture

of scallops (*Chlamys farreri*), Pacific oyster *Crassostrea gigas*, and seaweeds (*Saccharina japonica* and *Gracilaria lemaneiformis*) (Zhang et al. 2009). These species are grown in both monoculture and polyculture, from suspended longlines (Fang et al. 1996a) (Fig. 1). *S. japonica* monoculture occurs mainly near the mouth of the bay, bivalves are mainly cultured in the western part of the bay, and kelp and bivalve polyculture occurs in the middle part of the bay (Fig. 1). The co-cultivation of abalone *Haliotis discus hannai* with kelp (*S. japonica*) has also been developed, with the abalones held in lantern nets hanging vertically from the longlines. In 2012, production included approximately 84 500 t dry weight of *S. japonica*, 25 410 t wet weight of *G. lemaneiformis*, and approximately 15 000 and 60 000 t wet weight of *C. farreri* and *C. gigas*, respectively (data from Rongcheng Fishery Technology Extension Station). The main cultured species has shifted from scallop to oyster since 1996 because of reduced scallop production as a consequence of disease (Zhang et al. 2009).

To increase production, aquaculture has expanded from the bay to the open sea since the 1990s (Fang et al. 1996a). However, the total aquaculture production of kelp has not increased (Shi et al. 2011a). This may be related to a reduced supply of nutrients resulting from a decrease in the water exchange rate, which has been a consequence of reduced circulation because of the increase in aquaculture activities (Fang et al. 1996b). The hydrodynamic conditions have changed significantly because of the presence of suspended aquaculture (Shi et al. 2011a). Current speeds can be reduced by aquaculture facilities including rafts, and ropes impose drag (Grant & Bacher 2001, Duarte et al. 2003). The renewal of suspended particles for bivalve culture and nutrient regeneration for kelp have also been reduced (Grant & Bacher 2001, Duarte et al. 2003). Compared with the period of farming activities up to 1983, tidal currents had decreased by 50% by 1994 because of large-scale cultivation (Zhao et al. 1996). Based on a 2-dimensional model, Grant & Bacher (2001) estimated a reduction of 41% in the water exchange rate in SGB because of increased bottom friction with expansion of intensive suspended aquaculture. The vertical current has also changed because of suspended aquaculture (Fan & Wei 2010).

Sample collection

Sampling took place during 31 May to 4 June 2012 (early summer), 20 September to 2 October

2012 (early autumn), 22 to 25 April 2013 (spring), 21 to 25 July 2013 (summer), 16 to 17 October 2013 (autumn), and 15 to 17 January 2014 (winter) (Fig. 2). Two anchor stations for monitoring over complete tidal cycles of 25 h were established, one in April 2013 in the northern mouth of the bay (D1), and the other in October 2013 in the southern mouth (D2) (Fig. 2), respectively. At each station, surface water samples were collected by submersing a 1 l acid-cleaned polyethylene bottle from a boat, and bottom water samples were collected using a 5 l polymethyl methacrylate water sampler. River water samples were collected from the river edge in 0.5 l acid-cleaned polyethylene bottles, and groundwater was collected from wells around the bay (Fig. 2).

Water temperature and salinity were measured *in situ* using a WTW MultiLine F/Set3 multi-parameter probe. Each water sample was immediately filtered through a 0.45 μm pore size cellulose acetate filters (pre-cleaned with hydrochloric acid, pH = 2) into a polyethylene bottle that had previously been rinsed 3 times with some of the filtered water sample. The filtrates were fixed by the addition of saturated $HgCl_2$ solution (Liu et al. 2005), and the filters were dried at 45°C and weighed to determine the mass of suspended particulate matter (SPM).

Chemical analysis

Dissolved nutrient concentrations were measured in the laboratory using an Auto Analyzer 3 (Seal Analytical). Total dissolved nitrogen (TDN) and total dissolved phosphorus (TDP) were measured according to the methods of Grasshoff et al. (1999). The DIN

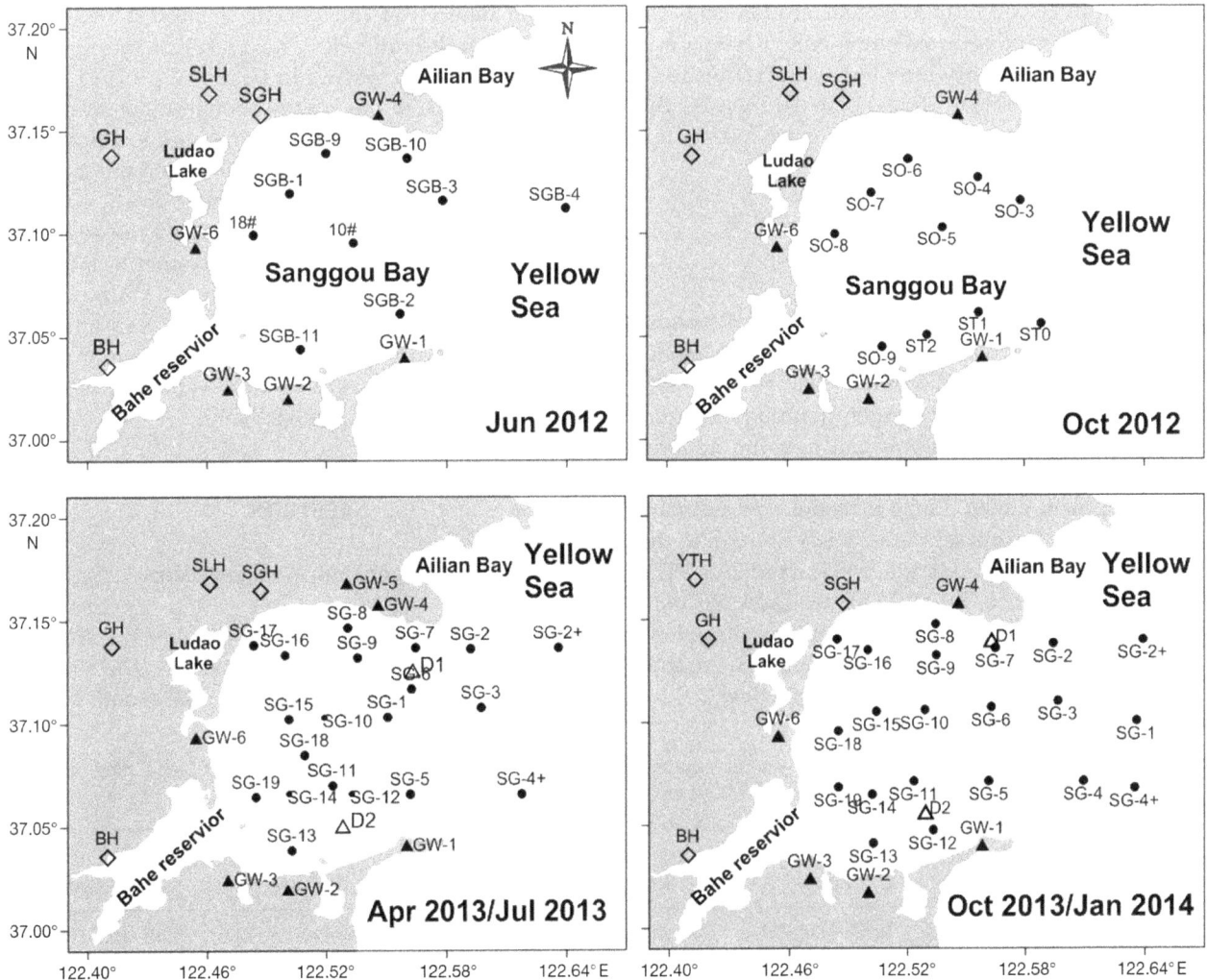

Fig. 2. Sampling stations in Sanggou Bay for the cruises during 2012 to 2014. (◇) River stations; (▲) groundwater stations (BH: Bahe; GH: Guhe; SLH: Shilihe; SGH: Sanggouhe; YTH: Yatouhe); (●) bay stations; (△) anchor stations

concentration was determined as the sum of the NO_3^-, NO_2^-, and NH_4^+ concentrations. The concentrations of dissolved organic nitrogen (DON) and dissolved organic phosphorus (DOP) were estimated by subtracting DIN from TDN and PO_4^{3-} from TDP, respectively. The analytical precision of NO_3^-, NO_2^-, NH_4^+, PO_4^{3-}, dissolved silicate (DSi), TDN, and TDP was <5%.

Statistical analysis

Statistical analyses were performed using the software SPSS 20.0 by IBM. One-way ANOVAs were used to analyze the individual effects of seasons and particular cultivation area on variations in SPM, and 2-way ANOVAs were used to analyze the combined effects of seasons and cultivation area on variations in SPM. Two-way ANOVAs were also used to analyze the effects of surface/bottom and seasons on variations in nutrient concentrations. Based on *a posteriori* homogeneity tests, Tukey's HSD or Tamhane's T2 comparisons were applied to assess the statistical significance of differences (p < 0.05) following ANOVA.

Nutrient budgets

Dissolved nutrient budgets for the study system were constructed based on the LOICZ box model (Gordon et al. 1996). This model has been widely used to construct nutrient budgets defining the internal biogeochemical processes and external nutrient inputs of estuarine and coastal ecosystems (Savchuk 2005, Liu et al. 2009). For our model, we assumed that the study system was in a steady state, and the bay was treated as a single well-mixed box. The water mass balance, salinity balance, and the nonconservative fluxes of nutrient elements based on nutrient concentrations and water budgets were estimated according to Eqs. (1) to (3), respectively:

$$V_R = V_{in} - V_{out} = -V_Q - V_P - V_G - V_W + V_E \quad (1)$$

$$V_X(S_1 - S_2) = S_R V_R \quad (2)$$

$$\Delta Y = \text{outflux} - \text{influx} = V_R C_R + V_X C_X - V_Q C_Q - V_P C_P - V_G C_G - V_W C_W \quad (3)$$

where V_R is the residual flow, and V_Q, V_P, V_G, V_W, V_E, V_{in}, V_{out}, V_X, and ΔY are the river discharge, precipitation, groundwater, wastewater, evaporation, inflow of water to the system of interest, outflow of water from the system of interest, the mixing flow between the 2 systems and nonconservative flux of nutrients, respectively. The volume of aquaculture effluent discharged directly into the system of interest was not considered, as the data were limited. We assumed that the salinity of fresh water (V_Q, V_P, and V_E) was 0. In Eq. (2), $S_R = (S_1+S_2)/2$, where S_1 and S_2 are the average salinity of the system of interest and the adjacent system, respectively. The total water exchange time (τ) of the system of interest was estimated from the ratio of V_S to ($V_R + V_X$), where V_S is the volume of the system. In Eq. (3), C_Q, C_P, C_G, C_W, C_R, and C_X are the average concentrations of nutrients in the river discharge, the precipitation, groundwater, wastewater, the residual flow, and the mixing flow, respectively. C_R and C_X equate to $(C_1 + C_2)/2$ and $(C_1 - C_2)$, respectively. C_1 and C_2 are the average concentrations of nutrients in the system of interest and the adjacent system, respectively. Outflux and influx are the total nutrient flux out of and into the system of interest, respectively. A negative or positive sign for ΔY indicates that the system of interest was a sink or a source, respectively.

RESULTS

Hydrographical chacteristics

The surface water temperature (Table 1) reflected the seasonality of this temperate system. The surface water temperature decreased from the mouth to the

Table 1. Seasonal variations in temperature, salinity, and suspended particulate matter (SPM) in Sanggou Bay, China, during the study. Mean values are given in parentheses

Season	Temperature (°C)		Salinity		SPM (mg l⁻¹)	
	Surface	Bottom	Surface	Bottom	Surface	Bottom
Spring	6.00–9.60 (7.60)	6.10–9.90 (7.80)	30.2–31.3 (30.8)	30.1–31.4 (30.7)	3.91–31.9 (13.6)	3.59–40.5 (14.9)
Summer	13.3–25.9 (20.0)	13.5–20.6 (17.0)	28.2–30.8 (30.0)	30.2–30.7 (30.4)	3.78–26.4 (13.9)	5.61–92.0 (37.6)
Autumn	17.7–25.0 (20.1)	16.6–23.3 (19.3)	29.1–30.0 (29.6)	29.3–29.9 (29.5)	5.75–29.3 (15.5)	11.9–67.8 (27.4)
Winter	1.80–5.70 (3.50)	0.90–5.30 (3.15)	29.2–30.6 (30.0)	29.2–30.4 (29.9)	2.27–54.0 (15.8)	3.04–54.7 (13.5)

west of the bay in spring and summer, but increased in this direction in autumn and winter. The horizontal distribution of temperature in the near-bottom layer was similar to that in surface water, but the temperatures were generally lower. The salinity of both surface and bottom water gradually increased from the west of the bay to mouth, except in winter. The salinity was lowest in autumn (Table 1).

The SPM concentrations varied considerably among seasons and cultivation areas, as evidenced by the large ranges shown in Table 1 and Fig. 3. The average concentration of SPM showed minor differences between surface and bottom waters in spring and winter, but was significantly less in surface water than in the bottom layer in both summer and autumn between different cultivation areas, especially those involving oyster and scallop monoculture (Fig. 3). A 1-way ANOVA indicated very significant differences in SPM concentration in bottom water of the bay in different seasons (p < 0.05). The subsequent post hoc Tamhane's T2 test showed that the concentrations of SPM in bottom water in summer and autumn differed significantly from those in spring and winter. In addition, a 1-way ANOVA indicated highly significant differences between different cultivation areas (p <

0.05). The subsequent post hoc Tamhane's T2 test showed that the values of SPM in both bottom and surface waters in the fish, oyster, and scallop cultivation areas differed significantly from those in the kelp, offshore, and bivalve and kelp areas.

Nutrients in rivers

Nutrient concentrations in rivers adjacent to SGB varied greatly during the study period (Table 2). The rivers were generally enriched with DIN relative to PO_4^{3-} (Table 2). The DIN was dominated by NO_3^-, which accounted for 73 to 98 % of DIN among all seasons. The NO_2^- concentrations in rivers were generally >2 µM except Bahe river (0.14–1.13 µM; Table 2). The PO_4^{3-} concentration ranged from 0.08 to 6.02 µM in the rivers, with an annual average of 1.45 µM. Seasonal variation of PO_4^{3-} in the Bahe river was similar to that in the Guhe river, and the PO_4^{3-} concentrations in the Bahe and Guhe rivers were lower than in the Shilihe and Sanggouhe rivers (Table 2). The DSi concentrations were high in our study rivers (average 182 µM; Table 2), indicating a high weathering rate associated with rivers adjacent

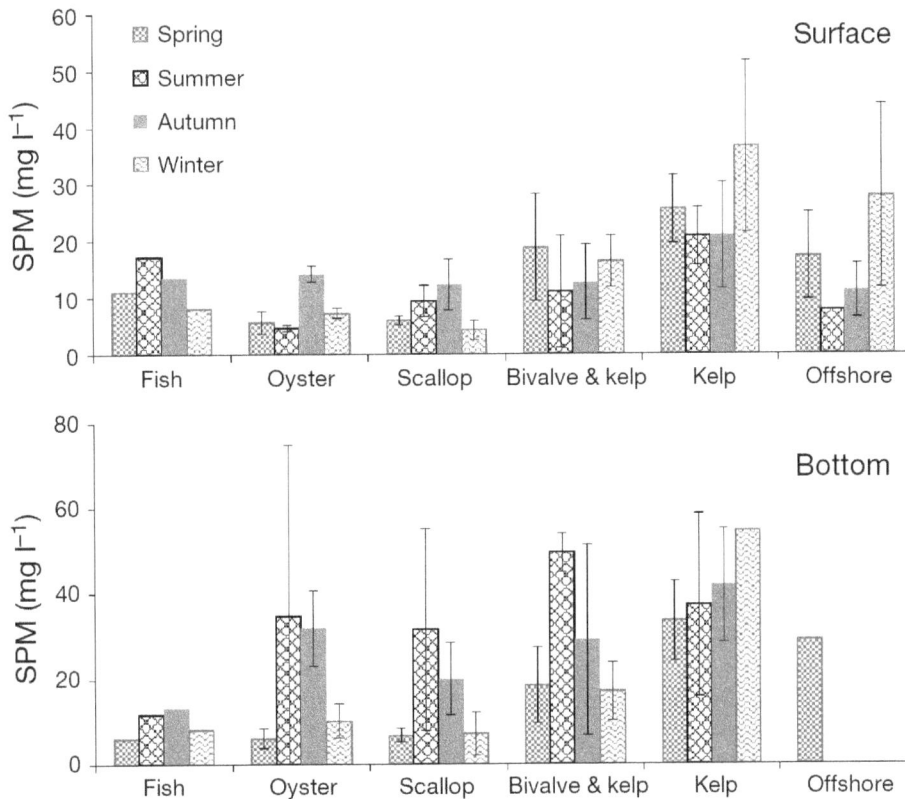

Fig. 3. Suspended particulate matter (SPM) concentrations (mg l^{-1}; mean ± SD) in various cultivation areas in different seasons during the study periods

Table 2. Nutrient concentrations (μM) and molar ratios in surface water in rivers adjacent to Sanggou Bay, China, in different seasons during the study. DSi: dissolved silicate. Dates are given as year-month

River	NH_4^+	NO_2^-	NO_3^-	PO_4^{3-}	DSi	N:P	Si:N
Bahe							
2012-06	2.53	0.15	0.48	0.17	29.0	18	9.0
2012-09	0.65	0.04	0.25	0.17	164	5.5	176
2013-04	2.91	0.32	28.9	2.23	47.9	14	0.4
2013-07	0.65	1.13	88.6	0.32	90.4	21	1.0
2013-10	1.41	0.15	10.5	0.56	246	21	20
2014-01	1.68	0.12	18.5	1.78	142	11	7.0
Guhe							
2012-06	36.3	17.5	309	0.19	72.3	1905	0.2
2012-09	9.86	10.4	530	0.08	102	7157	0.2
2013-04	22.5	5.18	288	1.80	55.6	175	0.2
2013-07	2.56	7.60	590	0.16	208	122	0.3
2013-10	12.6	6.40	240	–	143	–	0.6
2014-01	5.71	3.05	455	0.50	130	927	0.3
Shilihe							
2012-06	103	4.38	283	3.23	364	121	0.9
2012-09	17.8	6.70	600	0.22	282	2821	0.5
2013-04	93.0	17.0	503	2.89	243	212	0.4
2013-07	30.3	10.2	199	3.60	199	63	0.3
Sanggouhe							
2012-06	15.4	2.26	169	0.12	172	1505	0.9
2012-09	4.00	17.5	508	2.60	182	204	0.3
2013-04	8.23	10.2	351	1.44	166	256	0.5
2013-07	25.0	16.6	382	6.02	382	106	0.4
2013-10	2.73	4.40	362	–	318	–	0.9
2014-01	7.25	5.65	569	2.00	236	291	0.4
Yatouhe							
2013-10	4.68	3.20	420	–	189	–	–
2014-01	6.50	4.36	687	0.36	211	–	–

to the SGB. Except for Bahe river, the $DIN:PO_4^{3-}$ molar ratios in the rivers were significantly higher than the Redfield ratio (Table 2), indicating that phytoplankton might be limited by phosphorus despite high NO_3^- values, especially in summer in the Bahe and Guhe rivers. The high concentrations of DIN led to DSi:DIN ratios that were less than or approached a value of 1.

Spatial and temporal variations of nutrients in SGB

The concentrations of dissolved inorganic nutrients decreased gradually from offshore to the inner part of SGB in spring (April 2013; Fig. 4a), while the DON and DOP concentrations showed the opposite horizontal distribution (Fig. 4a). The concentrations of NO_3^- accounted for 53–92% and 56–89% of the DIN in surface and near-bottom layers, respectively. DON contributed 27–46% of TDN in surface water outside the bay, where kelp monoculture occurs, and ac-

counted for 46–87% of TDN inside of the bay. DON represented 40–84% of TDN in the near-bottom layer. For phosphorus compounds, PO_4^{3-} and DOP accounted for approximately 66 and 34% of TDP in the bay, respectively. The molar ratios of $DIN:PO_4^{3-}$ ranged from 7.8 to 31 (average 19 ± 7.9 SD) in surface water, and from 9.4 to 69 in the near-bottom layer, respectively. The average DSi:DIN ratio was higher than the Redfield ratio in both surface (1.3 ± 0.8) and bottom (1.2 ± 0.6) waters. Studies of nutrient uptake kinetics have shown that the threshold values for phytoplankton growth are 1.0 μM DIN and 0.1 μM PO_4^{3-} (Justi et al. 1995). In the western part of the bay, DIP concentrations were lower than the threshold values for phytoplankton growth (Fig. 4a). This suggests that phosphorus may be the most limiting element for phytoplankton growth in the following season.

During June 2012 (Fig. 4b), the levels of dissolved inorganic nutrients were lower than those in spring (Fig. 4a). The NO_3^-, NO_2^-, and NH_4^+ concentrations decreased gradually from offshore to the inner part of the bay, while PO_4^{3-} and DSi concentrations showed the opposite horizontal distribution. With respect to nitrogen compounds, NO_3^- comprised 24–78% of DIN in surface water and 34–72% in bottom water. Surface water was depleted in PO_4^{3-} (0.03–0.17 μM), which led to the $DIN:PO_4^{3-}$ ratios being significantly higher than the Redfield ratio. The DIN:DSi molar ratios ranged from 0.4 to 3.2 (average 1.6 ± 0.7). In July 2013, nutrient concentrations increased significantly from the mouth of the bay to the inner part (Fig. 4c), and were higher in the near-bottom layer than in surface water. The DIN was dominated by NH_4^+, which contributed 32–89% (mean 62%) and 32–69% (mean 52%) to DIN in surface water and the near-bottom layer, respectively. DON comprised 57–88% of the TDN in the entire bay, and DOP accounted for 34–75% and 46–81% of the TDP in surface water and the near-bottom layer, respectively. The molar ratios of $DIN:PO_4^{3-}$ were higher than the Redfield ratio in surface water, and the DSi:DIN ratios were higher than or comparable to the Redfield ratio. The PO_4^{3-} concentrations in surface water at 70% of the stations in June 2012 (Fig. 4b), and in the southeastern part of the bay in July 2013 (Fig. 4c), were lower than the threshold values. This suggested that phytoplankton growth might be limited by P in summer. In the western part of the bay (the main area for bivalve culture) the DIN concentrations were lower than or comparable to the threshold values, suggesting that N might be potentially limiting for phytoplankton growth in this part of the bay.

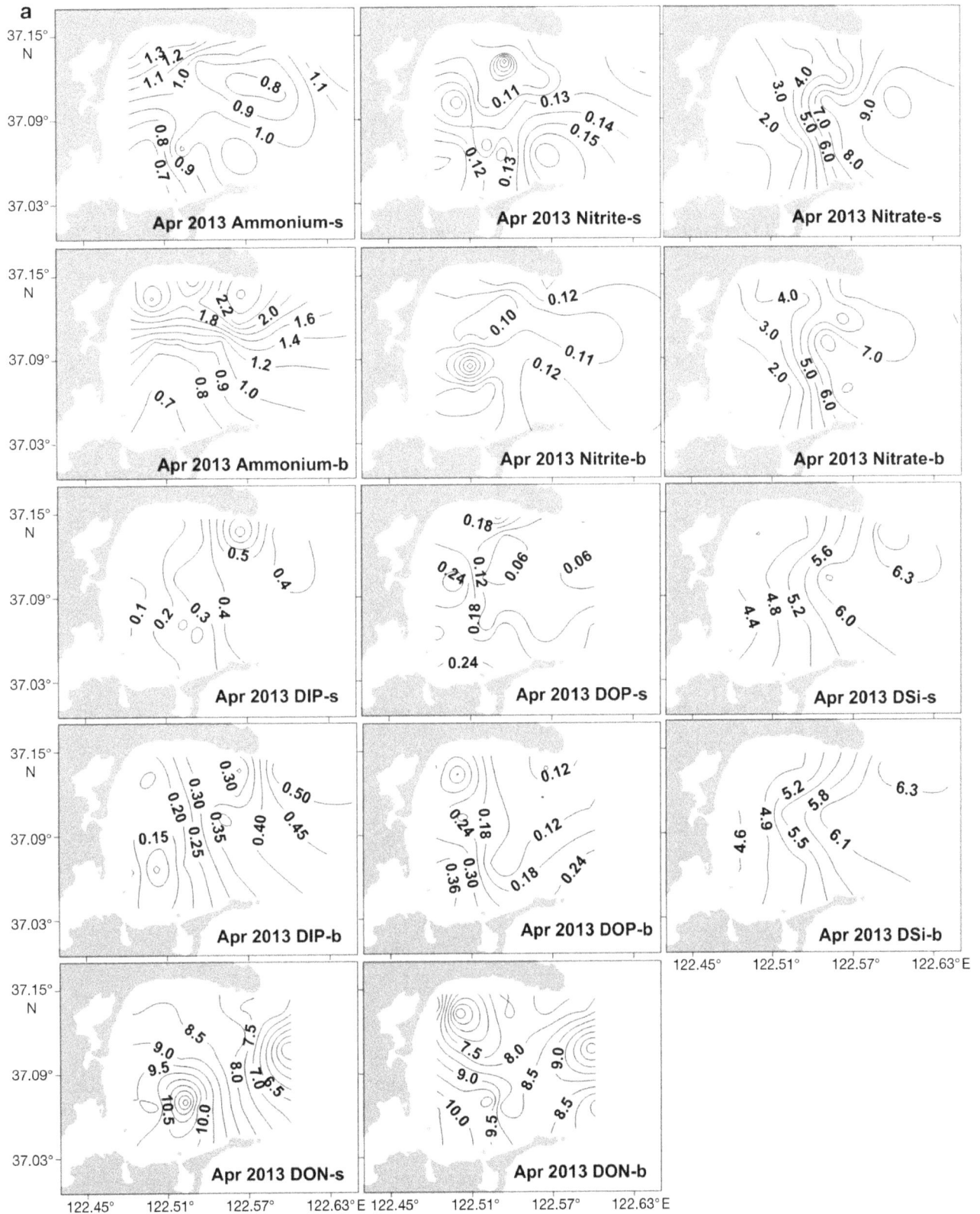

Fig. 4. Horizontal distributions of nutrients (μM) in Sanggou Bay: (a) April 2013; (b) June 2012; (c) July 2013; (d) October 2012; (e) October 2013; (f) January 2014. DIP (DOP): dissolved inorganic (organic) phosphorus, DSi: dissolved silicate, DON: dissolved organic nitrogen. s: surface; b: bottom

b

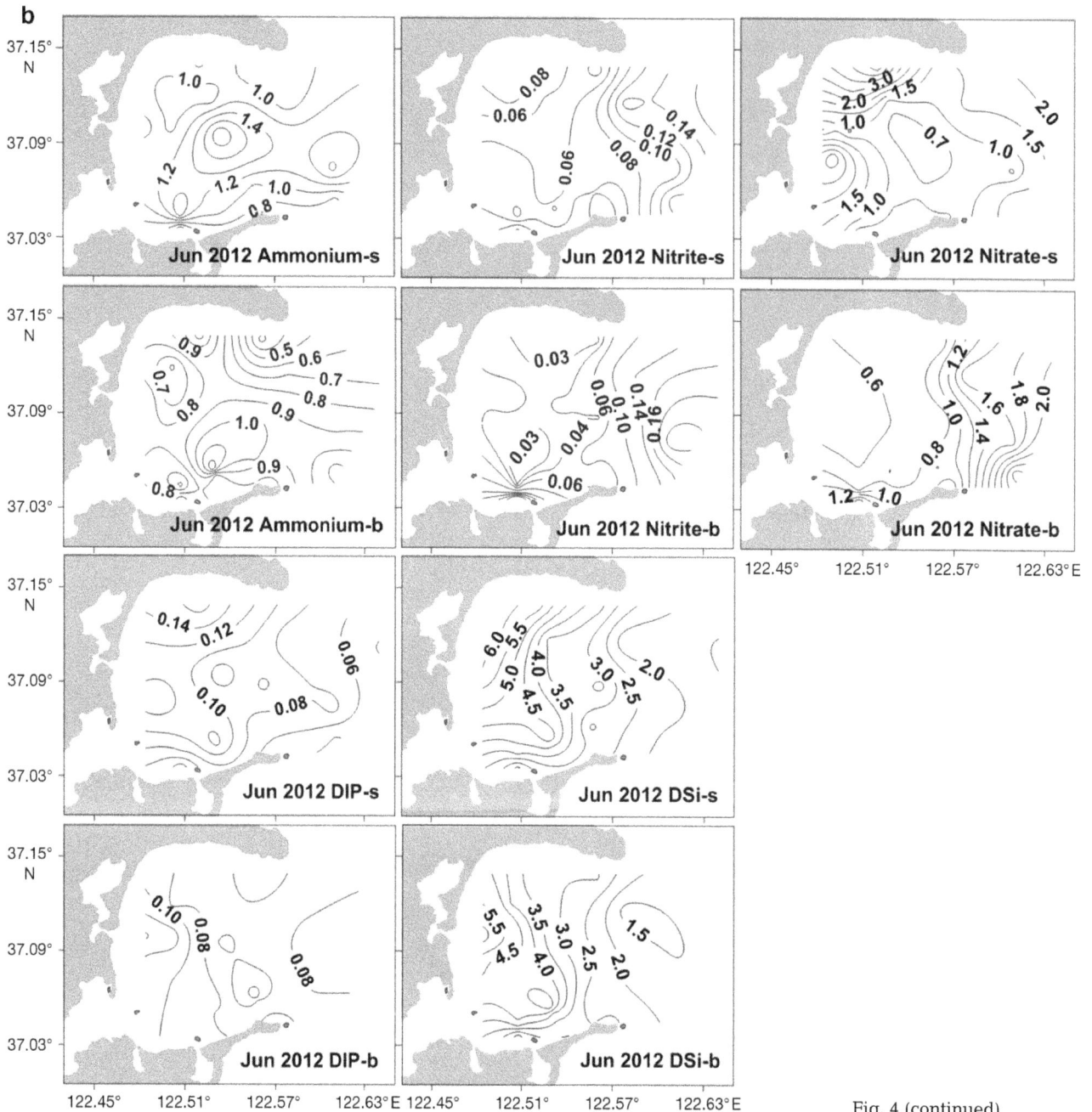

Fig. 4 (continued)

During the September–October 2012 study period, NO_3^- and NH_4^+ concentrations decreased from south to north in the bay; NO_2^-, DSi, and DOP increased gradually from west to east, and the PO_4^{3-} concentration increased from northeast to southwest (Fig. 4d). Throughout the entire bay, NO_3^- comprised 52–86 % of DIN, and NH_4^+ comprised 6–38 %. In October 2013, the NO_3^-, NO_2^-, DON, DIP, and DSi concentrations decreased from the mouth to the southwestern part of the bay (Fig. 4e). Throughout the entire bay, NO_3^- accounted for 55–84 % of DIN. DON comprised

27–48 % of TDN inside the bay, and 51–61 % in the kelp monoculture area. DOP contributed to 12–36 % and 16–50 % of TDP in surface water and the bottom layer, respectively. In autumn in both 2012 and 2013, the average $DIN:PO_4^{3-}$ ratios were higher than the Redfield ratio, while the DSi:DIN ratios in the water column were comparable to the Redfield ratio.

In winter, the horizontal distribution of nutrients was similar to that in spring (except for the NO_2^- and NH_4^+ concentrations), with higher concentrations in the near-bottom layer than in surface water (Fig. 4f).

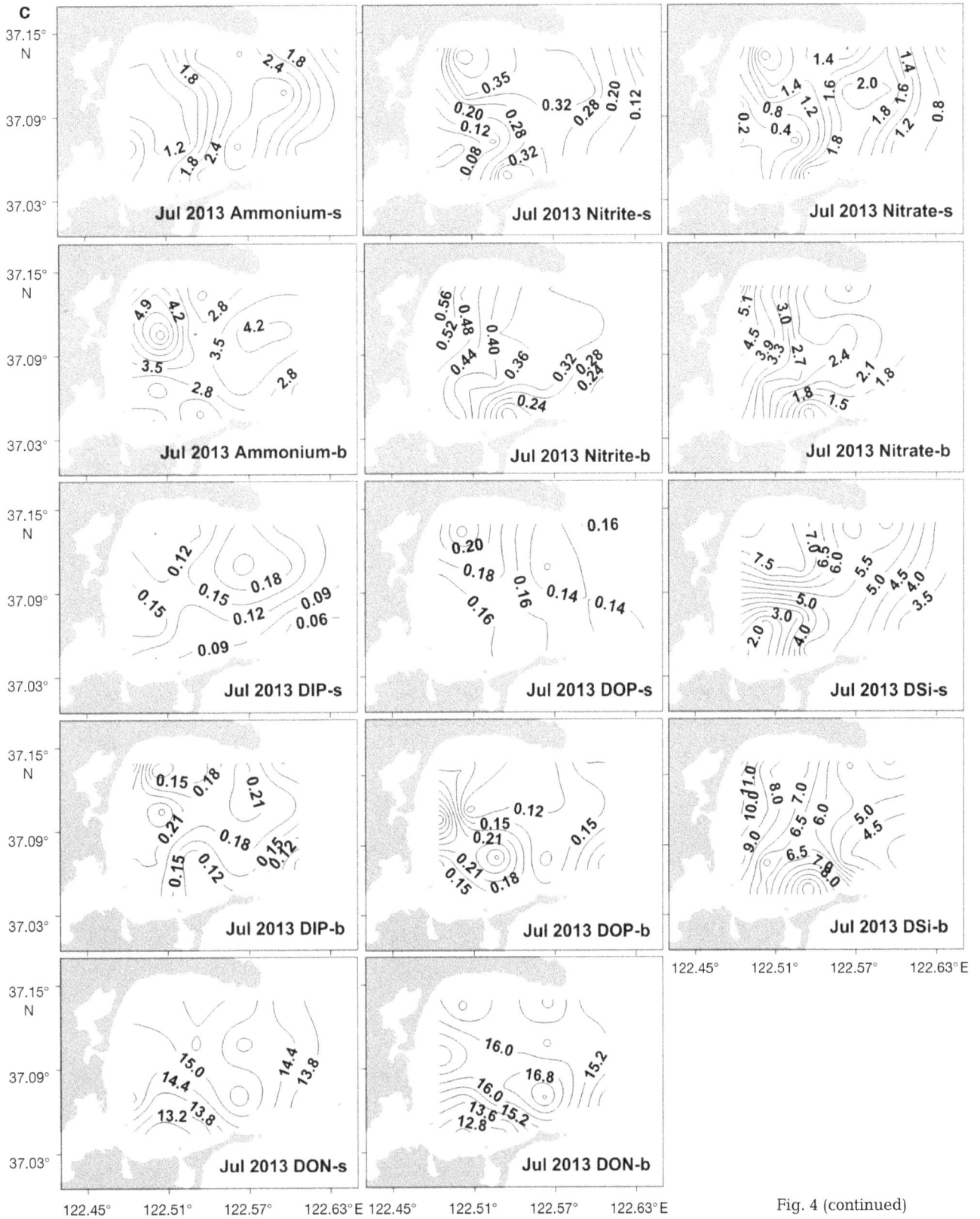

Fig. 4 (continued)

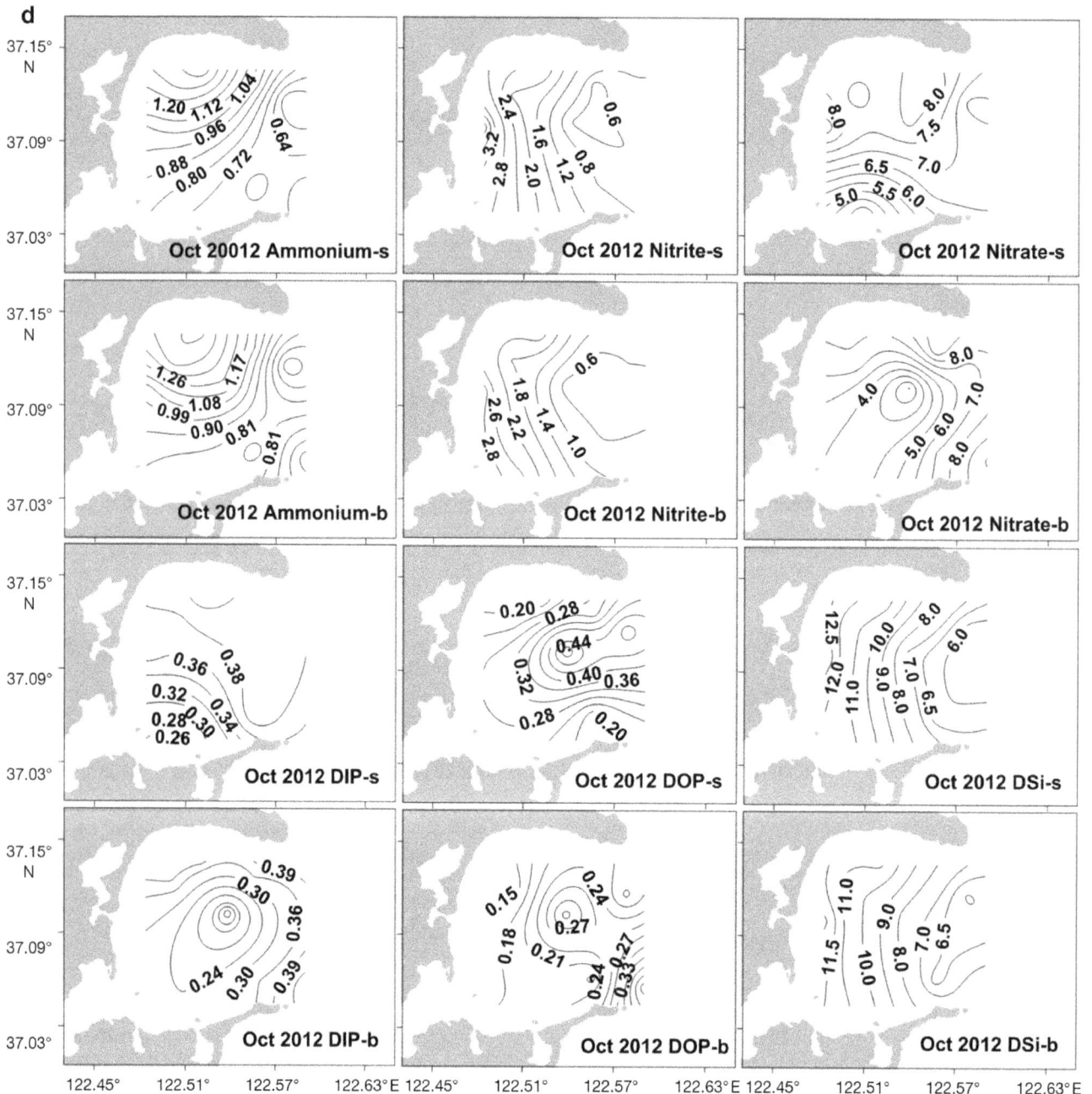

Fig. 4 (continued)

In the entire bay, NO_3^- accounted for 66–92% of DIN. DON was the dominant species of TDN, which represented 53–81% of TDN in the water column, and DOP represented 35–67% of TDP. The molar ratios of $DIN:PO_4^{3-}$ ranged from 20 to 62 and 17 to 46 in surface and bottom waters, respectively. The average DSi:DIN ratio in surface and bottom waters was comparable and significantly lower than the Redfield ratio. The results suggest that phosphorus may be a limiting element for phytoplankton growth in winter.

Seasonality in nutrient concentrations was evident in SGB (Figs. 4 & 5). At all sites, the NO_3^-, PO_4^{3-}, and DSi concentrations were significantly higher in autumn than in the other seasons. The average NO_3^- concentrations in surface (9.44 ± 4.00 µM) and bottom (9.72 ± 4.48 µM) waters in autumn exceeded those in summer by factors of 7.4 and 5.3, respectively. DIN was dominated by NO_3^-, except in summer. The DON concentrations in winter (16.0 ± 1.67 µM) were comparable to those in summer, and were signifi-

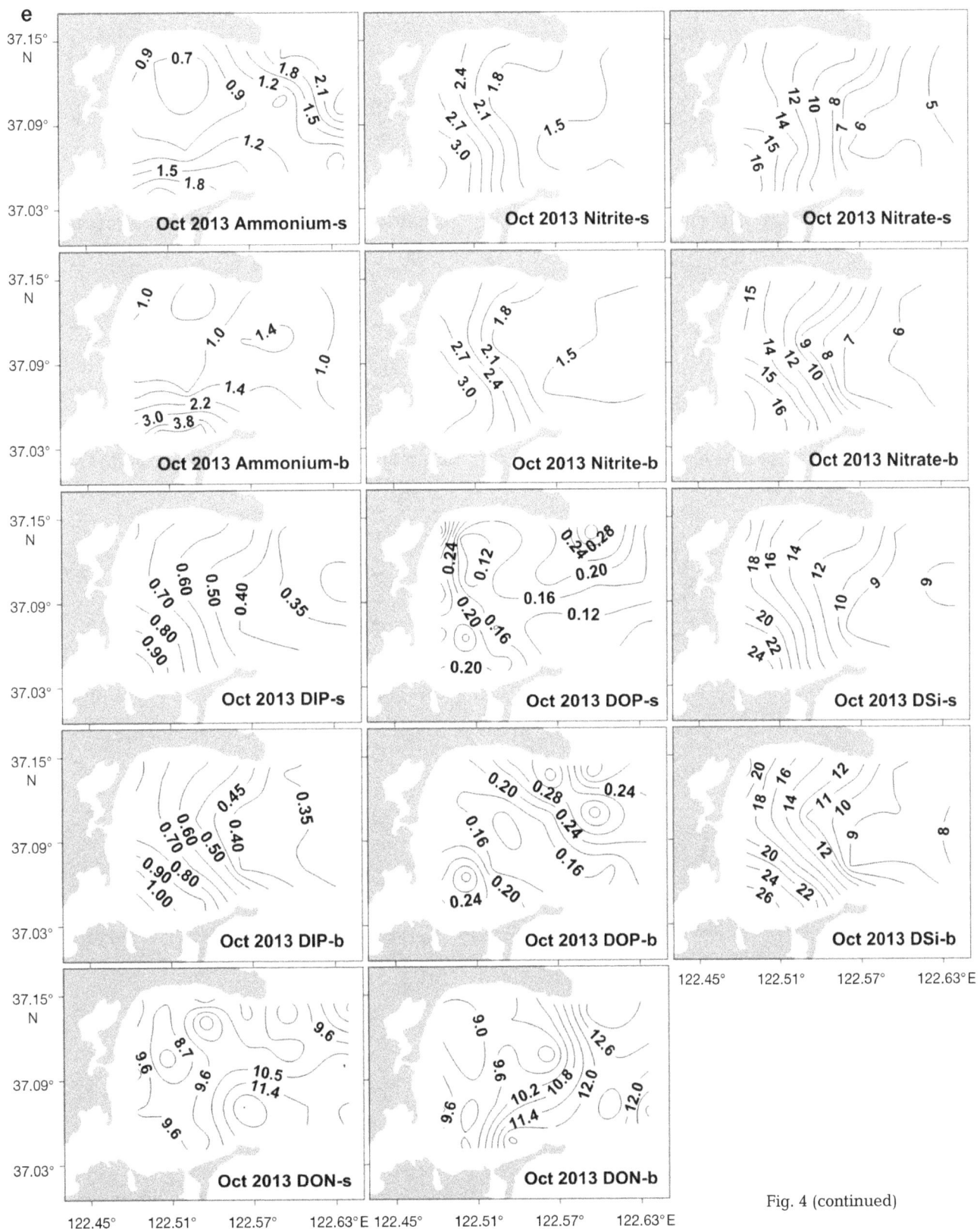

Fig. 4 (continued)

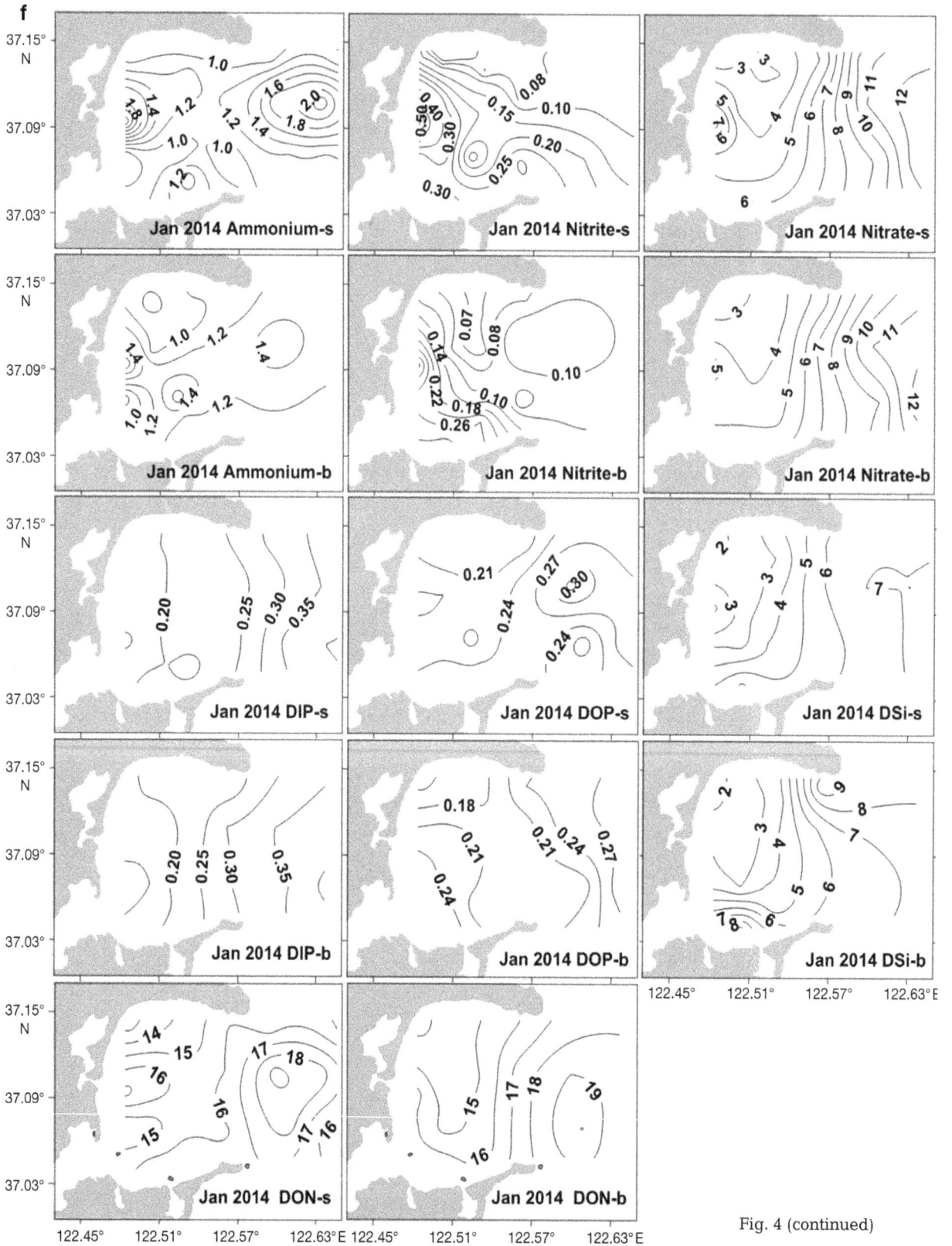

Fig. 4 (continued)

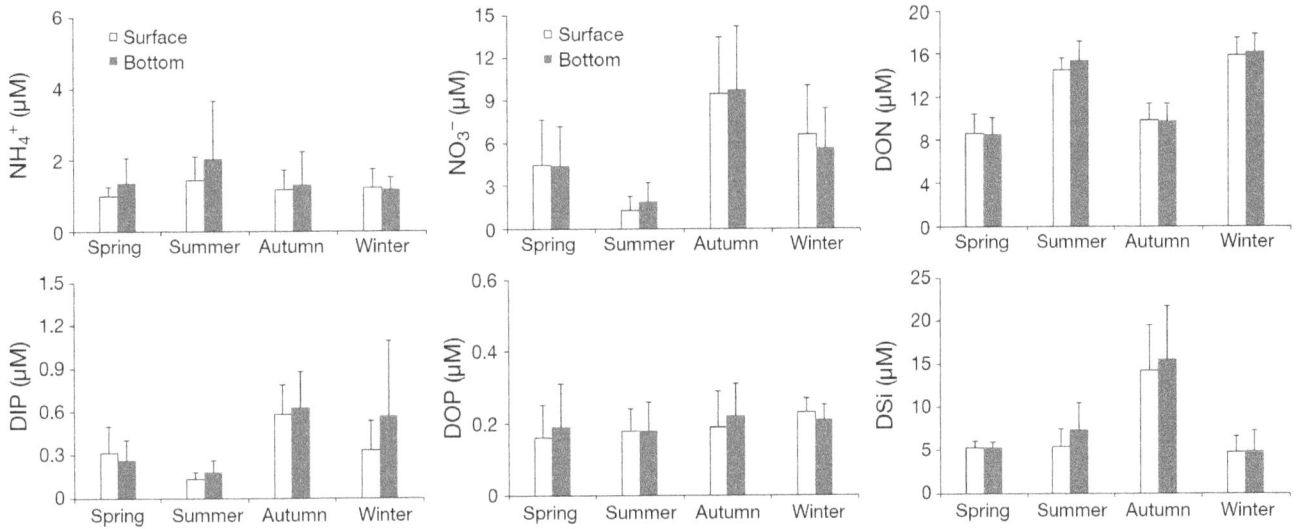

Fig. 5. Seasonal variations in nutrient concentrations (µM) in Sanggou Bay during the study period. DON: dissolved organic nitrogen, DIP (DOP): dissolved inorganic (organic) phosphorus, DSi: dissolved silicate

cantly higher than the concentrations in spring and autumn (Fig. 4). TDN was dominated by DON (59–82%), except in autumn (approximately 40%). Two-way ANOVA indicated highly significant differences in nutrient concentrations among seasons and layers ($p < 0.01$). The subsequent post hoc Tukey's HSD test showed that the nutrient concentrations in autumn differed significantly from those in other seasons ($p < 0.01$). Two-way ANOVA also indicated highly significant differences in nutrient concentrations among seasons and cultivation areas (Fig. 6; $p < 0.01$), suggesting that aquaculture activities significantly affect the nutrient composition in SGB.

Nutrients at the anchor stations

In April 2013, all nutrients changed during the tidal cycle at Stn D1 (Fig. 7a). The maximum concentrations usually occurred during high tide, indicating the outer bay as a nutrient source. The vertical profiles for concentrations of all dissolved inorganic nutrients at Stn D1 showed that the water column was well mixed (Fig. 7a). High concentrations of DON (9.01–13.8 µM) were found throughout the water column, and comprised up to 50% of TDN. The DIN:PO$_4^{3-}$ ratio ranged from 23 to 74 in surface water and from 30 to 132 in near-bottom water, and the DSi:DIN ratio ranged from 0.5 to 0.8 in surface water and from 0.4 to 0.9 in near-bottom water. At Stn D2, the nutrient concentrations were higher in near-bottom waters than in surface water, the exception being NH$_4^+$ and DOP (Fig. 7a). The DON (8.26–

10.5 µM) comprised 66–87% of TDN. The concentrations of DOP (0.08–0.35 µM) represented 25–73% of TDP, and indicated a well-mixed profile. The DIN:PO$_4^{3-}$ ratio increased from 8.0–20 in surface water to 11–37 in near-bottom water, while the DSi:DIN ratio decreased from 1.6–3.2 in surface water to 1.0–1.5 in near-bottom water. The nutrient concentrations at Stn D1 were higher than at D2.

Analysis of the concentrations of all nutrients during 18–19 October 2013 showed that the water column at Stn D1 was well mixed (Fig. 7b). No parameter showed significant differences between day and night, indicating that tidal mixing was the main factor affecting concentration changes. The concentrations of DON were 5.38–10.5 µM, which comprised 26–83% of TDN. The DOP concentrations were 0.05–0.34 µM, which represented 8–39% of TDP. The DIN:PO$_4^{3-}$ ratio was 23–36 (average 27) in surface water, and 22–51 (average 28) in bottom water. The DSi:DIN ratio was 0.7–1.0 (average 0.9) in surface water and 0.5–1.0 (average 0.8) in bottom water. At Stn D2, the concentrations of all nutrients in surface water showed a general decrease with increasing tide height. The DIN:PO$_4^{3-}$ and DSi:DIN ratios in surface water ranged from 22 to 32 and 0.8 to 1.0, respectively. The nutrient concentrations at Stn D1 were lower than at D2.

Water and nutrient budgets in SGB

Domestic wastewater is discharged directly into rivers adjacent to SGB, and so in developing a water

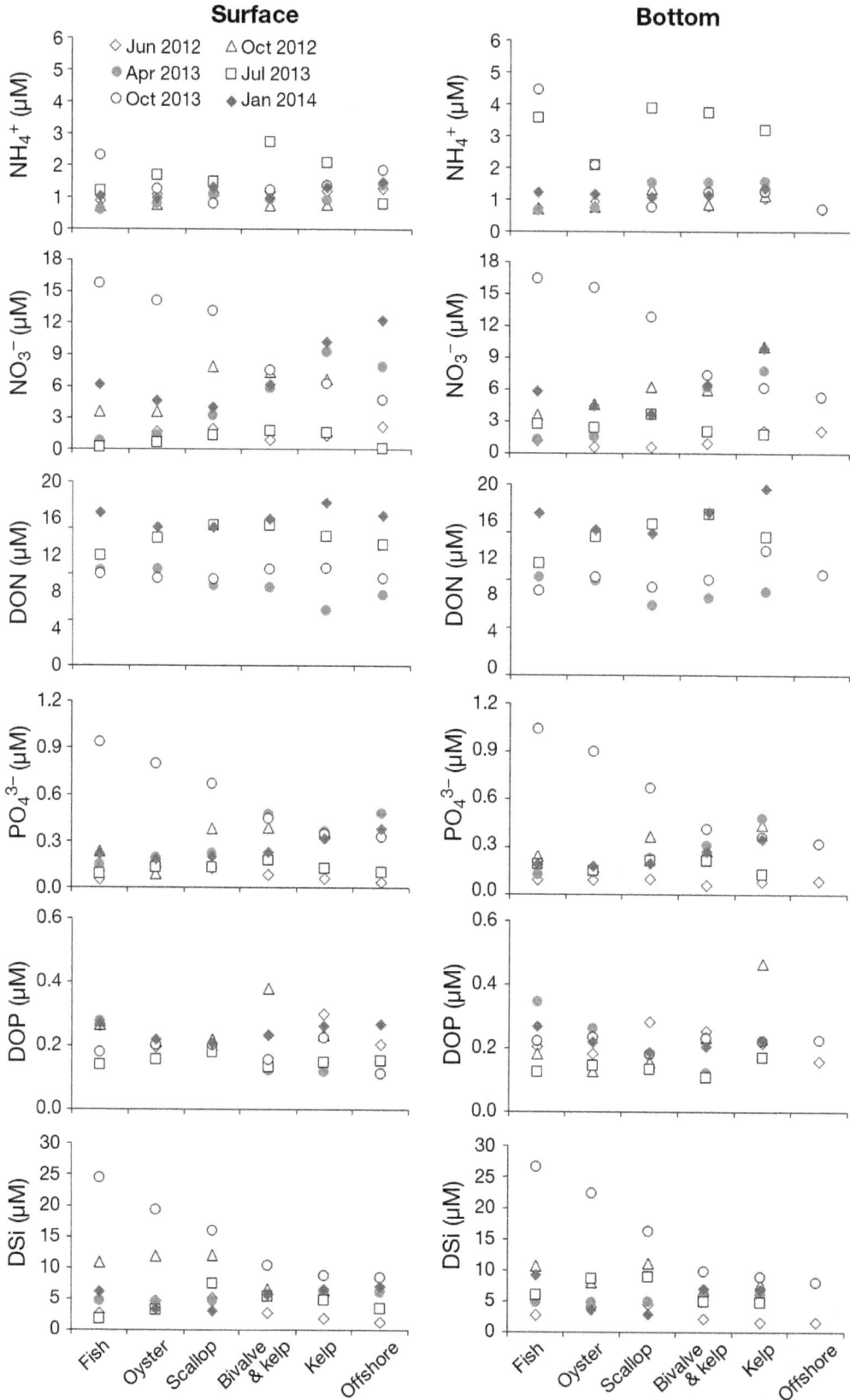

Fig. 6. Nutrient cycles, averaged for various aquaculture regions in Sanggou Bay Left: nutrients in surface water, right: nutrients in the near-bottom layer. DON: dissolved organic nitrogen, DOP: dissolved organic phosphorus, DSi: dissolved silicate

budget for the bay, sewage discharge was included in river discharges. The Guhe is the largest major river that directly empties into SGB. In developing the water budget (Fig. 8), we used the average discharge (V_Q) of the Guhe during 2011. The submarine groundwater discharge (SGD) was estimated based on submarine groundwater measurements made in June 2012. The groundwater discharge into SGB was calculated to be $(2.59–3.07) \times 10^7$ m^3 d^{-1}, based on the naturally occurring ^{228}Ra isotope (Wang et al. 2014). Generally, recirculated seawater accounts for 75 to 90% of total SGD (Moore 1996). Based on Ra isotopes, Beck et al. (2008) reported that recirculated seawater could account for approximately 90% of total SGD, and could increase as a consequence of precipitation (Guo et al. 2008). In our study, groundwater samples were collected during a summer in which substantial rainfall occurred. Based on the assumption that recirculated seawater could account for 90% of total SGD in SGB, the SGD was estimated to be $(2.59–3.07) \times 10^6$ m^3 d^{-1}. As the volume (V_S) of SGB is 10.8×10^8 m^3, the total water exchange time (τ) for SGB, estimated from the ratio $V_S/(V_R + V_X)$, was 22.4 d.

Scallop (*Chlamys farreri*) and oyster (*Crassostrea gigas*) are the main shellfish cultured in SGB. Aquaculture wastewater effluents are discharged directly into the bay. The minimum individual wet weight of oysters and scallops at harvest are 40 and 23 g (Nunes et al. 2003), respectively, and 60 000 t of oyster (wet weight) and 15 000 t of scallop are harvested annually from the bay (data from Rongcheng Fishery Technology Extension Station). Based on these data, we estimated that bivalve cultivation involved approximately 2.15×10^9 individuals during 2012. Based on excretion rates determined for bivalves and oysters in Sishili Bay (China) (Zhou et al. 2002a), the quantities of DIN and phosphate excreted by scallops were 3.84 and 0.21 µmol h^{-1} ind.$^{-1}$, respectively, and by oysters were 3.57 and 0.25 µmol h^{-1} ind.$^{-1}$, respectively. The bivalve growth

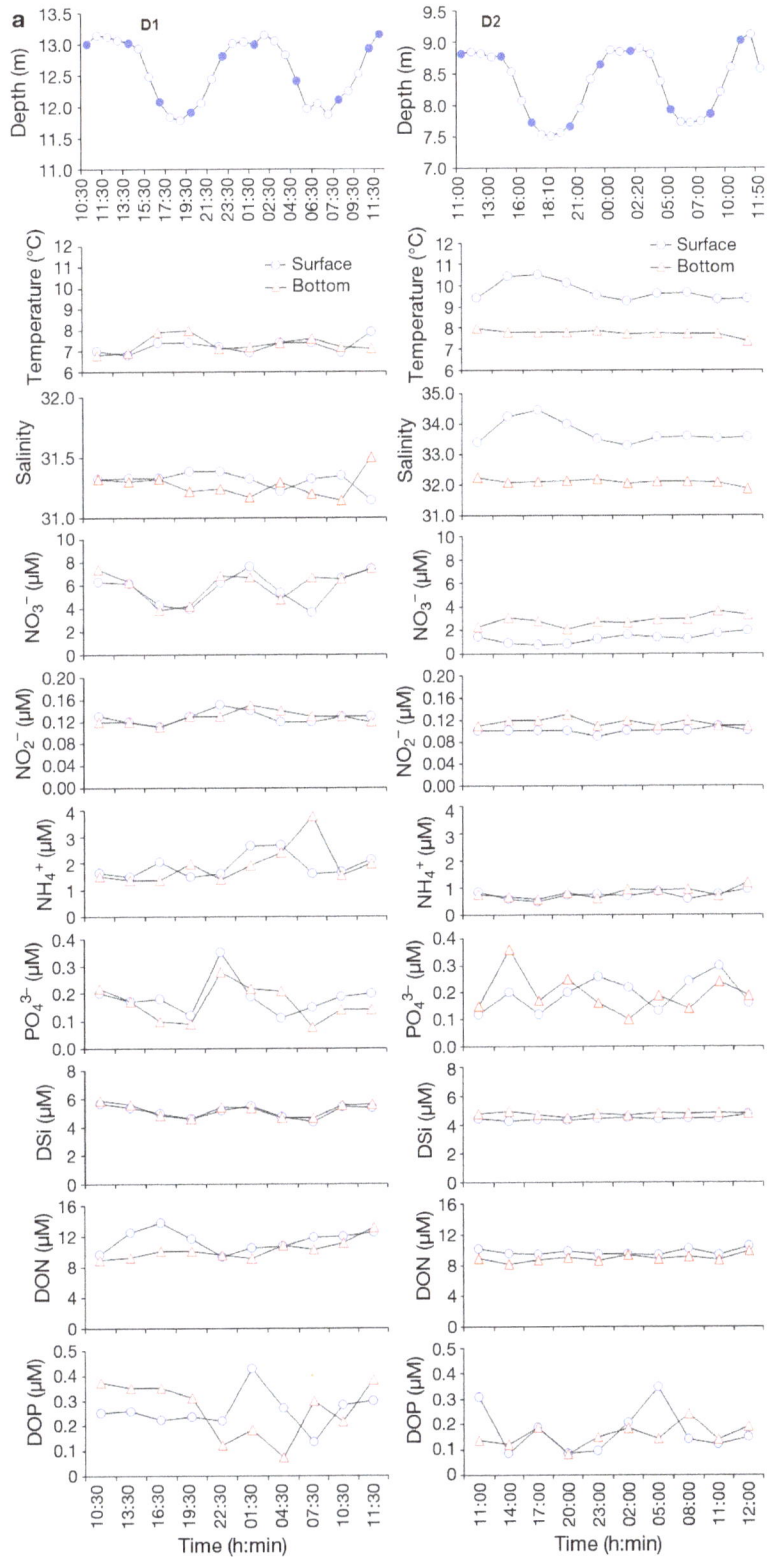

Fig. 7. Concentrations of nutrients (µM) at: (a) the anchor station in April 2013; (b; next page) the anchor station in October 2013. The water depth (m) in April 2013 and tide heights (cm) in October 2013 are provided, and the filled circles represent the nutrient sampling times. DSi: dissolved silicate, DON (DOP): dissolved organic nitrogen (phosphorus)

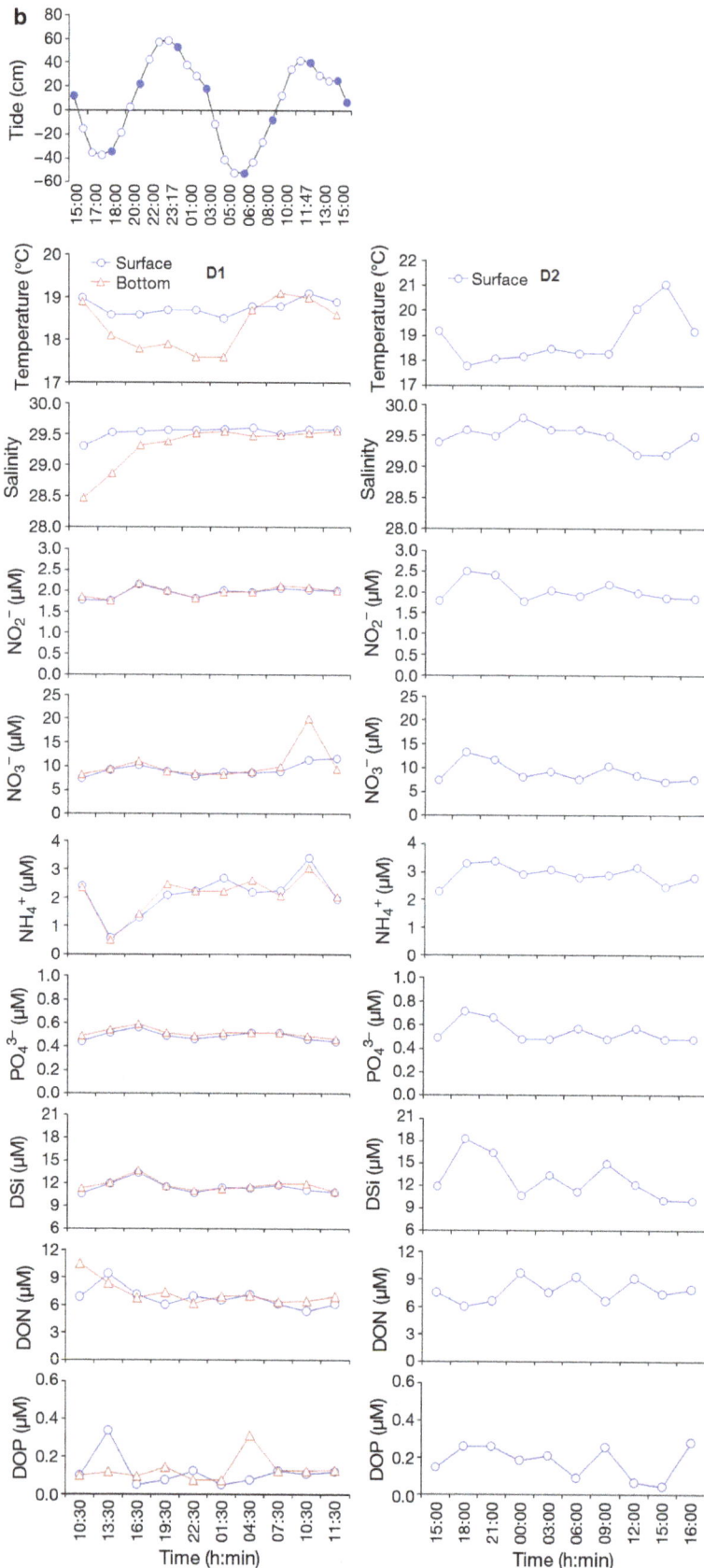

Fig. 7 (continued)

periods were mainly from May in one year to November in the following year (approximately 500 d). Hence, the total DIN and phosphate excreted by scallops and oysters in SGB amounted to 70.9×10^6 and 4.19×10^6 mol yr^{-1}, respectively. Nutrients are removed from the bay as a consequence of bivalve harvest. The dry weight nitrogen content of the soft tissue and shell of *C. gigas* is 8.19 and 0.12 % (Zhou et al. 2002b), respectively, while the phosphorus content is 0.379 and 62.1×10^{-4} % (Zhou et al. 2002b), respectively. The dry weight nitrogen and phosphorus content of the soft tissue of *C. farreri* is 12.36 and 0.839 % (Zhou et al. 2002b), respectively, and in the shell is 0.09 and 62.1×10^{-4} %, respectively. Therefore, in total the harvest of *C. farreri* and *C. gigas* removes 304 t of nitrogen and 16.7 t of phosphorus from the bay.

Saccharina japonica and *Gracilaria lemaneiformis* are the main algae cultivated in SGB. The weight of individual kelp plants at seeding is 1.2 g, and the cultivation area and density are 3331 ha and 12 ind. m^{-2}, respectively (Nunes et al. 2003). The dry weight:wet weight ratio of kelp is 1:10 (Tang et al. 2013). Hence, the dry weight of kelp at seeding is 48 t, while 87 040 t of dried kelp are produced annually in the bay (data from Rongcheng Fishery Technology Extension Station). The dry weight nitrogen and phosphorus content of kelp is 1.63 and 0.38 % (Zhou et al. 2002b), respectively. Hence, 1419 t of nitrogen and 331 t of phosphorus are removed from the bay as a consequence of kelp harvest. Similarly, 25 410 t wet weight of *G. lemaneiformis* are produced annually in the bay (data from Rongcheng Fishery Technology Extension Station). Therefore, 41.4 t of nitrogen and 9.66 t of phosphorus are removed from the bay as a consequence of *G. lemaneiformis* harvesting.

The nutrient transport fluxes from rivers and groundwater into SGB were determined from surveys undertaken during the period 2012 to 2014. The nutrient concentrations in rainwater were based on measurements at Qianliyan Island, in the western Yellow Sea (Han et al. 2013). Benthic fluxes in SGB were based on surveys undertaken during the same period

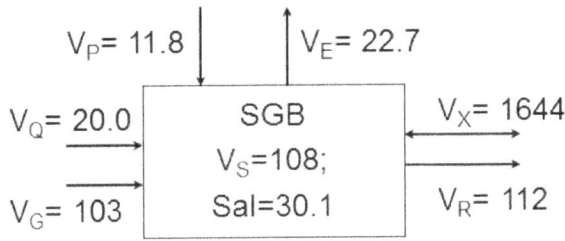

Fig. 8. Water and salt budgets for Sanggou Bay (SGB). Units: water volume, 10^7 m^3; water and salt fluxes, 10^7 m^3 and 10^7 psu m^3 mo^{-1}, respectively. V_Q, V_P, V_E, V_G, V_S, V_R, and V_X are the mean flow rate of river water, precipitation, evaporation, groundwater, the volume of the system of interest, the residual flow, and the mixing flow between the system of interest and the adjacent system, respectively. For comparison, salinity of the adjacent system = 32.23

(Ning et al. 2016, this Theme Section). For the nutrient budget, estimates of DSi removed through kelp and bivalve harvesting were not included, as no data were available.

The nutrient budgets showed that SGB behaved as a source of PO$_4$$^{3-}$ and as a sink of DSi and DIN (Table 3). The model results indicated that PO$_4$$^{3-}$ was mainly derived from bivalve excretion, which accounted for 65% of total influx, while benthic flux contributed 16% of total influx. Bivalve excretion may be an important source of PO$_4$$^{3-}$ when phytoplankton growth is phophorus-limited in the bay. The DSi load in the bay was mainly from river input and benthic flux, which contributed 47 and 34% of total influx (Table 3), respectively. Groundwater was the major source of DIN entering SGB, accounting for

41% of total influx. In addition, bivalve excretion accounted for 19% of total DIN influx. DIN and PO$_4$$^{3-}$ were mainly removed through kelp harvesting, which represented up to 64 and 81% of total outflux, respectively. The results show that aquaculture activities play an important role in nutrient cycling in SGB.

DISCUSSION

Nutrient transport in rivers

Nutrient levels in rivers varied widely (Table 2). The DIN concentrations in the rivers fell between those for polluted waters (110 µM) and severely polluted waters (350 µm) (Smith et al. 2003), except for the Bahe river. The DIN concentrations in the studied rivers were also higher than in most other small to medium-sized rivers in temperate China (Liu et al. 2009), and high relative to major Chinese rivers including the Yellow, Yangtze, and Pearl rivers (Liu et al. 2009). The extremely high DIN concentrations resulted in the high DIN:PO$_4$$^{3-}$ ratios in these rivers.

The DIN loading to streams is directly related to the extent of agriculture in the catchment (Heggie & Savage 2009). The high NO$_3$$^-$ concentrations, which dominated the DIN in rivers, is primarily attributable to anthropogenic nutrient sources, particularly to washout of fertilizers not used by target plants (Bellos et al. 2004). Rivers in the study area flow through villages and Rongcheng City, then discharge directly into SGB. Untreated industrial and domestic sewage is also discharged directly into rivers. The drainage

Table 3. Nutrient budgets for Sanggou Bay, China. $V_R C_R$: residual nutrient transport out of the system of interest (Eq. 1); $V_X C_X$: mixing exchange flux of nutrients (Eq. 2); influx (outflux): total nutrient flux into (out of) the system of interest. Δ (=$\sum$outflux − $\sum$influx) is the non-conservative flux of nutrients. Negative and positive signs of Δ indicate that the system is a sink or a source, respectively. DIP (DIN): dissolved inorganic phosphorus (nitrogen), DSi: dissolved silicate (units: 10^6 mol)

	DIP	DSi	DIN	Reference
River input ($V_Q C_Q$)	0.29	22.4	83.2	Present study
Atmospheric deposition ($V_P C_P$)	0.41	0.87	14.6	Han et al. (2003)
Groundwater discharge ($V_G C_G$)	0.55	8.27	155	Wang et al. (2014)
Benthic fluxes	1.05	16.3	57.8	Ning et al. (2016)
Bivalve excretion	4.19		70.9	Zhou et al. (2002a,b)
Influx	**6.49**	**47.8**	**382**	
Kelp harvest	−10.7		−101	Zhou et al. (2002a,b)
Gracilaria lemaneiformis harvest	−0.32		−2.96	Zhou et al. (2002a,b)
Bivalve harvest	−1.19		−21.7	Zhou et al. (2002a,b), Zhang et al. (2013)
Residual flow ($V_R C_R$)	−0.38	−8.26	−7.31	Present study
Mixing exchange ($V_X C_X$)	−0.65	−16.1	−26.2	Present study
Outflux	**13.2**	**24.4**	**159**	
ΔY (=$\sum$outflux − $\sum$influx)	**6.71**	**−23.4**	**223**	

areas of the Yatouhe, Sanggouhe, and Shilihe rivers are small (<30 km^2) and are therefore readily affected by human activities. We conclude that the high NO$_3^-$ concentrations in rivers are derived from agriculture, urban, and industrial wastewater in their drainage basins, as well as surface runoff from Rongcheng City.

The concentrations of PO$_4^{3-}$ in the Bahe and Guhe rivers were between those for pristine (0.5 μM) and clean (1.4 μM) water, and apparently lower than in the Shilihe and Sanggouhe rivers (Table 2). The high PO$_4^{3-}$ concentration (up to 6.02 μM) in the Sanggouhe, and industrial and domestic sewage, might be the most important sources of PO$_4^{3-}$ to water bodies. DSi is little affected by human activities (Jennerjahn et al. 2009) and mainly originates from natural sources. The high DSi levels in rivers adjacent to SGB may be related to the underlying rock types and weathering rates.

Rain events can result in nutrient inputs derived from hinterland areas. Approximately 73.3% of annual precipitation occurs during summer (June to September), and the annual rainfall in Rongcheng City is 819.6 mm. River discharges can be enhanced by rainfall, and weathering rates are affected by precipitation and temperature (Liu et al. 2011), which can lead to higher nutrient values during the wet seasons. High nutrient concentrations (especially dissolved silicate) but low salinities were found in the bay (Fig. 4), suggesting that rainfall might be an important factor affecting nutrient supply to SGB in summer.

Nutrient fluxes from the bay to the Yellow Sea

In this study, nutrient budgets were developed to provide an overview of nutrient cycles under the impact of aquaculture activities. Despite some uncertainties, the nutrient budgets indicated that large quantities of nitrogen and silicate would probably be buried in the sediment or transformed into other forms in the bay (Table 3). Seaweeds can absorb large amounts of nutrients from the water column, resulting in the removal of these nutrients from the system when the plants are harvested (Schneider et al. 2005). The budgets indicated that a large proportion of DIN and DIP were removed during seaweed and bivalve harvesting (Table 3), demonstrating that aquaculture activities are a significant sink for nutrients in the bay.

Based on the budgets, nutrient fluxes from SGB to the Yellow Sea were estimated as the sum of the net residual flux ($V_R C_R$) and mixing flux ($V_X C_X$) (Table 3). With the exception of DIN, nutrient fluxes to the Yellow Sea were 1.1 to 3.6 times the riverine input ($F_{model} = VC_Q$), indicating that nutrient cycling in the bay (including regeneration, aquaculture effluents) may magnify the riverine fluxes, especially bivalve excretion, which contributed to 65% of the total DIP influx. Additionally, the molar ratios of DIN:PO$_4^{3-}$ and DSi:DIN were approximately 49 and 0.2 in all external nutrient inputs to the studied system, respectively, while the corresponding flux ratios in the output waters to the Yellow Sea were approximately 35 and 0.7. These ratios deviated significantly from the Redfield ratio, indicating that aquaculture activities have significantly influenced nutrient cycling in the bay.

Wang et al. (2014) estimated that approximately 4.76×10^7 mol mo^{-1} of DIN and 5.58×10^6 mol mo^{-1} of PO$_4^{3-}$ are input from fertilizer and feed, based on protein data of shellfish and kelp in the bay during summer being used to construct a mass balance. Based on their data, fertilizer and feed would be the major source of nutrients in the bay. By visiting local farming households, we confirmed that fertilizers were used; however, fertilizer and feed are only used in fish farming during summer in SGB, thus the amounts might be far below the estimated values. If fertilizer and feed for fish farming were taken into account, the uncertainty might rise. Hence, nutrient input from feed was ignored in the box model. Furthermore, aquaculture effluents were not taken into account. Consequently, more studies on nutrient cycling in relation to aquaculture activities in SGB are needed to improve our understanding of the nutrient sink or source function of the bay.

Effects of aquaculture activities on nutrient biogeochemical cycles

The nutrient concentrations varied significantly among seasons in SGB. The dissolved inorganic nutrient levels in SGB in summer were quite low compared with other seasons; they increased from summer to autumn and reached the highest values in October (Figs. 4 & 5), indicating a shift from consumption to autumn accumulation. These seasonal variations corresponded with aquaculture activities in the bay, and this was confirmed by statistical analysis. Zhang et al. (2012) reported that nutrient biogeochemical processes and cycles were significantly affected by intensive kelp and bivalve aquaculture activities in SGB. Shi et al. (2011a) also reported that

Saccharina japonica assimilates substantial nutrients in spring. During the growth period of kelp from November to May, the NO_3^- and PO_4^{3-} concentrations decreased rapidly because of assimilation by kelp (Fig. 6). Nitrogen removed through kelp harvesting accounted for 64% of total outflux (Table 3). Kelp was a net sink for nutrients during winter and spring, and competed with phytoplankton for nutrient utilization during kelp seeding; as a consequence, phytoplankton growth was restrained. Following the kelp harvest in late May, phytoplankton could grow fast because of adequate solar radiation and temperature. As a result, the dissolved inorganic nutrient concentrations continued to decrease (Figs. 4–6).

Shellfish aquaculture generally commences in May, during the period when kelp is harvested. Bivalves in turn become another source of nutrients through excretion. During early summer, the bivalves are in the early growth stage, and produce only low levels of nutrients. The dissolved nutrients released through bivalve excretion have the potential to stimulate phytoplankton production at local scales and promote the risk of harmful algal blooms (Pietros & Rice 2003, Buschmann et al. 2008). The highest concentrations of chlorophyll *a* have been reported in summer (Hao et al. 2012). The dissolved nutrients in aquaculture effluents, coupled with high solar radiation, result in high phytoplankton production in summer (Shpigel 2005). At this time, *Gracilaria lemaneiformis* replaces kelp, and is cultivated from June to October in SGB; because it can use available nitrogen efficiently (Buschmann et al. 2008), it absorbs nutrients from seawater and probably reduces the nutrient levels in summer. This probably leads to the nutrient levels dropping rapidly to the lowest level in summer (Fig. 6).

In September, the bivalves are in active growth stages and generate large quantities of metabolic byproducts. The maximum metabolic rates for oysters are recorded in July and August (Mao et al. 2006), and lead to high nutrient concentrations in seawater (Fig. 5). Bivalves filter phytoplankton larger than 3 μm in size, thereby reducing their biomass in the water column (Newell, 2004). Phytoplankton growth is also limited by the level of solar radiation (Shi et al. 2011b). Thus, as nutrient utilization by phytoplankton decreased, the dissolved inorganic nutrient concentrations increased as a result, and increased to a greater extent in regions where bivalve monoculture occurred. Based on the nutrient budget in our study, phosphorus released from bivalve excretion could account for 65% of total influx to SGB. Hence, from June to October, prior to kelp seeding, bivalves and

fish excretion may constitute an important nutrient source in SGB, leading to increased nutrient levels. Particulate waste material (feces or pseudofeces) from bivalves and phytoplankton are consumed by bivalves, and the nutrients involved may be removed through bivalve harvesting (Shpigel 2005, Troell et al. 2009). As top-down grazers, bivalves filter phytoplankton, which results in a reduction in the nutrient turnover time and speeds up nutrient cycling.

Nutrients can be produced indirectly via remineralization and subsequent release from enriched sediments (Forrest et al. 2009). Nutrient release from sediment is also a common phenomenon occurring beneath bivalve farms in SGB (Cai et al. 2004, Sun et al. 2010). The nutrient budgets also show that benthic flux is another important source of nutrients in SGB, especially for DIP and DSi (Table 3), and that this is significantly affected by aquaculture activities in the bay (Ning et al. 2016). Based on studies of other bivalve culture systems and natural or restored oyster reefs, it is evident that benthic fluxes are determined by processes involving filter feeding and excretion of dissolved nutrients, as well as biodeposition and sediment remineralization of nutrients (Newell 2004, Forrest et al. 2009). The TDN in SGB was dominated by DON in both summer and winter (Figs. 4 & 5), as observed in landbased aquaculture (Jackson et al. 2003, Herbeck et al. 2013). Burford & Williams (2001) reported that most of the dissolved nitrogen leaching from feed and shrimp feces was in organic rather than in inorganic forms. Hence, DON leaching from feces or pseudofeces might be an important source of DON in the bivalve cultivation regions in SGB (Fig. 6). Furthermore, increased sedimentation of organic matter from feces and pseudofeces underneath mussel farms can have significant ecosystem effects on the biogeochemical cycles of nitrogen and phosphorus (Stadmark & Conley 2011).

Biogeochemical cycling of DSi can be affected by diatom dissolution, sediment resuspension, and terrigenous input. In our study, the average concentrations of DSi increased by 9.0 μM from July to October, and decreased rapidly from 14.2 to 4.76 μM in January. Phytoplankton abundance was tightly controlled by filter feeding of oysters (Hyun et al. 2013), so the high metabolic rates of oysters may result in a reduction of diatom biomass, leading to high levels of DSi in autumn. In addition, as the water depth in SGB is ≤20 m, sediment resuspension and diatom dissolution might be important sources of DSi during the summer to autumn period. The dissolution of diatom frustules depends on a variety of factors, including microbial activity (Olli et al. 2008). Bacteria can attack the organic matrix protecting the diatom frustule, exposing biogenic silica, and substantially increase the dissolution rate (Bidle & Azam 1999). The maximum biomass in SGB occurred in autumn (Chen 2001), and diatoms dominated in the bay in summer. Consequently, dissolution of diatom frustules may be an important source of DSi in the bay.

Although the aquaculture area and quantities of effluents released in SGB were high (Table 3), nutrient levels in the bay were not significantly elevated compared with other bays used for aquaculture, including Jiaozhou Bay (Liu et al. 2007) and Sishili Bay (Zhou et al. 2002b). This is attributed to the fact that nutrients released from shellfish are taken up by seaweeds during their growth periods. Large-scale kelp cultivation plays an important role in keeping nutrients at low levels and maintaining relatively good water quality.

Effects of physical factors on nutrient changes

The marine IMTA culture system used in SGB is suspended aquaculture. Water exchange between SGB and the Yellow Sea could be hindered by kelp (*S. japonica*), especially during kelp harvesting (Zeng et al. 2015). Our depth study showed that nutrien changes over the tidal cycle generally closely followed changes in water depth at Stn D2 (Fig. 7), indicating that water exchange is greater at Stn D1 (in the northern mouth of SGB), and weaker at Stn D2. Furthermore, in April 2013, the nutrients were well mixed at Stn D1, while at Stn D2, the nutrient concentrations were higher in bottom water than in the surface water (Fig. 7). This indicates that the current was affected by the aquaculture facilities and kelp at Stn D2, which may have led to higher nutrient concentrations in the bottom water than in the surface water. These results are consistent with the *in situ* measurements of Zeng et al. (2015), which showed that the vertical tidal flux at the northern entrance of SGB was much larger than at the southern entrance. In addition, the current structure in SGB has been significantly changed by the presence of aquaculture activities (Shi et al. 2011a). The tidal current in the surface layer is only half that in the middle layer when kelp is at its maximum length (Shi et al. 2011a). As a result, particulate matter and nutrients in bottom waters are constrained from entering the upper water layers because of the influence of aquaculture facilities and species (Wei et al. 2010).

The current flow generally tends to decrease in suspended aquaculture areas because of the extra drag caused by the presence of aquaculture facilities. In SGB, bivalves and fish are grown in cages, nets, or other containers hung from floats or rafts. Based on a 3-dimensional physical–biological coupled aquaculture model (Shi et al. 2011a), the average current flow speed can be reduced by approximately 63% by aquaculture facilities and cultured species. Moreover, Grant & Bacher (2001) reported a 20% reduction in current speed in the main navigation channel in SGB, and a 54% reduction in the middle of the culture area because of the effects of suspended aquaculture. Nutrients are likely to be retained in the bay because of the weaker current in the bivalve culture areas. The nutrient budgets showed that bivalve excretion was an important source of nutrients (Table 3). Large quantities of nutrients could accumulate in the west of the bay, and red tides have occurred in SGB in recent years (Zhang et al. 2012). The effects of consequent shading and competition pressure from the increased algae biomass on the valuable habitats involved may negatively affect the seagrass meadows in the southwest of the bay, and the production of bivalves may be reduced. To conserve the natural services provided by the bay, aquaculture effluents should be treated before they are released into natural water bodies.

Water exchange can also cause differences in nutrient species inside and outside SGB. Wei et al. (2010) observed that the flow speed declined by approximately 70% from the mouth to the southwestern part of the bay, and the outflow was slowed by the increased aquaculture activities and infrastructure (Fan & Wei 2010). Thus, movement of nutrients from the southwest of the bay to the open sea may be impeded, which was suggested by the high concentrations of nutrients found in this part of the bay in summer and autumn (Fig. 4).

Long-term trends of nutrients in SGB

Fig. 9 shows compiled data for DIN, DSi, and PO_4^{3-} in SGB, based on historical data and our observations (Song et al. 1996, P. Sun et al. 2007, S. Sun et al. 2010, Zhang et al. 2010, 2012, this study), reflecting the long-term variations for the period 1983 to 2014. No trends in the PO_4^{3-} concentrations were evident because of the high variability in this parameter (Fig. 9). In contrast, the DIN concentrations increased over time and were significantly higher in 2003 to 2011 than in previous years (Fig. 9). Prior to the 1980s, kelp was the main aquaculture species, and the DIN concentration was low in the bay (Fang et al. 1996a,

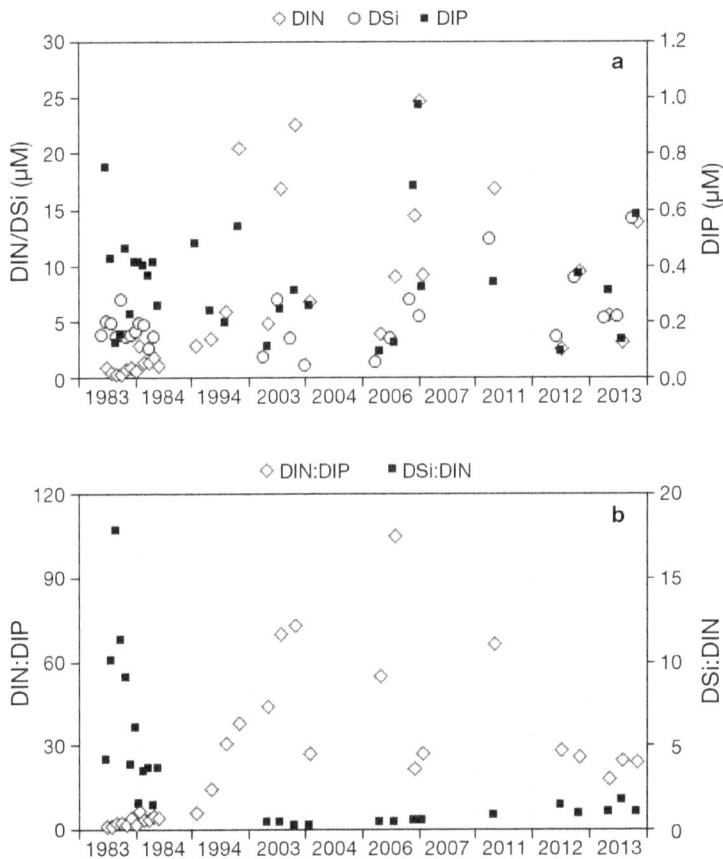

Fig. 9. Long-term changes in (a) the concentrations of dissolved inorganic nitrogen (DIN), dissolved inorganic phosphorus (DIP), and dissolved silicate (DSi) concentrations, and (b) the DIN:PO$_4$$^{3-}$ and DSi:DIN ratios for the period 1983 to 2013

bays used for aquaculture, including Chesapeake Bay in the US (Tango et al. 2005), and Jiaozhou (Shen 2002, Sun et al. 2011) and Daya Bays (Wang et al. 2009) in China. Turner et al. (1998) reported that the risk of harmful algal blooms increases with shifts in the DSi:DIN ratio to values <1, when phytoplankton becomes dominated by non-diatom species. Molar ratios of DSi:DIN in SGB changed from 1.4–18 in 1983 to <1 during the 2003 to 2011 period. Red tides were observed in April 2011 (Zhang et al. 2012), and were apparent in small areas in 2013. In addition, an increase in the DIN concentration will lower the DSi:DIN ratio, and could change ecosystem structure of the bay (Billen & Garnier 2007).

Because of its combination of environmental, economic, and social benefits (Allsopp et al. 2008, Nobre et al. 2010), IMTA has been gaining recognition as a sustainable approach to aquaculture, and the water quality in SGB has remained in good condition compared with other bays affected by aquaculture activities. Environmental management strategies will need to include both reduction of nutrient pollution and monitoring of the relative abundance of nutrients. The ecological and economic health of SGB should be tightly monitored to ensure a rapid response to critical changes.

CONCLUSION

We have reported on the nutrient dynamics of SGB, which represents a typical watershed for IMTA. The results of our investigation show that aquaculture activities play an important role in nutrient cycling in SGB. Nutrients showed considerable seasonal variation in the bay, and nutrient composition and distribution were also affected by the cultured species in the bay. The nutrient budgets showed that SGB behaved as a source of PO$_4$$^{3-}$ and as a sink of DSi and DIN. The model results indicated that PO$_4$$^{3-}$ was mainly derived from bivalve excretion. Bivalve excretion may be an important source of PO$_4$$^{3-}$ when phytoplankton growth is phosphorus-limited in the bay. Seaweed and bivalve harvesting play an important role in removing DIN and PO$_4$$^{3-}$ from the bay. Under the combined effects of natural processes and aquaculture activities, nutrient biogeochemistry in the bay has been affected.

Ning et al. 2016). Polyculture was introduced into the bay for economic reasons (Fang et al. 1996a), and its rapid development may have been responsible for increasing levels of nutrients in the bay, and resulted in long-term alterations to the nutrient conditions (Shi et al. 2011a, Zhang et al. 2012). In SGB, nutrient-rich aquaculture effluents are released into the natural water body without prior treatment. The high concentrations of nitrogen in aquaculture effluents mainly originate from excess feed or from excretion from the farmed animals (Burford & Williams 2001).

As a result of the increased nitrogen levels, the DIN:DIP ratios in SGB shifted from severe nitrogen limitation in 1983 to the ecologically desirable Redfield ratio (16) in summer 1994, and continued to increase until summer 2006, when the DIN:PO$_4$$^{3-}$ ratio reached 105; phytoplankton growth is now limited by phosphorus in summer. The increase in the DIN:PO$_4$$^{3-}$ ratios in SGB is a common phenomenon observed in long-term studies of estuarine and coastal areas affected by human activities, and also in semi-closed

Acknowledgements. This research was funded by the Chinese Ministry of Science and Technology (2011CB409802), the Natural Sciences Foundation of China (NSFC: 41221004), and the Taishan Scholars Programme of Shandong Province. We thank colleagues at the Yellow Sea Fisheries Research Institute, Chinese Academy of Fishery Sciences, East China Normal University, and the Ocean University of China for their help during field investigations.

LITERATURE CITED

Allsopp M, Johnston P, Santillo D (2008) Challenging the aquaculture industry on sustainability. Greenpeace International, Amsterdam

ä Bacher C, Grant J, Hawkins AJS, Fang JG, Zhu MY, Besnard M (2003) Modelling the effect of food depletion on scallop growth in Sungo Bay (China). Aquat Living Resour 16:10–24

ä Beck AJ, Rapaglia JP, Kirk Cochran J, Bokuniewicz HJ, Yang SH (2008) Submarine groundwater discharge to Great South Bay, NY, estimated using Ra isotopes. Mar Chem 109:279–291

ä Bellos D, Sawidis T, Tsekos I (2004) Nutrient chemistry of River Pinios (Thessalia, Greece). Environ Int 30:105–115

Bidle KD, Azam F (1999) Accelerated dissolution of diatom silica by marine bacterial assemblages. Nature 39:508–511

ä Billen G, Garnier J (2007) River basin nutrient delivery to the coastal sea: assessing its potential to sustain new production of non-siliceous algae. Mar Chem 106:148–160

ä Bostock J, McAndrew B, Richards R, Jauncey K and others (2010) Aquaculture: global status and trends. Philos Trans R Soc Lond B Biol Sci 365:2897–2912

ä Bouwman AF, Pawłowski M, Liu C, Beusen AHW, Shumway SE, Glibert PM, Overbeek CC (2011) Global hindcasts and future projections of coastal nitrogen and phosphorus loads due to shellfish and seaweed aquaculture. Rev Fish Sci 19:331–357

ä Burford MA, Williams KC (2001) The fate of nitrogenous waste from shrimp feeding. Aquaculture 198:79–93

ä Buschmann AH, Varela DA, Hernández-González MC, Huovinen P (2008) Opportunities and challenges for the development of an integrated seaweed-based aquaculture activity in Chile: determining the physiological capabilities of *Macrocystis* and *Gracilaria* as biofilters. J Appl Phycol 20:571–577

Cai LS, Fang JG, Dong SL (2004) Preliminary studies on nitrogen and phosphorus fluxes between seawater and sediment in Sungo Bay. Mar Fish Res 25:57–64 (in Chinese with English abstract)

Chen HW (2001) Correlation analysis on bacteriological indexes and environmental parameters for surface water of Sanggou Bay. Mar Environ Sci 20:29–33 (in Chinese with English abstract)

ä Duarte P, Meneses R, Hawkins AJS, Zhu M, Fang J, Grant J (2003) Mathematical modelling to assess the carrying capacity for multi-species culture within coastal waters. Ecol Model 168:109–143

Fan X, Wei H (2010) Modeling studies on vertical structure of tidal current in a typically coastal raft-culture area. Prog Fish Sci 31:78–84 (in Chinese with English abstract)

ä Fang JG, Sun HL, Yan JP, Kuang SH, Li F, Newkirk GF, Grant J (1996a) Polyculture of scallop *Chlamys farreri* and kelp *Laminaria japonica* in Sungo Bay. Chin J Oceanol Limnol 14:322–329

Fang JG, Kuang SH, Sun HL, Li F, Zhang AJ, Wang XZ, Tang TY (1996b) Mariculture status and optimizing measurements for the culture of scallop *Chlamys farreri* and kelp *Laminaria japonica* in Sungou Bay. Mar Fish Res 17:95–102 (in Chinese with English abstract)

FAO (Food and Agriculture Organization of the United Nations) (2010) The state of world fisheries and aquaculture 2010. Fisheries and Aquaculture Department, FAO, Rome

ä Fei XG (2004) Solving the coastal eutrophication problem by large scale seaweed cultivation. Hydrobiologia 512: 145–151

ä Forrest BM, Keeley NB, Hopkins GA, Webb SC, Clement DM (2009) Bivalve aquaculture in estuaries: review and synthesis of oyster cultivation effects. Aquaculture 298: 1–15

ä Fu MZ, Pu XM, Wang ZL, Liu XJ (2013) Integrated assessment of mariculture ecosystem health in Sanggou Bay. Acta Ecol Sin 33:238–316 (in Chinese with English abstract)

Gordon DC, Boudreau PR, Mann KH, Ong JE and others (1996) LOICZ biogeochemical modeling guidelines. LOICZ Reports and Studies (5). Land–Ocean Interactions in the Coastal Zone, Texel

ä Grant J, Bacher C (2001) A numerical model of flow modification induced by suspended aquaculture in a Chinese Bay. Can J Fish Aquat Sci 58:1–9

Grasshoff K, Kremling K, Ehrhardt M (1999) Methods of seawater analysis. In: Hansen HP, Koroleff F (eds) Determination of nutrients. Wiley-VCH, Weinheim, p 159–228

Guo ZR, Huang L, Liu HT, Yuan XJ (2008) The estimation of submarine inputs of groundwater to a coastal bay using radium isotopes. Acta Geosci Sin 29:647–652 (in Chinese with English abstract)

Han LJ, Zhu YM, Liu SM, Zhang J, Li RH (2013) Nutrients of atmospheric wet deposition from the Qianliyan Island of the Yellow Sea. China Environ Sci 33:1174–1184 (in Chinese with English abstract)

Hao LH, Sun PX, Hao JM, Du BB, Zhang XJ, Xu YS, Bi JH (2012) The spatial and temporal distribution of chlorophyll-a and its influencing factors in Sanggou Bay. Ecol Environ Sci 21:338–345 (in Chinese with English abstract)

ä Heggie K, Savage C (2009) Nitrogen yields from New Zealand coastal catchments to receiving estuaries. N Z J Mar Freshw Res 43:1039–1052

ä Herbeck LS, Unger D, Wu Y, Jennerjahn TC (2013) Effluent, nutrient and organic matter export from shrimp and fish ponds causing eutrophication in coastal and back-reef waters of NE Hainan, Tropical China. Cont Shelf Res 57: 92–104

ä Hyun JH, Kim SH, Mok JS, Lee JS, An SU, Lee WC (2013) Impacts of long-line aquaculture of Pacific oysters (*Crassostrea gigas*) on sulfate reduction and diffusive nutrient flux in the coastal sediments of Jinhae-Tongyeong, Korea. Mar Pollut Bull 74:187–198

ä Jackson C, Preston N, Thompson PJ, Burford M (2003) Nitrogen budget and effluent nitrogen components at an intensive shrimp farm. Aquaculture 218:397–411

ä Jennerjahn TC, Nasir B, Pohlenga I (2009) Spatio-temporal variation of dissolved inorganic nutrients related to hydrodynamics and land use in the mangrove-fringed Segara Anakan Lagoon, Java, Indonesia. Reg Environ Change 9:259–274

ä Jiang ZJ, Fang JG, Zhang JH, Mao YZ, Wang W (2007) Forms and bioavailability of phosphorus in surface sediments from Sungo Bay. Huan Jing Ke Xue (Environ Sci) 28:2783–2788 (in Chinese with English abstract)

Justić D, Rabalais NN, Turner RE, Dortch Q (1995) Changes in nutrient structure of river-dominated coastal waters: stoichiometric nutrient balance and its consequences. Estuar Coast Shelf Sci 40:339–356

Kuang S, Fang J, Sun H, Li F (1996) Seston dynamics in Sanggou Bay. Mar Fish Res 17:60–67 (in Chinese with English abstract)

ä Liu SM, Zhang J, Chen HT, Zhang GS (2005) Factors influencing nutrient dynamics in the eutrophic Jiaozhou Bay, North China. Prog Oceanogr 66:66–85

ä Liu SM, Li XN, Zhang J, Wei H, Ren JL, Zhang GL (2007) Nutrient dynamics in Jiaozhou Bay. Water Air Soil Pollut Focus 7:625–643

ä Liu SM, Hong GH, Zhang J, Ye XW, Jiang XL (2009) Nutrient budgets for large Chinese estuaries. Biogeosciences 6:2245–2263

▶ Liu SM, Li RH, Zhang GL, Wang DR and others (2011) The impact of anthropogenic activities on nutrient cycling dynamics in the tropical Wenchanghe and Wenjiaohe Estuary and Lagoon system in East Hainan, China. Mar Chem 125:49–68

ä Mao YZ, Zhou Y, Yang HS, Wang RC (2006) Seasonal variation in metabolism of cultured Pacific oyster, *Crassostrea gigas*, in Sanggou Bay, China. Aquaculture 253:322–333

ä Marinho-Soriano E, Nunes SO, Carneiro MAA, Pereira DC (2009) Nutrients' removal from aquaculture wastewater using the macroalgae *Gracilaria birdiae*. Biomass Bioenergy 33:327–331

ä Moore WS (1996) Large groundwater inputs to coastal waters revealed by ^{226}Ra enrichments. Nature 380:612–614

ä Neori A, Chopin T, Troell M, Buschmann AH and others (2004) Integrated aquaculture: rationale, evolution and state of the art emphasizing seaweed biofiltration in modern mariculture. Aquaculture 231:361–391

Newell RIE (2004) Ecosystem influences of natural and cultivated populations of suspension-feeding bivalve molluscs: a review. J Shellfish Res 23:51–61

▶ Newell RIE, Koch EW (2004) Modeling seagrass density and distribution in response to changes in turbidity stemming from bivalve filtration and seagrass sediment stabilization. Estuaries 27:793–806

ä Ning Z, Liu S, Zhang G, Ning X and others (2016) Impacts of an integrated multi-trophic aquaculture system on benthic nutrient fluxes: a case study in Sanggou Bay, China. Aquacult Environ Interact 8:221–232

▶ Nobre AM, Robertson-Andersson D, Neori A, Sankar K (2010) Ecological-economic assessment of aquaculture options: comparison between abalone monoculture and integrated multi-trophic aquaculture of abalone seaweeds. Aquaculture 306:116–126

ä Nunes JP, Ferreira JG, Gazeau F, Lencart-Silva J, Zhang XL, Zhu MY, Fang JG (2003) A model for sustainable management of shellfish polyculture in coastal bays. Aquaculture 219:257–277

▶ Olli K, Clarke A, Danielsson Å, Aigars J, Conley D, Tamminen T (2008) Diatom stratigraphy and long-term dissolved silica concentrations in the Baltic Sea. J Mar Syst 73:284–299

ä Pietros JM, Rice MA (2003) The impacts of aquacultured oysters, *Crassostrea virginica* (Gmelin, 1791) on water column nitrogen and sedimentation: results of a mesocosm study. Aquaculture 220:407–422

ä Savchuk OP (2005) Resolving the Baltic Sea into seven sub-basins: N and P budgets for 1991–1999. J Mar Syst 56: 1–15

ä Schneider O, Sereti V, Eding EH, Verreth JAJ (2005) Analysis of nutrient flows in integrated intensive aquaculture systems. Aquacult Eng 32:379–401

Shen ZL (2002) Long-term changes in nutrient structure and its influences on ecology and environment in Jiaozhou Bay. Oceanol Limnol Sin 33:322–331 (in Chinese with English abstract)

ä Shi J, Wei H, Zhang L, Yuan Y, Fang JG, Zhang JH (2011a) A physical-biological coupled aquaculture model for a suspended aquaculture area of China. Aquaculture 318: 412–424

Shi HH, Fang GH, Hu L, Zheng W (2011b) Analysis on response of pelagic ecosystem to kelp mariculture within coastal waters. J Waterway Harbor 32:213–218

Shpigel M (2005) Bivalves as biofilters and valuable by-products in land-based aquaculture systems. In: Dame RF, Olenin S (eds) The comparative roles of suspension feeders in ecosystems. Springer-Verlag, Dordrecht, p 183–197

ä Smaal A, van Stralen M, Schuiling E (2001) The interaction between shellfish culture and ecosystem processes. Can J Fish Aquat Sci 58:991–1002

ä Smith SV, Swaney DP, Talaue-McManus L, Bartley JD and others (2003) Humans, hydrology, and the distribution of inorganic nutrient loading to the ocean. Bioscience 53: 235–245

Song HJ, Li RX, Wang ZL, Zhang XL, Liu P (2007) Interannual variations in phytoplankton diversity in the Sanggou Bay. Adv Mar Sci 25:332–339 (in Chinese with English abstract)

Song XL, Yang Q, Sun Y, Yin H, Jiang SL (2012) Study of sedimentary section records of organic matter in Sanggou Bay over the last 200 years. Acta Oceanol Sin 34: 120–126 (in Chinese with English abstract)

Song YL, Cui Y, Sun Y, Fang JG, Sun HL, Kuang SH (1996) Study on nutrient state and influencing factors in Sanggou Bay. Mar Fisher Res 17(2):41–51 (in Chinese with English abstract)

ä Stadmark J, Conley DJ (2011) Mussel farming as a nutrient reduction measure in the Baltic Sea: consideration of nutrient biogeochemical cycles. Mar Pollut Bull 62:1385–1388

Sun PX, Zhang ZH, Hao LH, Wang B and others (2007) Analysis of nutrient distributions and potential eutrophication in seawater of the Sanggou Bay. Adv Mar Sci 25: 436–445 (in Chinese with English abstract)

Sun S, Liu SM, Ren JL, Zhang JH, Jiang ZJ (2010) Distribution features of nutrients and flux across the sediment-water interface in the Sanggou Bay. Acta Oceanol Sin 32: 108–117 (in Chinese with English abstract)

Sun S, Li CL, Zhang GT, Sun XX, Yang B (2011) Long-term changes in the zooplankton community in the Jiaozhou Bay. Oceanol Limnol Sin 42:625–631 (in Chinese with English abstract)

Tang QS, Fang JG, Zhang JH, Jiang ZJ, Liu HM (2013) Impacts of multiple stressors on coastal ocean ecosystems and integrated multi-trophic aquaculture. Prog Fish Sci 34:1–11 (in Chinese with English abstract)

▶ Tango PJ, Magnien R, Butler W, Luckett C, Luckenbach M, Lacouture R, Poukish C (2005) Impacts and potential effects due to *Prorocentrum minimum* blooms in Chesapeake Bay. Harmful Algae 4:525–531

The People's Republic of China Ministry of Agriculture Fisheries Bureau (2013) China Fishery Statistical Yearbook.

China Agriculture Press, Beijing (in Chinese)

Troell M, Joyce A, Chopin T, Neori A, Buschmann AH, Fang JG (2009) Ecological engineering in aquaculture-potential for integrated multi-trophic aquaculture (IMTA) in marine offshore systems. Aquaculture 297:1–9

▶ Turner RE, Qureshi N, Rabalais NN, Dortch Q, Justi D, Shaw RF, Cope J (1998) Fluctuating silicate: nitrate ratios and coastal plankton food webs. Proc Natl Acad Sci USA 95:13048–13051

▶ Wall CC, Peterson BJ, Gobler CJ (2008) Facilitation of seagrass Zostera marina productivity by suspension-feeding bivalves. Mar Ecol Prog Ser 357:165–174

Wan L (2012) Effect of shellfish farming on nutrient salts of seawater in Sanggou Bay in spring. Environ Sci Manag 2012(6):62–64 (in Chinese with English abstract)

▶ Wang XL, Du JZ, Ji T, Wen TY, Liu SM, Zhang J (2014) An estimation of nutrient fluxes via submarine groundwater discharge into the Sanggou Bay—a typical multi-species culture ecosystem in China. Mar Chem 167:113–122

▶ Wang ZH, Zhao JG, Zhang YJ, Cao Y (2009) Phytoplankton community structure and environmental parameters in aquaculture areas of Daya Bay, South China Sea. J Environ Sci (China) 21:1268–1275

Wei H, Zhao L, Yuan Y, Shi J, Fan X (2010) Study of hydrodynamics and its impact on mariculture carrying capacity of Sanggou Bay: observation and modeling. Prog Fish Sci 31:65–71 (in Chinese with English abstract)

▶ Wu H, Peng RH, Yang Y, He L, Wang WQ, Zheng TL, Lin GH (2014) Mariculture pond influence on mangrove areas in south China: significantly larger nitrogen and phosphorus loadings from sediment wash-out than from tidal water exchange. Aquaculture 426–427:204–212

Yang YF, Li CH, Nie XP, Tang DL, Chuang EK (2005) Development of mariculture and its impacts in Chinese coastal waters. Rev Fish Biol Fish 14:1–10

▶ Yu HY, Bao LJ, Wong CS, Hu Y, Zeng EY (2012) Sedimentary loadings and ecological significance of polycyclic aromatic hydrocarbons in a typical mariculture zone of South China. J Environ Monit 14:2685–2691

▶ Yuan XT, Zhang MJ, Liang YB, Liu D, Guan DM (2010) Self-pollutant loading from a suspension aquaculture system of Japanese scallop (Patimopecten yessoensis) in the Changhai sea area, Northern Yellow Sea of China. Aquaculture 304:79–87

▶ Zeng DY, Huang DJ, Qiao XD, He YQ, Zhang T (2015) Effect of suspended kelp culture on water exchange as estimated by in situ current measurement in Sanggou Bay, China. J Mar Syst 149:14–24

▶ Zhang JH, Hansen PK, Fang JG, Wang W, Jiang ZJ (2009) Assessment of the local environmental impact of intensive marine shellfish and seaweed farming—application of the MOM system in the Sungo Bay, China. Aquaculture 287:304–310

▶ Zhang JH, Shang DR, Wang W, Jiang ZJ, Xue SY, Fang JG (2010) The potential for utilizing fouling macroalgae as feed for abalone Haliotis discus hannai. Aquacult Res 41:1770–1777

▶ Zhang JH, Wang W, Hang TT, Liu DH and others (2012) The distributions of dissolved nutrients in spring of Sungo Bay and potential reason of outbreak of red tide. J Fish China 36:132–138 (in Chinese with English abstract)

Zhang JH, Fang JG, Tang QS, Ren LH (2013) Carbon sequestration rate of the scallop Chlamys farreri cultivated in different areas of Sanggou Bay. Prog Fish Sci 34:12–16 (in Chinese with English abstract)

Zhang LH, Zhang XL, Li RX, Wang ZL and others (2005) Impact of scallop culture in dinoflagellate abundance in the Sanggou Bay. Adv Mar Sci 23:342–346 (in Chinese with English abstract)

Zhao J, Zhou SL, Sun Y, Fang JG (1996) Research on Sanggou bay aquaculture hydro-environment. Mar Fish Res 17:68–79

Zhou Y, Mao YZ, Yang HS, He YZ, Zhang FS (2002a) Clearance rate, ingestion rate and absorption efficiency of the scallop Chlamys farreri measured by in situ biodeposition method. Acta Ecol Sin 22:1455–1462 (in Chinese with English abstract)

Zhou Y, Yang HS, Liu SL, He YZ, Zhang FS (2002b) Chemical composition and net organic production of cultivated and fouling organisms in Sishili Bay and their ecological effects. J Fish China 26:21–27 (in Chinese with English abstract)

7

Temporal variations in suspended particulate waste concentrations at open-water fish farms in Canada and Norway

Lindsay M. Brager[1], Peter J. Cranford[1,*], Henrice Jansen[2], Øivind Strand[3]

[1]Department of Fisheries and Oceans, Bedford Institute of Oceanography, 1 Challenger Drive, Dartmouth, Nova Scotia B2Y 4A2, Canada

[2]Wageningen IMARES–Institute for Marine Resources and Ecosystem Studies, Korringaweg 5, 4401 NT Yerseke, The Netherlands

[3]Institute of Marine Research, Nordnesgaten 50, 5005 Bergen, Norway

ABSTRACT: The co-cultivation of finfish and bivalve filter feeders, with the purpose of recycling solid waste effluents and enhancing aquaculture revenues, has stimulated efforts to characterize particulate waste dynamics at open-water fish farms. Temporal variability in waste concentrations in the water column was studied at Atlantic salmon *Salmo salar* farms in eastern Canada and Norway, and at a sablefish *Anoplopoma fimbria* farm in western Canada. Turbidity and chlorophyll *a* (chl *a*) sensors were used to continuously monitor suspended particulate matter (SPM) and chl *a* concentrations at various depths and distances to the net-pens. Time-series analysis of these data indicated that SPM fluctuations at the study sites corresponded largely with tidal periodicity and variations in phytoplankton biomass (chl *a*). ANOVA comparisons of mean SPM levels in water flowing from the direction of net-pens (potential farm influence) and towards the farm (control) generally indicated insignificant effects of fish wastes on SPM levels at the farms ($p > 0.05$). A significant effect of the farm was detected by an ANOVA comparison of SPM concentrations collected at 5 m depth at farm and reference stations in an oligotrophic fjord, but the calculated level of waste enhancement over the sampling period was extremely small (0.02 mg l^{-1}). Results indicate that temporal variations in SPM concentrations around the open-water fish farms were largely driven by natural processes and that the addition of fish wastes had a negligible effect. Consequently, there is little rationale for introducing commercial extractive species (e.g. bivalves) in open-water integrated multi-trophic aquaculture systems to mitigate the horizontal flux of particulate fish wastes.

KEY WORDS: Aquaculture wastes · Suspended particulate matter · Integrated multi-trophic aquaculture · *Salmo salar* · *Anoplopoma fimbria*

INTRODUCTION

Aquaculture has been responsible for the continuing growth in global fish production since capture production levelled off in the mid-1990s. Global aquaculture is the fastest growing food-producing sector, with nearly half of the world's seafood supply now sourced from aquaculture (FAO 2012). A main challenge facing this industry is sustaining a continued increase in fish production while minimizing the impact on the environment (Sugiura et al. 2006, Navarrete-Mier et al. 2010, Taranger et al. 2015). The open nature of many fish culture systems allows for continuous exchange of water between the cages and the surroundings (Beveridge 1984, Gowen & Bradbury 1987, Folke & Kautsky 1989, Beveridge et

*Corresponding author: peter.cranford@dfo-mpo.gc.ca

al. 1991, 1994, Phillips et al. 1991) and the rapid expansion of caged fish culture has raised a general concern about the release of solid and dissolved waste products to the environment (Troell & Norberg 1998, Perez et al. 2002, Cheshuk et al. 2003, Whitmarsh et al. 2006, Valdemarsen et al. 2012, Taranger et al. 2015). Conspicuous environmental effects can stem from the increased organic matter loading on the seafloor, and consequential changes in benthic habitat and communities (Strain & Hargrave 2005, Kutti et al. 2007). Large faecal particles and uneaten feed sink rapidly and accumulate in sediments below and within the vicinity of the net-pens (Cromey et al. 2002, Olsen et al. 2008, Nickell et al. 2009). Fine waste particles can remain in suspension for long periods and there is some evidence that suspended particulate matter (SPM) levels are enhanced in waters adjacent to fish farms (Jones & Iwama 1991, Lefebvre et al. 2000, MacDonald et al. 2011, Lander et al. 2013, Brager et al. 2015, Bannister et al. 2016). However, particle enhancement around fish pens has not been consistently observed (Buschmann et al. 1996, Pridmore & Rutherford 1992, Lander et al. 2013, Brager et al. 2015) and the reasons for these different conclusions on waste enhancement of the natural particle field are not fully understood.

Integrated multi-trophic aquaculture (IMTA) involves the co-culture of several commercial species in a system designed to facilitate the conversion of wastes produced at each trophic level into additional aquaculture revenue (Chopin et al. 2001, Troell et al. 2003). Open-water IMTA systems generally include the primary finfish component, a dissolved inorganic waste extractive component, such as seaweeds, and filter- and/or deposit-feeding species to capture, assimilate and extract particulate organic wastes. IMTA has been practiced for centuries in Asia (Li 1987, Fang et al. 1996, Qian et al. 1996), and the cultivation of scallops, kelp and abalone in Sungo Bay, China, has been commercially successful at industrial scales (Fang et al. 1996, Troell et al. 2009). IMTA has more recently been explored in Canada, Scotland, Australia and Norway, where several pilot experiments have been conducted (Stirling & Okumus 1995, Cheshuk et al. 2003, Chopin et al. 2001, Handå et al. 2012a, 2012b). Development of the particulate organic extractive component of IMTA has largely focused on the introduction of filter-feeding bivalve molluscs inside and/or adjacent to fish net-pens (Soto 2009). However, the expected enhancement of bivalve growth in polyculture and IMTA systems has not consistently been realized (Jones & Iwama 1991, Taylor et

al. 1992, Stirling & Okumus 1995, Gryska et al. 1996, Cheshuk et al. 2003, Lander et al. 2004, Sarà et al. 2009, Navarrete-Mier et al. 2010, Handå et al. 2012a, Lander et al. 2012, Jiang et al. 2013). The effectiveness of bivalves to extract fish wastes under open-water conditions also appears to be much lower than anticipated (Troell & Norberg 1998, Cranford et al. 2013). Improving interactions between fish wastes and extractive species through refinements in IMTA system design may be expected to improve crop yields and the intended ecological benefits of the IMTA approach. However, this optimization of farm design requires knowledge on particulate waste dynamics and transport pathways under relevant hydrographic conditions.

Lander et al. (2013) examined the characteristics of suspended particles released from salmon farms and concluded that particle concentrations within 5 m of fish cages are significantly enhanced with highly organic particles over the long term, and that reliable pulses of this material exits fish farms in the upper water column over daily feeding cycles. Brager et al. (2015) expanded on this work by conducting high-resolution 3-D spatial surveys of SPM concentrations around 4 fish farms in Canada. This latter study also showed that fish waste enhancement was highly localized, but concluded that there was no evidence of a waste plume from the net-pens and that natural seston patchiness can confound the identification of waste particle enhancement. Although these spatial surveys were conducted during fish feeding periods when particle concentrations have been reported to be elevated (e.g. Lander et al. 2013), any spatial sampling represents a snapshot in time and it is possible that the sampling periods did not coincide with the export of fish wastes. The present study expanded on these past studies by focusing on temporal variations in particulate waste concentrations adjacent to open-water fish net-pens. The primary objective was to quantify the contribution of particulate fish wastes to natural food resources available to filter feeders in IMTA systems. Open-water net-pen farms in Norway and Canada were selected for this research to include environments with different trophic states and ambient suspended particle characteristics. The Norwegian sites are characterized as oligotrophic environments with the seston dominated by phytoplankton, while the mesotrophic Canadian sites typically contain a more complex and variable particle regime. According to 2012 statistics, Norway and Canada contributed 79 and 3%, respectively, of the 1.66 M tonnes of cultured fish produced in the North Atlantic region (FAO 2012).

MATERIALS AND METHODS

Study sites

Time-series data on suspended particle concentrations were collected at 5 fish farms (Fig. 1). On the Canadian east coast, field studies were conducted at 2 Atlantic salmon *Salmo salar* farms in the macrotidal Passamaquoddy Bay region of the Bay of Fundy: Navy Islands (45° 1.8' N, 67° 0.3' W) and Charlie Cove (45° 1.8' N, 66° 52.0' W). On the Canadian west coast,

data were collected at a site culturing *Anoplopoma fimbria* (black cod) in Kyuquot Sound on the northwest coast of Vancouver Island (50° 2.8' N, 127° 17.8' W). The Navy Islands site contained 15 circular net pens (25 m diameter), while the Charlie Cove and Kyuquot Sound farms had 8 (25 m diameter) and 2 (25 m square) stocked pens at the time of sampling, respectively. Both east coast sites (Navy Islands and Charlie Cove) are 23 m in depth, and subject to strong tidal currents and mixing owing to the 6 m average diurnal tidal range (Trites & Garrett 1983,

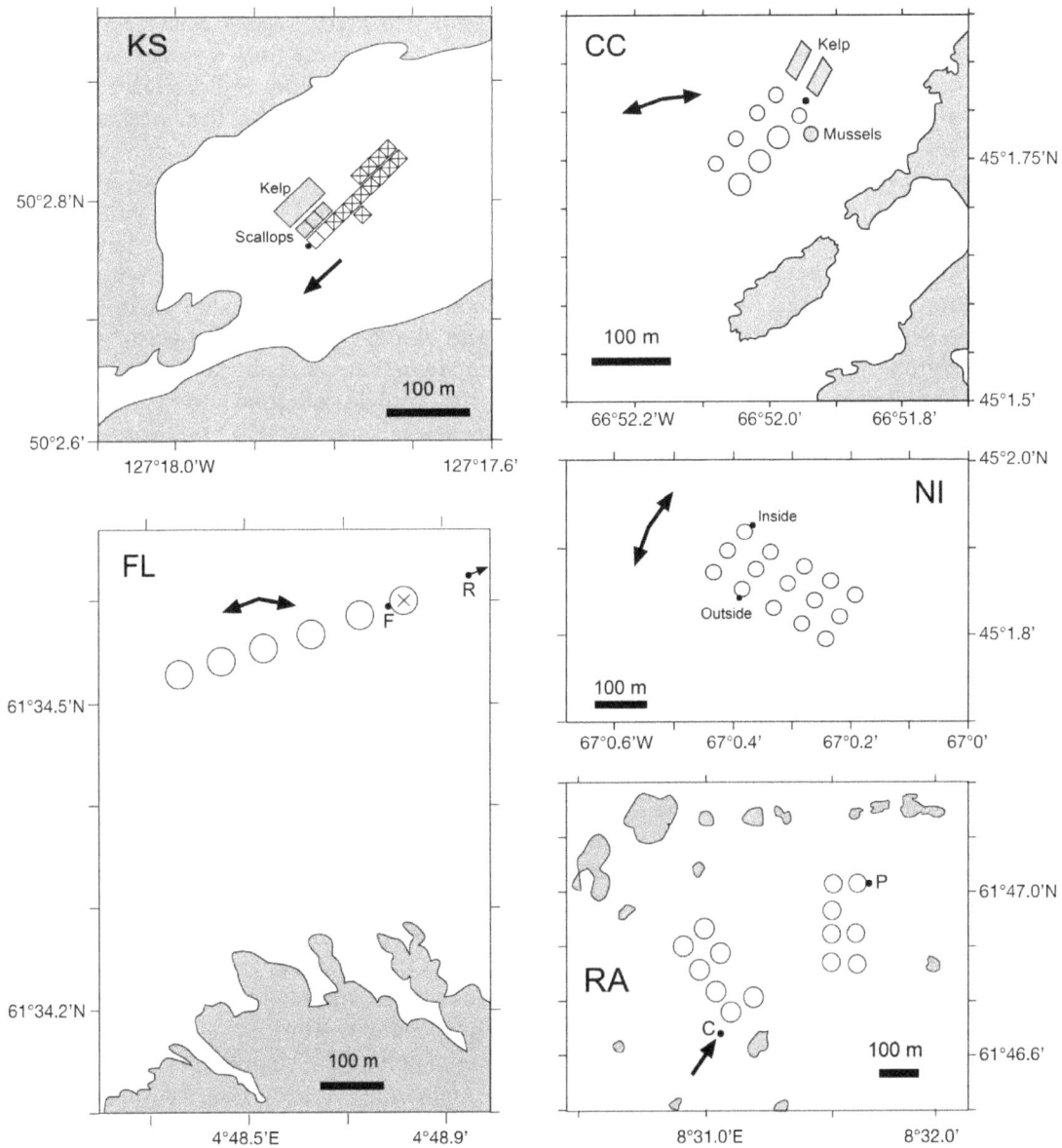

Fig. 1. Navy Islands (NI), Charlie Cove (CC) and Kyuquot Sound (KS) study sites in Canada, and the Rataran (RA) and Flåtegrunnen (FL) study sites in Norway. Fish net-pens are represented by open symbols; shaded symbols indicate net-pens for other taxa; closed circles show the locations of instrument moorings or profiling stations. Pens not containing fish are indicated by an 'x'. The arrows show prevailing current directions during flood and ebb tides. The reference site (R) at the FL farm was located 1 km to the east of the farm site (F). Profiling (P) and current meter (C) locations are identified for the RA farm

Thompson et al. 2002). The Kyuquot Sound site is approximately 28 m in depth, and subject to a 3 m average semi-diurnal tidal range. All Canadian sites contained fish in their final season of growth and feeding was underway during sampling periods. Each salmon cage is known to contain between 30 000 to 50 000 fish, giving a minimum of 450 000 and 240 000 fish at the Navy Islands and Charlie Cove farms, respectively. The Kyuquot Sound farm contained approximately 100 000 fish of harvest size in 2 square steel frame net-pens. Other pens at this farm contained small quantities of scallops and kelp.

The 2 Atlantic salmon farms sampled in Norway were the Flåtegrunnen farm in the Florø area (61° 34.5' N, 4° 48.6' E) and the Rataran farm in the Frøya area (63° 46.8' N, 8° 31.0' E; Fig. 1). The Flåtegrunnen farm was located in relatively deep water (75 to 200 m depth) and contained 8 circular pens (50 m diameter), with 7 pens containing fish. During the sampling period, the farm contained a total of 3050 tonnes of 13-mo-old salmon and feed was supplied at 20 tonnes per day. The Rataran farm contained 14 net-pens (50 m diameter), with each pen containing an average of 620 tonnes of 18-mo-old salmon. During sampling at this farm, feed was being pumped to the net-pens at 80 tonnes per day. The floating net-pens at the Canadian farms extended from the surface down to 15 m depth while the Norwegian net-pens were 25 m deep. All 5 fish farms studied were located at least 1 km away from other aquaculture activities.

Particle-sensing instrumentation and experimental design

Times-series of SPM were obtained using moored instruments or through repeated water column profiling at a set location. Particle sensors attached to moorings consisted of optical back-scatter (OBS) sensors and chlorophyll a (chl a) fluorometers (FLNTUSB ECOMeter, WET Labs). The OBS sensors provided information on total SPM concentrations, including both natural seston and particulate farm wastes, while the chl a fluorometers provided simultaneous data on phytoplankton biomass. Chl a was monitored to control for the potential influence of natural phytoplankton variations on SPM levels. The instruments contained internal batteries and a data logger for autonomous recording. For profiling purposes, a sensor platform was employed that consisted of a CTD (MicroCTD, AML Oceanographic), a chl a fluorometer (Cyclops 7, Turner Designs) and a transmis-

someter with a 25 cm optical path length for measuring SPM concentrations (c-Rover CRV5, WET Labs). The instrument frame was designed to be compact and light for manual profiling operations, with the sensors encased within a streamlined polyethylene jacket that allowed water to flow past the instruments while preventing instrument entanglement with mooring ropes and nets. Current meters were deployed at each farm site.

A control/impact experimental design was utilized at each farm site, with the exception of Rataran (see below), to detect the presence of particulate farm wastes. However, results from previous studies on spatial variability in SPM levels at each farm dictated the use of 2 different approaches for detecting the presence of particulate fish wastes. Brager et al. (2015) showed that natural seston patchiness at the mesotrophic Canadian farm sites results in natural differences in SPM levels at potential reference and farm mooring sites that would confound the interpretation of any site comparisons. Consequently, sensor moorings were only located in the vicinity of the farms and the mean effect of waste effluents on particulate matter concentrations was assessed by comparing particle data from periods when the current at the mooring site was flowing towards the farm (reference condition with no farm influence) and from the direction of net-pens (potential SPM enhancement from farm wastes). Preliminary work at the Flåtegrunnen farm showed a highly uniform spatial distribution in SPM levels in this oligotrophic fjord (P. Cranford unpubl. data). Reference and farm mooring sites were therefore used for this site to detect the mean effect of farm wastes on particulate matter concentrations, in addition to the directional approach described above for the other farms.

At the Navy Islands farm, particle sensors were hung at 5 m depth from the side of fish cages on opposite sides of the farm (Fig. 1). These 2 mooring sites are referred to as the 'inside' and 'outside' moorings, which reflect their position relative to the direction of flood tide. Two acoustic Doppler velocimeters (Argonaut ADV; SonTeK/YSI) were also attached to the same instrument frames to measure variations in current speed and direction. The instruments were programmed for burst sampling (1 Hz sampling for 10 s) at 2 min intervals from 16 to 18 November 2010. Single moorings were deployed at the Charlie Cove and Kyuquot Sound farms. At Charlie Cove, the mooring was deployed 20 m away from a net-pen (Fig. 1) with ECOMeters attached at 5 and 15 m depth on the same mooring. An Infinity-EM current meter (JFE Advantech) was attached to the mooring at 1.5 m depth and an Argonaut ADV at 15 m depth. The 10 s instrument data bursts were recorded every 10 min between 5 and 15 July 2011. At Kyuquot Sound, the mooring was hung from the side of the net pen containing the largest number of fish (Fig. 1), with ECOMeters attached at 1 and 20 m depth and an Infinity current meter located at 19 m. The 10 s data bursts were logged at 1 min intervals between 24 and 28 July 2011. Two moorings were deployed at the Flåtegrunnen farm (Fig. 1). An ECOMeter and an Infinity current meter were suspended at 5 m depth from the edge of an empty fish pen. This location was 50 m from the nearest pen containing fish. A second ECOMeter was deployed at the same depth at a site located 1000 m away (Fig. 1). The instruments at this site were programmed for 10 s data bursts every 5 min between 10 and 12 September 2013. For ease of reference, sensor deployment details for each farm are summarized in Table 1.

Water column profiling was conducted 2 m away from the side of a net-pen at the Rataran farm (63° 47.02′ N, 8° 31.69′ E; Fig. 1). Temporal variations in SPM, chl a, salinity and temperature between 0.5 and 30 m depth were recorded by continuously lowering and raising the profiler while logging data at 1 s intervals using Windmill 7 data acquisition and visualization software (Windmill Software). Sampling was conducted between 13:38 and 15:00 h on 6 August 2014. During profiling operations, currents at 20 m depth were measured at 5 min intervals with an Aanderra ADCP current meter moored to the southwest of the farm (63° 46.74′ N, 8° 31.02′ E; Fig. 1).

The WET Labs ECOMeters and Cyclops fluorometer were calibrated in the laboratory using chl a and total SPM standards (n = 14). Standards were pre-pared by suspending known concentrations of algal cells (mixed flagellates) or fine sediment (natural clay/silt mixture pre-sieved through a 64 μm mesh) in filtered (0.45 μm) seawater. SPM and chl a concentrations in all standards were determined using routine gravimetric and fluorometric procedures as described in Brager et al. (2015). The c-Rover transmissometer was calibrated in the laboratory using ground fish feed standards (n = 16) and under field conditions with natural seston samples collected in triplicate at the same depth as the cRover (n = 101). The laboratory calibration was conducted to confirm sensor output linearity with a potential fish farm waste. However, the seston calibration was used for calculating SPM concentrations at the farm site. Comparisons of data obtained using the different SPM (ECOMeter and cRover) and chl a (ECOMeter and Cyclops) sensors were conducted under field conditions at the study sites to confirm that they provided comparable concentrations. All calibration equations and curve fit data are summarized in Table 2.

Data analysis

SPM (mg l^{-1}) and chl a (μg l^{-1}) data for each site were averaged to obtain a single value for each 10 s sample burst. No further data smoothing was performed prior to statistical analysis; however, a 3-sample running mean was used for plotting time-series data. Tidal elevation predictions for the Canadian farms were obtained for sampling periods using the WebTide tidal prediction model (version 7.0.1) developed by Fisheries and Oceans Canada (www.bio.gc.

Table 1. Deployment types at each location included fixed depth instrument moorings (number of mooring sites indicated) or water-column profiling. Details of particle and current meter sensor deployments at the 5 fish farms (see Fig. 1)

Farm	Type (no.)	Dates	Instruments	Distance to pen (m)	Depth (m)	Sampling frequency (min)
Navy Islands	Mooring (2)	16–18 Nov 2010	ECOMeter, Argonaut ADV	1	5	2
Charlie Cove	Mooring (1)	5–15 Jul 2011	ECOMeter	20	5	10
			Infinity-EM		1.5	10
			ECOMeter		15	10
			Argonaut ADV		15	10
Kyuquot	Mooring (1)	24–28 Jul 2011	ECOMeter	1	1	1
			Infinity-EM		19	1
			ECOMeter		20	1
Flåtegrunnen	Mooring (2)	10–12 Sep 2013	ECOMeter	50	5	5
			Infinity-EM	50	5	5
			ECOMeter	1000	5	5
Rataran	Profiling	6 Aug 2014	Cyclops, cRover, MicroCTD	2	1–30	1 s
Rataran	Mooring (1)	4–15 Aug 2014	Aanderra ADCP	100	20	5

Table 2. Results of regression analysis of suspended particulate matter (SPM) and chlorophyll a (chl a) instrument responses (counts or mV) to calibration standards prepared as described in the 'Materials and methods'. SEE: standard error of estimate

Parameter	Instrument	Equation	r^2	p	SEE
SPM (mg l^{-1})	ECOMeter 2018	SPM = (0.005 × counts) + 0.930	0.998	0.0001	0.28
	ECOMeter 2017	SPM = (0.004 × counts) + 0.815	0.997	0.0001	0.34
	cRover (seston)	SPM = −12.173 × ln(counts) + 18.439	0.892	0.0001	1.80
	cRover (feed)	SPM = −15.911 × ln(counts) + 23.217	0.999	0.0001	0.30
Chl a (µg l^{-1})	ECOMeter 2018	Chl a = (0.013 × counts) − 2.949	0.991	0.0045	0.25
	ECOMeter 2017	Chl a = (0.013 × counts) − 2.949	0.955	0.0041	0.26
	Cyclops 7	Chl a = (4.938 × mV) − 0.221	0.999	0.0001	0.18

ca/science/research-recherche/ocean/webtide/index-en.php). Tidal elevation measurements from monitoring stations located near the Norway farms were obtained from the Norwegian Hydrographic Service (Norwegian Mapping Authority). Spectral analysis was conducted on the SPM time-series data to identify any regularly repeating patterns. All SPM data were linearly detrended prior to using the periodogram function of MATLAB. Cyclic periodicities were obtained from peak periods (hour) in the periodogram plots. Contour plots of chl a and SPM concentrations measured at the Rataran farm were produced with Surfer 9 software (Golden Software) using the ordinary kriging interpolation method.

Hypotheses on the potential enhancement of SPM concentrations with particulate wastes from fish net-pens at each sampling location were tested using one-way ANOVA (Systat version 13, SPSS). SPM data collected at single mooring locations were pooled by current direction, with 12 direction levels representing 30° compass bearing increments. The hypothesis tested was that mean SPM concentrations were equal for all flow directions. The alternate hypothesis was accepted if the p-value was ≤0.05. In the event of a significant effect, a single Tukey post hoc comparison was performed comparing differences in mean SPM levels in water flowing to the sampling site directly from the farm and from the opposite direction. SPM enhancement with waste particles was calculated from differences in mean SPM concentrations for these same 2 groups of pooled data (mean SPM from direction of farm minus mean SPM in water flowing towards the farm). Data collected at the Flåtegrunnen salmon farm included sampling locations near the farm (50 m) and at a reference site located 1 km away. SPM concentrations measured at 5 m depth at both locations were compared using ANOVA, and the degree of SPM enhancement was calculated from the difference in

mean concentrations. Prior to performing each ANOVA, the data were screened for normality and homoscedasticity by examining normal probability and residual plots. The independence assumption was tested based on calculation of the Durbin–Watson D-statistic and the first-order autocorrelation. The results of statistical tests were only reported when these assumptions were not violated.

RESULTS

Moored instruments

Time series of SPM and chl a concentrations and current speed data for the different mooring locations and fish farms are shown in Figs. 2 to 5. Mean (±SD) SPM concentrations measured at 5 m depth on the inner- and outer-bay sides of the Navy Islands farm were equal (2.4 ± 0.2 mg l^{-1} for both sites) and displayed similar temporal variability (Fig. 2). Relatively high winds on the afternoon of 17 November (data from Environment Canada), which reached 25 km h^{-1}, preceded peaks in current speed and SPM and chl a. Chl a and SPM concentrations were not correlated at this site (r^2 = 0.01). Predominant current directions at both locations were 35° and 200° on the flood and ebb tide, respectively. Although data collection was limited to 4 tidal cycles, spectral analysis indicated 12.3 h periodicity in SPM at this farm. ANOVA indicated a significant effect of current direction on mean SPM concentrations at both the inside and outside mooring locations (Table 3). However, Tukey post hoc comparisons showed no significant difference in SPM concentrations in water flowing from the direction of the net-pens, compared with water flowing from the opposite direction (Table 3).

Measurements made at 2 depths over a 10 d period at the Charlie Cove farm showed temporal variability

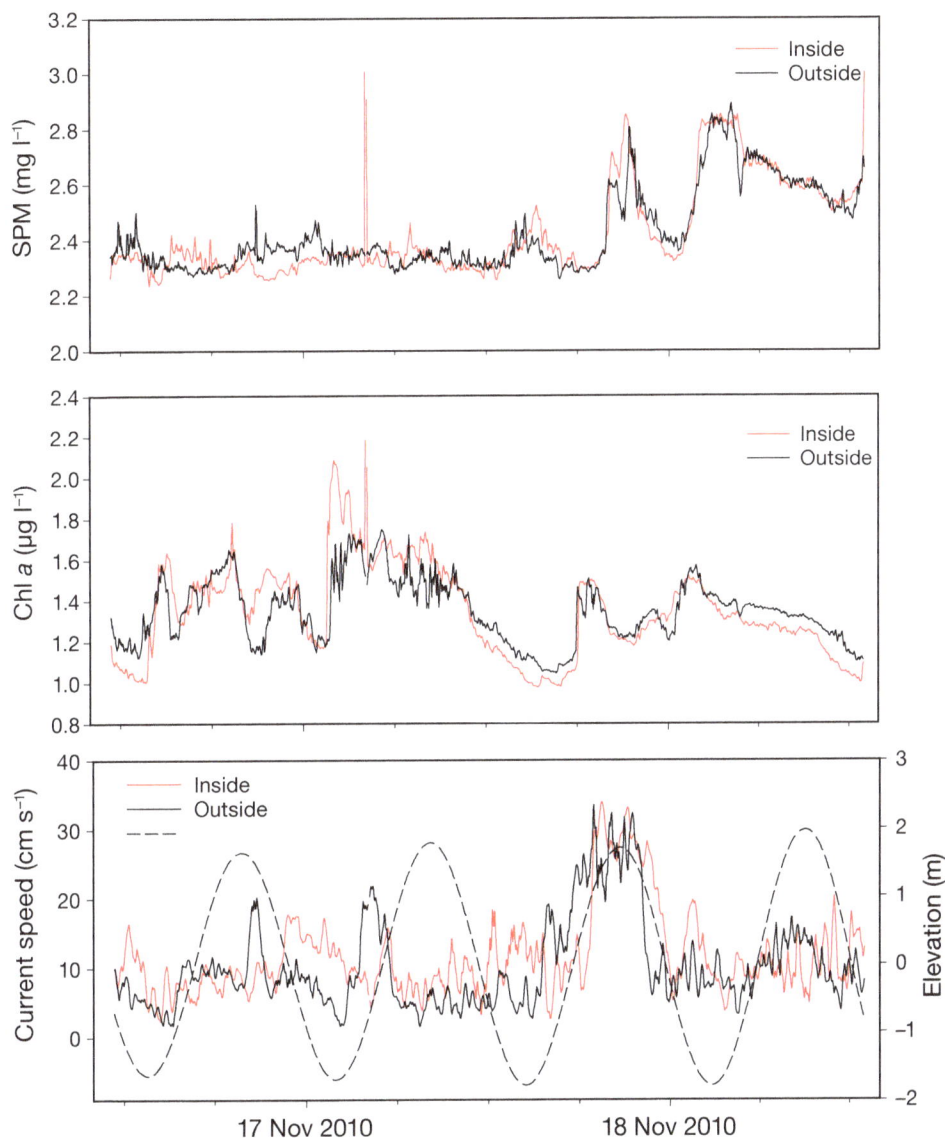

Fig. 2. Suspended particulate matter (SPM) and chlorophyll *a* (chl *a*) concentrations and current speeds, collected at 2 min intervals at 5 m depth, at the Navy Islands salmon farm. Data are from the 'inside' and 'outside' stations shown in Fig. 1. All instruments were located within 3 m of a net-pen. The dashed line is the predicted tidal elevation

in both SPM and chl *a* that corresponded to a periodicity of 12.2 to 12.7 h (Fig. 3). SPM was higher at 15 m depth (4.2 ± 0.5 mg l^{-1}) compared with values measured at 5 m (3.4 ± 0.3 mg l^{-1}), and SPM and chl *a* at both depths were not correlated ($r^2 < 0.03$). The current meter moored at 15 m depth stopped recording early, but changes in current speed and direction were similar for both depths during the period when both instruments were operating (Fig. 3). The predominant current directions at this site were 160° during flood tide and 250° on the ebb. ANOVA comparisons of SPM concentrations in water flowing from different directions (based on the 5 m depth current data) indicated a significant effect of current direction on mean SPM concentrations at both depths. However, the Tukey comparisons did not indicate a significant enhancement of SPM in water flowing from the direction of the salmon net-pens (Table 3).

Currents at the Kyuquot Sound farm were generally slower than measured at the other farm sites and were dominated by flows to the southwest on both flood and ebb tides. Mean SPM concentrations were greater at 1 m depth (2.7 ± 0.9 mg l^{-1}) than at 20 m (2.0 ± 0.1 mg l^{-1}), and SPM and chl *a* values were poorly correlated at both depths ($r^2 < 0.25$). SPM lev-

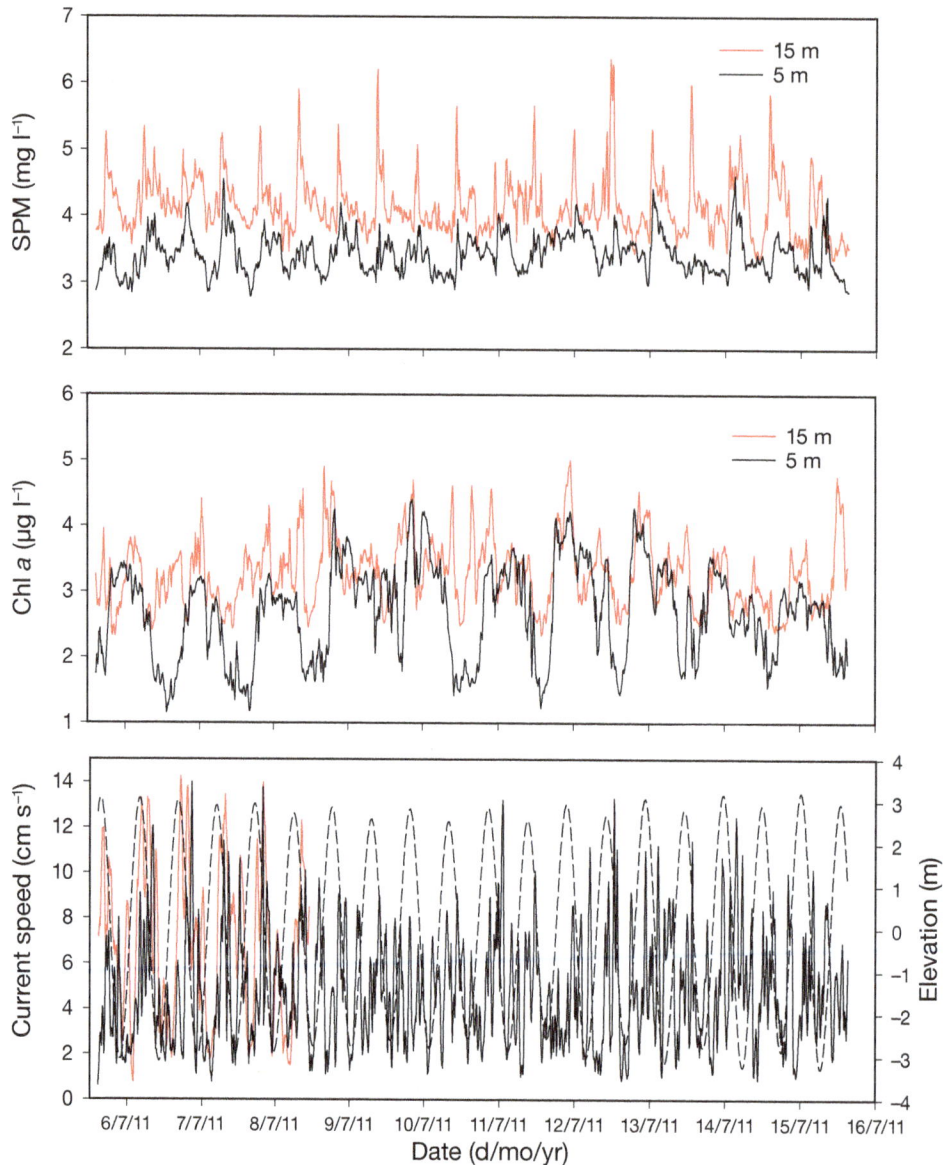

Fig. 3. Suspended particulate matter (SPM) and chlorophyll *a* (chl *a*) concentrations and current speeds, collected at 10 min intervals, at the Charlie Cove salmon farm. Data are from 5 and 15 m depths at a single mooring located 20 m from a net-pen (Fig. 1). The dashed line is the predicted tidal elevation. The instrument mooring was 20 m from the nearest salmon cage

els exhibited strong periodicity at 12.1 h and between 22 and 26 h. The periods of elevated SPM at 1 m depth (Fig. 4) corresponded with periods when water was flowing from the southeast. ANOVA comparisons indicated significant directional changes in mean SPM concentrations at both depths, but that there was no significant SPM enhancement during periods when water was flowing from the direction of the net-pens (Table 3).

Currents measured at 5 m depth adjacent to the Flåtegrunnen salmon farm were strongly bidirectional, with predominant flow directions during flood

and ebb tides at 105° and 255°, respectively (Fig. 1). SPM concentrations measured at 5 m depth at the farm and reference locations (Fig. 5) were lower than measured at the 3 Canadian farms and were correlated with chl *a* concentrations (Fig. 6). SPM and chl *a* levels at the farm and reference sites showed similar temporal variability, with slightly higher SPM concentrations measured at the farm mooring (Fig. 5). Spectral analysis of these data was not conducted owing to the relatively short sampling period. ANOVA comparisons showed no significant differences in mean SPM levels in water flowing to and

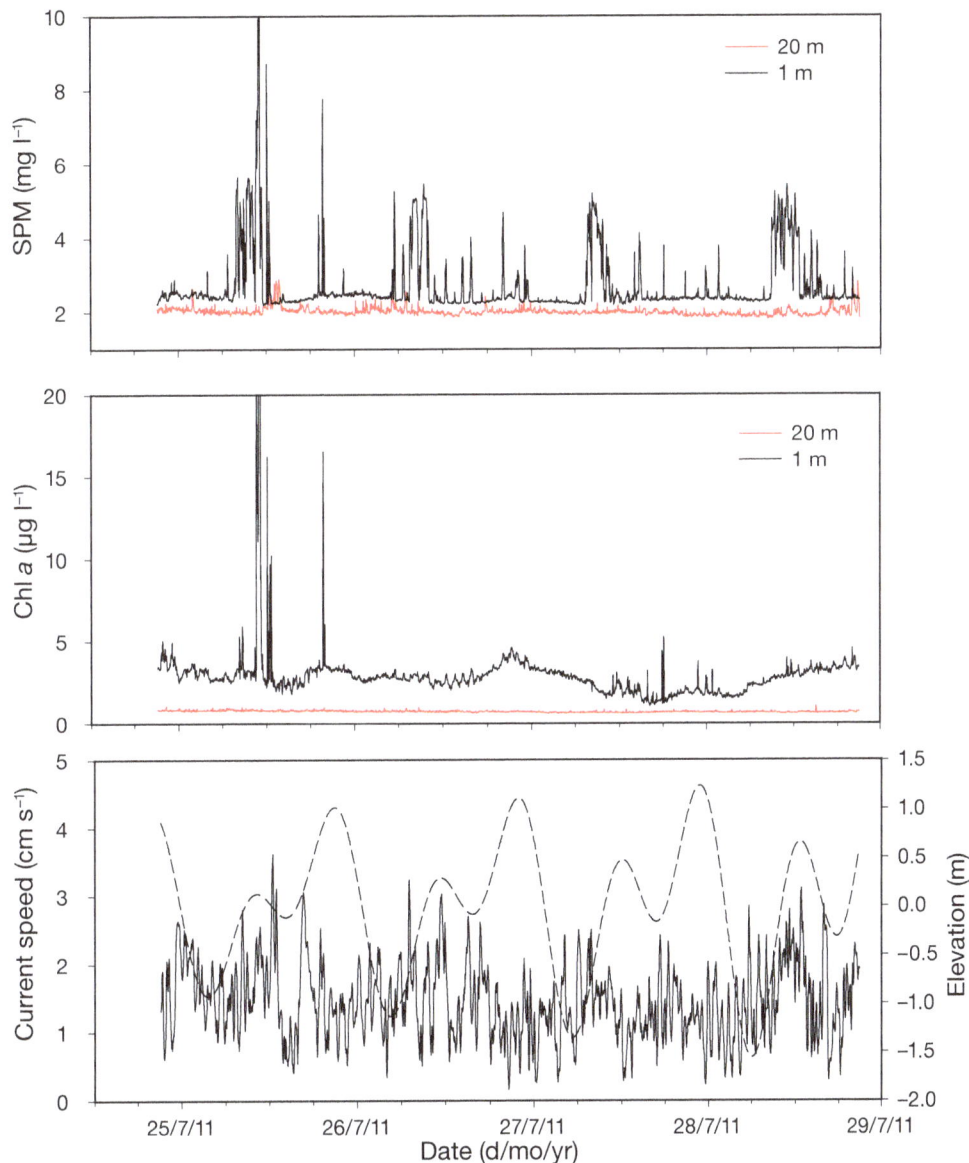

Fig. 4. Suspended particulate matter (SPM) and chlorophyll *a* (chl *a*) concentrations and current speeds, collected at 1 min intervals, at the Kyuquot Sound black cod farm. Data are from 1 and 20 m depths (19 m for current speed) at a mooring located 1 m from a net-pen (Fig. 1). The dashed line is the predicted tidal elevation

from the farm (i.e. no significant direction effect), but detected a significant difference in SPM concentrations at the farm site compared with the reference site (Table 3). The correlation between SPM and chl *a* at the farm site was lower than for the reference site (Fig. 6).

Calculations of the mean increase in SPM concentrations at each mooring location, assumed to originate from the addition of particulate farm wastes, are given in Table 3. In all cases, the degree of particle enhancement was small (0.01 to 0.24 mg l⁻¹). The only significant level of waste enhancement was de-

tected for the mooring located 50 m away from for the Flåtegrunnen salmon farm (Table 1). The mean increase in SPM at this site, compared with the reference site, was 0.02 mg l⁻¹.

Time-series profiling

Time-series contour plots of chl *a* and SPM concentrations in the water column beside the Rataran salmon pen are shown in Fig. 7. During the 85 min sampling period, a total of 25 profiles (from 0.5 to

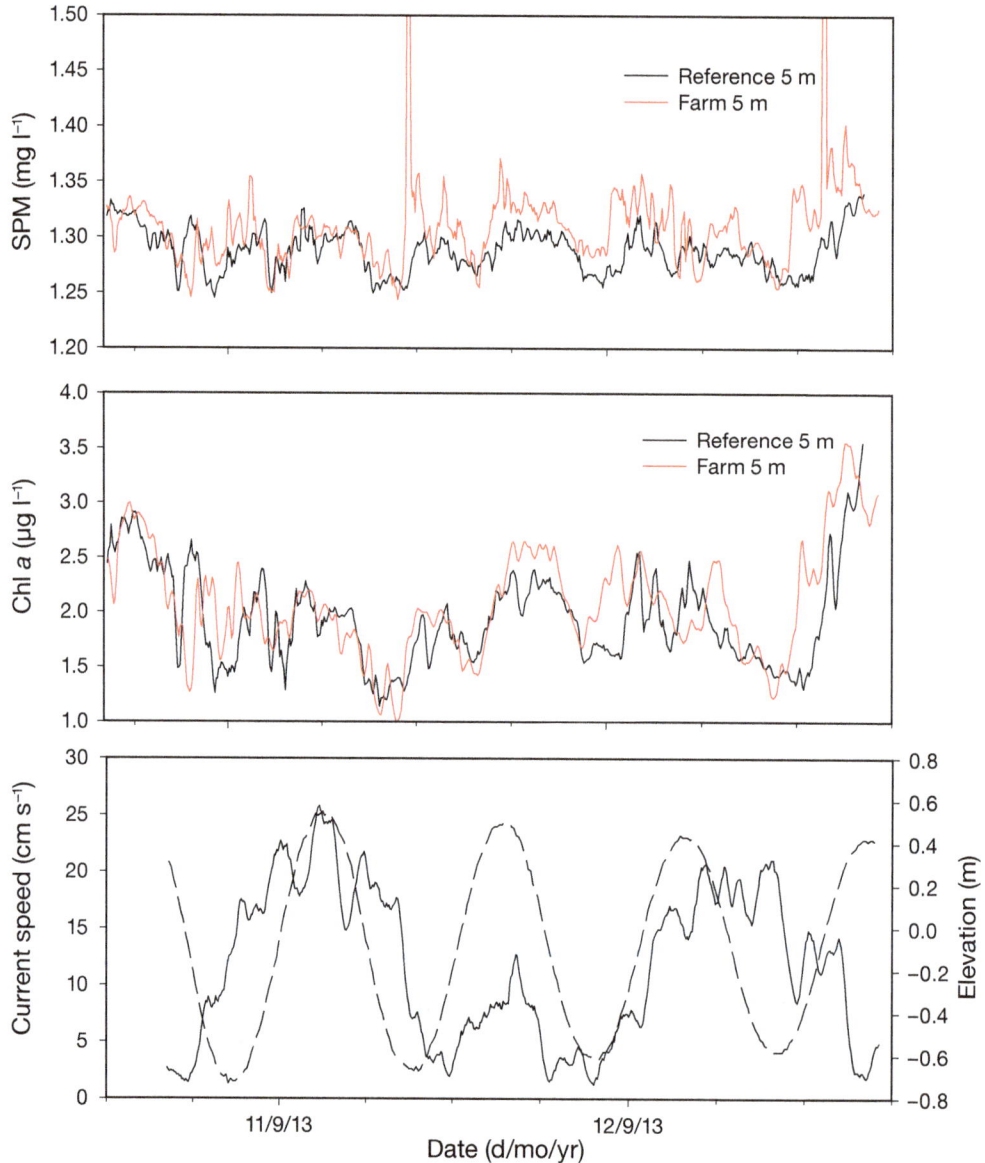

Fig. 5. Suspended particulate matter (SPM) and chlorophyll *a* (chl *a*) concentrations and current speeds, collected at 5 min intervals at a depth of 5 m, at the farm (20 m from nearest salmon cage) and reference stations at the Flåtegrunnen salmon farm (Fig. 1). The dashed line is the measured tidal elevation

30 m depth) were conducted. The water depth at the profiling location was 40 m. A chl *a* maximum was observed between 8 and 15 m depth and was observed to move upward during the sampling period (Fig. 7) in a manner similar to that of the pycnocline (data not shown). SPM concentrations were highest near the surface and generally declined with depth. However, particle concentrations below 25 m depth occasionally increased above a background level of approximately 1.9 mg l^{-1} (Fig. 7). The relationship between chl *a* and SPM concentrations at this site (Fig. 8) was more complex than for the Flåtegrunnen

farm (Fig. 6). SPM in surface water (0 to 10 m depth) was negatively related with chl *a* levels, while the opposite trend was observed for deeper water. Water between 25 and 30 m depth generally contained relatively low SPM and chl *a* levels, but SPM occasionally increased by as much as 1 mg l^{-1} (Fig. 8). Current meter data collected to the southwest of the farm indicated a highly unidirectional flow (averaged 33° between 4 and 15 August 2014) towards the farm on both flood and ebb tides (Fig. 1). During the sampling period, the mean direction was 30.9° (SD = 7.5°), with a mean current speed of 10.3 m s^{-1} (SD = 3.9 m s^{-1}).

Table 3. Summary of ANOVA results comparing mean suspended particulate matter concentrations with current direction (12 direction levels representing 30 degree increments) or between farm and reference locations (Flåtegrunnen farm only). Significant ANOVA results (in **bold**) for the direction comparisons were followed by a single Tukey post hoc test comparing mean suspended particulate matter (SPM) values in the 2 direction bins that represent water flowing directly towards and away from the closest net-pen. The mean SPM enhancement at the mooring site was calculated from these 2 mean values

Location	Source	df	Mean squares	F-ratio	p-value	Mean enhancement $(mg\ l^{-1})$
Navy Islands Inside	Direction	11	0.399	16.888	**<0.001**	
	Error	1362	0.024			
	Tukey				0.072	−0.29
Navy Islands Outside	Direction	11	0.227	11.318	**<0.001**	
	Error	1363	0.02			
	Tukey				0.052	0.05
Charlie Cove 5 m	Direction	11	0.403	4.082	**<0.001**	
	Error	1535	0.099			
	Tukey				1.000	0.04
Charlie Cove 15 m	Direction	11	1.108	4.855	**<0.001**	
	Error	1535	0.228			
	Tukey				1.000	0.05
Kyuquot Sound 1 m	Direction	11	2.459	44.559	**<0.001**	
	Error	4302	0.728			
	Tukey				0.096	0.24
Kyuquot Sound 20 m	Direction	11	0.407	26.846	**<0.001**	
	Error	4302	0.015			
	Tukey				1.000	0.02
Flåtegrunnen Farm	Direction	11	0.002	0.643	0.792	
	Error	555	0.003			
	Tukey				–	0.01
Flåtegrunnen Reference	Direction	11	0.003	8.54	**<0.001**	
	Error	538	0			
	Tukey				0.917	0.01
Flåtegrunnen	Location	1	0.142	73.895	**<0.001**	0.02
Farm versus reference	Error	1114	0.002			

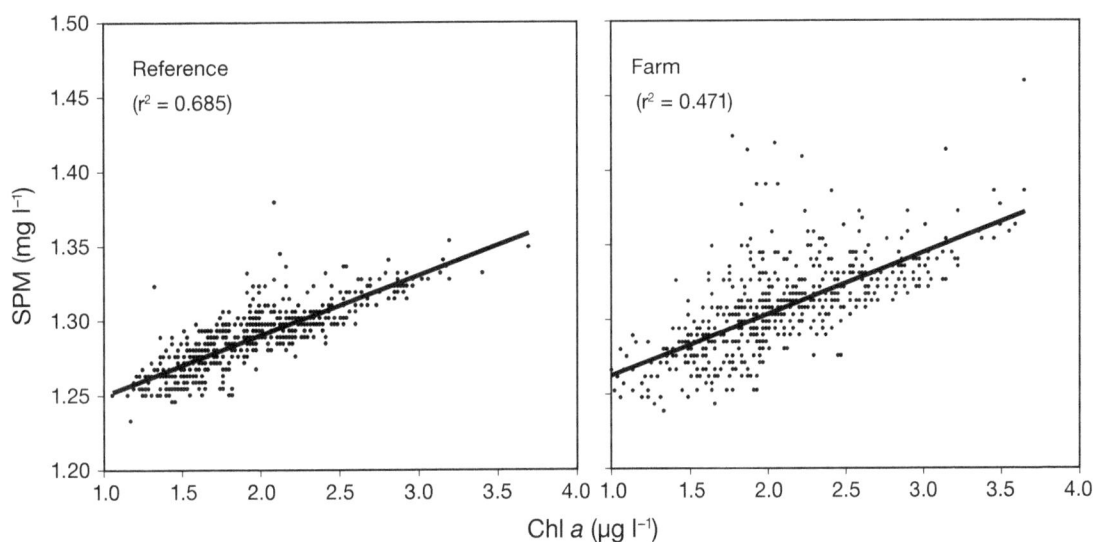

Fig. 6. Relationships between chlorophyll a (chl a) and suspended particulate matter (SPM) concentrations at 5 m depth at the Flåtegrunnen farm and reference stations (Fig. 1)

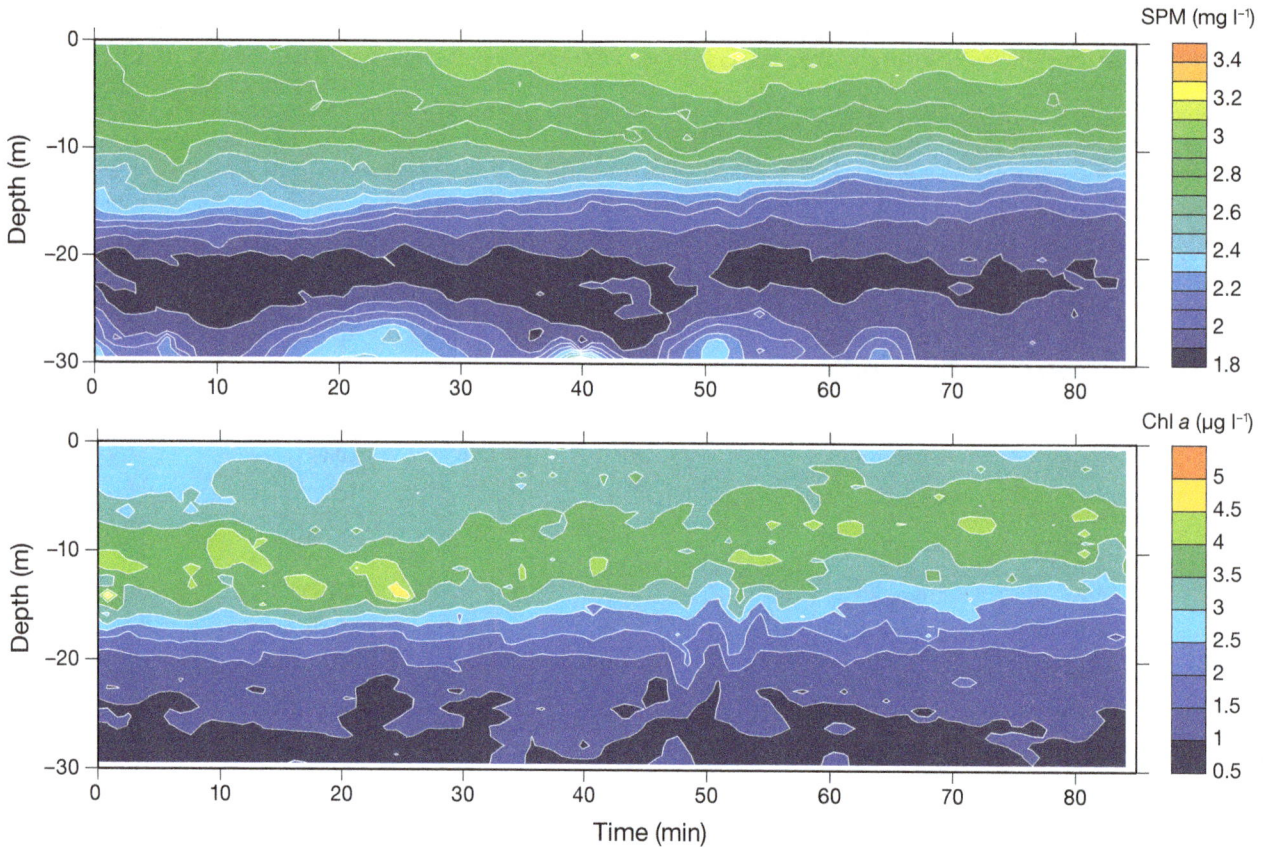

Fig. 7. Time-series contour plots of total suspended particulate matter (SPM) and chlorophyll *a* (chl *a*) concentrations in the water column at 2 m distance from a salmon net-pen on the outflow side of the Rataran farm (Fig. 1). The plots are based on *in situ* sensor profiles from 0.5 to 30 m depth, conducted at 3 min intervals, starting at 13:38 h on 6 August 2014

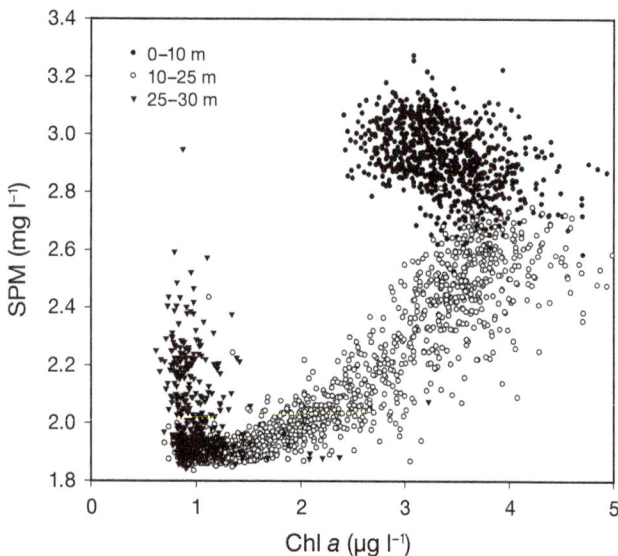

Fig. 8. Relationships between chlorophyll *a* (chl *a*) and suspended particulate matter (SPM) concentrations over 3 depth ranges beside a salmon net-pen at the Rataran farm

DISCUSSION

Studies on SPM loading from open-water fish farms have not previously examined whether there is any persistent enhancement of the ambient seston in estuarine/coastal systems. The results of the present study showed the dynamic nature of the seston around 5 coastal fish farms, and demonstrate that fish farming has an overall negligible effect on SPM concentrations in the vicinity of the farms. These conclusions are based on periodogram power spectral density estimates, statistical comparisons of mean SPM levels flowing to and from the direction of net-pens, and comparisons between farm and reference sites. Temporal fluctuations in SPM levels at the study sites predominantly corresponded with tidal constituents. Frequency peaks corresponding to the M2 principal lunar semi-diurnal tidal constituent (12.4 h) were evident in the SPM time series collected in this investigation. The Kyuquot Sound data from 1 m depth also indicated a strong 24 h cycle (i.e. P1 principal solar diurnal tide). In tidally driven environments, variability of SPM concentration at tidal frequencies has been well documented (Cloern et al. 1989, Velegrakis et al. 1997, McCandliss et al. 2002). Previous studies have identified horizontal SPM concentration gradients along the tidal excursion (Weeks et al. 1993, Velegrakis et al. 1997) that will result in tidal cycle SPM variations at a given location. In the macro-tidal Passamaquoddy Bay, the tides have a significant effect on the seston (Dowd 2003). Although tidal advection has often been characterized as the predominant mechanism for short-term variability of SPM, other mechanisms (e.g. river sources, terrestrial inputs, wind, anthropogenic impacts) contribute to SPM variability observed at the tidal time scale (Fegley et al. 1992, Velegrakis et al. 1999, McCandliss et al. 2002, Shi 2010, Strohmeier et al. 2015). The recurring peaks in SPM concentrations at 1 m depth at the Kyuquot Sound mooring corresponded with periods when the predominantly southwest currents briefly switch to the southeast. This current direction indicates a terrestrial source of particles from a stream located to the west of the farm.

The companion study by Brager et al. (2015) concluded that the export of fish wastes from farms was highly localized and episodic. Consequently, it is important to consider the possible consequences of this spatial variability on the capacity of the present study to detect the presence of particulate wastes. The analysis of the particle sensor data focused specifically on periods when the current was flowing from the nearest net-pen towards the particle sensors. This approach was taken specifically to minimize the importance of the geographic location and the limited number of sensor moorings deployed. Based on this approach, the probability of missing waste discharges at all 5 farm sites is believed to be extremely low. The high sampling frequency of electronic particle sensors allows for the collection of large sample sizes and, consequently, results in high statistical power to detect small differences between temporal data sets. Hypothesis testing provided no indication of a significant increase in mean SPM levels at any of the farms during periods when water was flowing to the sensors from the direction of the net-pen. The only indication of particle enhancement from a fish farm was for Flåtegrunnen, where SPM levels at 5 m depth near the farm were found to be significantly different than for the reference site located 1 km away. However, the mean level of SPM enhancement at this site was only 0.02 mg l^{-1}. Given that the precision of the ECOmeter SPM calibration was approximately 0.3 mg SPM l^{-1} (standard error of estimate, Table 2),

the actual degree of seston enhancement at this farm cannot be accurately quantified other than to state that the contribution of fish wastes to the natural particle field was very low. The detection of such a small anthropogenic effect at the Flåtegrunnen farm was facilitated by the low seston conditions and dominance of phytoplankton over non-living particulate matter in this oligotrophic fjord (Fig. 6).

Water column profiling with particle sensors allows for greater depth coverage that can be achieved with moored instruments, albeit with a reduced sampling period. Continuous profiling immediately adjacent to the Rataran salmon pen on 6 August 2014 detected relatively high SPM levels, compared with chl a, in the 0.5–10 m and 25–30 m depth ranges. During sampling, the current was flowing from the direction of a net-pen that contained harvest-sized salmon that were actively feeding. The elevated SPM levels in the upper water column appear to occur naturally, as opposed to resulting from the presence of fish feed 'fines', and may consist of detrital material from the abundant macrophyte beds in this region. Spatial sampling around this farm showed a similar vertical distribution of SPM in the upper water column regardless of the profiling location (data not shown). SPM levels in the 25–30 m depth range occasionally increased by as much as 1.0 mg l^{-1} above the background concentration. This layer is below the depth of the salmon pen and the periodic increases in SPM likely indicate from the presence of salmon faeces.

The presence of finfish aquaculture sites has previously been predicted and shown to impact the surrounding suspended particle field. Troell & Norberg (1998) calculated that farm waste production, including uneaten whole feed pellets, feed 'fines' and faecal wastes, could increase suspended solids concentrations inside a fish cage anywhere from 3- to 30-fold. That study assumed 10% feed wastage. Early estimates of feed loss were approximately 20% (Beveridge 1987), but have since been reduced through improved feeding control mechanisms (Reid et al. 2009). The most recent feed wastage estimates are below 5% (Cromey et al. 2002, Perez et al. 2002, Strain & Hargrave 2005, Stucchi et al. 2005). Waste concentrations exiting the cages will rapidly decline through dilution and the rapid sedimentation of feed pellets and faeces (Troell & Norberg 1998). MacDonald et al. (2011) reported significantly elevated SPM concentrations at 2 m depth within 3 salmon farms in the Bay of Fundy, including the Charlie Cove farm studied herein. However, Brager et al. (2015) concluded from extensive spatial sampling that any farm-induced effect on the surrounding particle field would be highly localized and episodic. This is consistent with the results of Lander et al. (2013), who reported elevated SPM levels at 5 m depth inside the Charlie Cove salmon cages during sampling in 2004, but only during salmon feeding periods. These authors reported that waste concentrations dropped to ambient levels between 5 and 10 m distance from the cage. These conclusions are consistent with results from the present study that showed a lack of a waste particle signal at 20 m from this farm, and the extremely low waste signal detected 50 m away from the Flåtegrunnen farm. However, sampling within 2 m of fish pens at the Navy Islands, Kyuquot Sound and Rataran farms did not detect a significant increase in fish wastes in the water column. The latter farm was the largest salmon farm studied and sampling was conducted continuously over an 85 min period as feed was being pumped into the cage at approximately 0.5 tonnes h^{-1}. Faecal matter appeared to be intermittently present in the lower water column, but SPM levels beside the pen remained at ambient concentrations. This indicates that feed wastage at this farm was very low. Water profiling with SPM sensors inside a Kyuquot Sound fish pen before, during and after feeding also showed no increase in SPM from the addition of feed 'fines' (Brager et al. 2015). Feed is the highest single cost in salmon farming, and maximizing utilization and minimizing loss due to pellet breakage and dust formation strongly motivates continued progress in the formulation of food pellets. Improvements over the past decade in pellet durability during transport at the fish farm and stability in water may have contributed to the low degree of feed wastage detected in the present study.

The stimulus for the present study on the temporal dynamics of particulate wastes at coastal fish farms, as well as the companion study on spatial dynamics (Brager et al. 2015), was the desire to optimize interactions between fish wastes and extractive species in IMTA systems (see Introduction). The IMTA concept strives to increase global food production and reduce environmental impacts from intensive fish culture. Bivalve culture has been the primary focus of studies on particulate waste extraction in IMTA systems. However, the results of the present study, along with those from other studies, provide a weight of evidence against the practical incorporation of commercial bivalve operations in IMTA systems. In brief, SPM and bivalve growth enhancement in waters around fish farms tend to be absent, miniscule or highly localized (see above), and model predictions of fish waste capture efficiency by bivalve populations

in IMTA systems are low owing to the limited time available for the bivalves to intercept wastes in the horizontal particle flux (Cranford et al. 2013). The establishment of commercial-scale bivalve culture at fish farms requires strong evidence of a persistent and elevated food supply over a scale sufficient for the bivalves to effectively capture the wastes. Previous observations that fish wastes can comprise an important fraction of the bivalve diet have been limited to areas with low seston concentrations and organic content, inside or very close to net-pens and/or during winter, when natural food is relatively scarce (reviewed in Cranford et al. 2013). Although the intensive salmon farming operations at the Rataran and Flåtegrunen sites take place within oligotrophic environments, the presence of a consistent and significantly elevated supply of waste particles to help support bivalve co-culture was not supported by the results of the present study. Environmental concerns regarding the discharge of particulate fish wastes are directed primarily at the benthic habitat and community effects that stem from the rapid deposition of unused feed pellets and faeces below the net-pens (e.g. Bannister et al. 2016). Even if some faeces were to reach bivalves held beside the pens, this material does not appear to be effectively utilized (Handå et al. 2012b), and any benefit may be negated by faeces deposition by the bivalves themselves (Cranford et al. 2013). Future open-water IMTA research should focus on the development of systems that rely on extractive species held below fish cages.

Acknowledgements. This work was supported by the Department of Fisheries and Oceans Canada through contributions to the strategic Canadian Integrated Multi-Trophic Aquaculture Network (CIMTAN), and the Norwegian Research Council trough funding of the ERA (Environmental responses to organic and inorganic effluents from fin-fish aquaculture; project no. 228871) and EXPLOIT (Exploitation of nutrients from salmon aquaculture; project no. 216201) projects. The authors thank Drs. Jon Grant, Shawn Robinson, Steve Cross, Raymond Bannister, Paul Hill and Mike Dowd for their many contributions, advice and assistance throughout this project.

LITERATURE CITED

Bannister RJ, Johnsen IA, Kupka-Hansen P, Kutti T, Asplin L (in press) (2016) Near- and far-field dispersal modelling of organic waste from Atlantic salmon aquaculture in fjord systems. ICES J Mar Sci doi:10.1093/icesjms/fsw027
Beveridge MCM (1984) Cage aquaculture. Blackwell Science, Oxford
Beveridge MCM (1987) Cage aquaculture. Blackwell Science, Oxford
Beveridge MCM, Phillips MJ, Clarke RM (1991) A quantita-

tive and qualitative assessment of wastes from aquatic animal production. In: Brune D, Tomasso JR (eds) Aquaculture and water quality. Advances in world aquaculture 3. The World Aquaculture Society, Baton Rouge, LA, p 506–533

Beveridge MCM, Ross GL, Kelly AL (1994) Aquaculture and biodiversity. Ambio 23:497–502

ä Brager LM, Cranford PJ, Grant J, Robinson SMC (2015) Spatial distribution of suspended particulate wastes at open-water Atlantic salmon and sablefish aquaculture farms in Canada. Aquacult Environ Interact 6:135–149

ä Buschmann A, Troell M, Kautsky N, Kautsky L (1996) Integrated tank cultivation of salmonids and *Gracilaria chilensis* (Gracilariales, Rhodophyta). Hydrobiologia 326:75–82

ä Cheshuk BW, Purser GJ, Quintana R (2003) Integrated open-water mussel (*Mytilus planulatus*) and Atlantic salmon (*Salmo salar*) culture in Tasmania, Australia. Aquaculture 218:357–378

ä Chopin T, Buschmann AH, Halling C, Troell M and others (2001) Integrating seaweeds into marine aquaculture systems: a key towards sustainability. J Phycol 37:975–986

ä Cloern JE, Powell TM, Huzzey LM (1989) Spatial and temporal variability in South San Francisco Bay II. Temporal changes in salinity, suspended sediments, and phytoplankton biomass and productivity over tidal time scales. Estuar Coast Shelf Sci 28:599–613

ä Cranford PJ, Reid GK, Robinson SMC (2013) Open water integrated multi-trophic aquaculture: constraints on the effectiveness of mussels as an organic extractive component. Aquacult Environ Interact 4:163–173

ä Cromey CJ, Nickell TD, Black KD (2002) DEPOMOD — modeling the deposition and biological effects of waste solids from marine cage farms. Aquaculture 214:211–239

ä Dowd M (2003) Seston dynamics in a tidal inlet with shellfish aquaculture: a model study using tracer equations. Estuar Coast Shelf Sci 57:523–537

ä Fang JG, Sun HL, Yan JP, Kuang SH, Li F, Newkirk G, Grant J (1996) Polyculture of scallop *Chlamys farreri* and kelp *Laminaria japonica* in Sungo Bay. Chin J Oceanology Limnol 14:322–329

FAO (2012) The state of world fisheries and aquaculture 2012. FAO, Rome

ä Fegley SR, MacDonald BA, Jacobsen TR (1992) Short-term variations in the quantity and quality of seston available to benthic suspension feeders. Estuar Coast Shelf Sci 34:393–412

Folke C, Kautsky N (1989) The role of ecosystems for a sustainable development of aquaculture. Ambio 18:234–243

Gowen RJ, Bradbury NB (1987) The ecological impact of salmonid farming in coastal waters: a review. Oceanogr Mar Biol Annu Rev 25:563–575

Gryska A, Parsons J, Shumway SE, Geib K, Emery I, Kuenster S (1996) Polyculture of sea scallops suspended from salmon cages. J Shellfish Res 15:481 (Summary)

ä Handå A, Min H, Wang X, Broch OJ, Reitan KI, Helge R, Olsen Y (2012a) Incorporation of fish feed and growth of blue mussels (*Mytilus edulis*) in close proximity to salmon (*Salmo salar*) aquaculture: implications for integrated multi-trophic aquaculture in Norwegian coastal waters. Aquaculture 356–357:328–341

ä Handå A, Ranheim A, Olsen AJ, Altin D, Reitan KI, Olsen Y, Reinertsen H (2012b) Incorporation of salmon fish feed and feces components in mussels (*Mytilus edulis*): implications for integrated multi-trophic aquaculture in Nor-

wegian coastal waters. Aquaculture 370–371:40–53

ä Jiang Z, Wang G, Fang J, Mao Y (2013) Growth and food sources of Pacific oyster *Crassostrea gigas* integrated culture with sea bass *Lateolabrax japonicus* in Ailian Bay, China. Aquacult Int 21:45–52

ä Jones TO, Iwama GK (1991) Polyculture of the Pacific oyster, *Crassostrea gigas* (Thunberg), with Chinook salmon, *Oncorhynchus tshawytscha*. Aquaculture 92:313–322

ä Kutti T, Hansen PK, Ervik A, Høisæter T, Johannessen P (2007) Effects of organic effluents from a salmon farm on a fjord system. II. Temporal and spatial patterns in infauna community composition. Aquaculture 262:355–366

Lander T, Barrington K, Robinson S, MacDonald B, Martin J (2004) Dynamics of the blue mussel as an extractive organism in an integrated aquaculture system. Bull Aquacult Assoc Can 104:19–28

ä Lander TR, Robinson SMC, MacDonald BA, Martin JD (2012) Enhanced growth rates and condition index of blue mussels (*Mytilus edulis*) held at integrated multi-trophic aquaculture (IMTA) sites in the Bay of Fundy. J Shellfish Res 31:997–1007

ä Lander TR, Robinson SMC, MacDonald BA, Martin JD (2013) Characterization of the suspended organic particles released from salmon farms and their potential as a food supply for the suspension feeder, *Mytilus edulis* in integrated multi-trophic aquaculture (IMTA) systems. Aquaculture 406-407:160–171

ä Lefebvre S, Barille L, Clerc M (2000) Pacific oyster (*Crassostrea gigas*) feeding responses to a fish-farm effluent. Aquaculture 187:185–198

ä Li S (1987) Energy structure and efficiency of a typical Chinese integrated fish farm. Aquaculture 65:105–118

ä MacDonald BA, Robinson SMC, Barrington KA (2011) Feeding activity of mussels (*Mytilus edulis*) held in the field at an integrated multi-trophic aquaculture (IMTA) site (*Salmo salar*) and exposed to fish food in the laboratory. Aquaculture 314:244–251

ä McCandliss RR, Jones SE, Hearn M, Latter R, Jago CF (2002) Dynamics of suspended particles in coastal waters (southern North Sea) during a spring bloom. J Sea Res 47:285–302

ä Navarrete-Mier F, Sanz-Lazaro C, Marin A (2010) Does bivalve mollusc polyculture reduce marine fin fish farming environmental impact? Aquaculture 306:101–107

ä Nickell TD, Cromey CJ, Borja A, Black KD (2009) The benthic impacts of a large cod farm — Are there indicators for environmental sustainability? Aquaculture 295:226–237

Olsen LM, Holmer M, Olsen Y (2008) Perspectives of nutrient emission from fish aquaculture in coastal waters: literature review with evaluated state of knowledge. Final report. Fishery and Aquaculture Industry Research Fund (FHF), Oslo

▶ Perez OM, Telfer TC, Beveridge MCM, Ross LG (2002) Geographical Information Systems (GIS) as a simple tool to aid modeling of particulate waste distribution at marine fish cage sites. Estuar Coast Shelf Sci 54:761–768

Phillips MJ, Beveridge MCM, Clarke RM (1991) Impact of aquaculture on water resources. In: Brune D, Tomasso JR (eds) Aquaculture and water quality. Advances in world aquaculture 3. The World Aquaculture Society, Baton Rouge, LA, p 568–591

Pridmore RD, Rutherford JC (1992) Modelling phytoplankton abundance in a small enclosed bay used for salmon farming. Aquacult Fish Manage 23:525–542

ä Qian PY, Wu CY, Wu M, Xie YK (1996) Integrated cultivation of the red alga *Kappaphycus alvarezii* and the pearl oyster *Pinctada martensi*. Aquaculture 147:21–35

Reid GK, Liutkus M, Robinson SMC, Chopin TR and others (2009) A review of the biophysical properties of salmonid faeces: implications for aquaculture waste dispersal models and integrated multi-trophic aquaculture. Aquacult Res 40:257–273

► Sarà G, Zenone A, Tomasello A (2009) Growth of *Mytilus galloprovincialis* (*Mollusca bivalvia*) close to fish farms: a case of integrated multi-trophic aquaculture in the Tyrrhenian sea. Hydrobiologia 636:129–136

► Shi JZ (2010) Tidal resuspension and transport processes of fine sediment within the river plume in the partially mixed Changjiang River estuary, China: a personal perspective. Geomorphology 121:133–151

Soto D (2009) Integrated mariculture: a global review. FAO Fisheries and Aquaculture Technical Paper 529. FAO, Rome

► Stirling HP, Okumus I (1995) Growth and production of mussels (*Mytilus edulis* L.) suspended at salmon cages and shellfish farms in two Scottish sea lochs. Aquaculture 134:193–210

Strain PM, Hargrave BT (2005) Salmon aquaculture, nutrient fluxes and ecosystem processes in southwestern New Brunswick. In: Hargrave BT (ed) Environmental effects of marine finfish aquaculture. Springer, Berlin, p 29–57

► Strohmeier T, Strand Ø, Alunno-Bruscia M, Duinker A and others (2015) Response of *Mytilus edulis* to enhanced phytoplankton availability by controlled upwelling in an oligotrophic fjord. Mar Ecol Prog Ser 518:139–152

Stucchi D, Sutherland TA, Levings C, Higgs D (2005) Nearfield deposition model for salmon aquaculture waste. In: Hargrave BT (ed) Handbook of environmental chemistry 5, Part M. Springer-Verlag, Berlin, p 157–179

► Sugiura SH, Marchant DD, Kelsey K, Wiggins T, Ferraris RP (2006) Effluent profile of commercially used low-phosphorus fish feeds. Environ Pollut 140:95–101

► Taranger GL, Karlsen Ø, Bannister RJ, Glover KA and others (2015) Risk assessment of the environmental impact of Norwegian Atlantic salmon farming. ICES J Mar Sci 72: 997–1021

► Taylor BE, Jamieson G, Carefoot TH (1992) Mussel culture in British Columbia: the influence of salmon farms on growth of *Mytilus edulis*. Aquaculture 108:51–66

► Thompson KR, Dowd M, Shen Y, Greenberg DA (2002) Probabilistic characterization of tidal mixing in a coastal embayment: a Markov chain approach. Cont Shelf Res 22:1603–1614

Trites RW, Garrett CJR (1983) Physical oceanography of the Quoddy Region. Can Spec Publ Fish Aquat Sci 64:9–34

► Troell M, Norberg J (1998) Modelling output and retention of suspended solids in an integrated salmon-mussel culture. Ecol Model 110:65–77

► Troell M, Halling C, Neori A, Chopin T, Buschmann AH, Kautsky N, Yarish C (2003) Integrated mariculture: asking the right questions. Aquaculture 226:69–90

Troell M, Joyce A, Chopin T, Neori A, Buschmann AH, Fang JG (2009) Ecological engineering in aquaculture — potential for integrated multi-trophic aquaculture (IMTA) in marine offshore systems. Aquaculture 297:1–9

► Valdemarsen TB, Bannister RJ, Hansen PK, Holmer M, Ervik A (2012) Biogeochemical malfunctioning in sediments beneath a deep-water fish farm. Environ Pollut 170:15–25

Velegrakis AF, Bishop C, Lafite R, Oikonomou EK, Lecouturier M, Collins MD (1997) Investigation of meso- and macro-scale sediment transport, hydrodynamics biogeochemical processes and fluxes in the channel. FLUX-MANCHE II Final Report, MAST II:128–143

► Velegrakis AF, Collins MD, Lafite MB, Oikonomou EK and others (1999) Sources, sinks and resuspension of suspended particulate matter in the eastern English Channel. Cont Shelf Res 19:1933–1957

► Weeks AR, Simpson JH, Bowers D (1993) The relationship between concentrations of suspended particulate material and tidal processes in the Irish Sea. Cont Shelf Res 13:1325–1334

► Whitmarsh DJ, Cook EJ, Black KD (2006) Searching for sustainability in aquaculture: an investigation into the economic prospects for an integrated salmon-mussel production system. Mar Policy 30:293–298

8

Carbon, nitrogen and phosphorus budgets of silver carp *Hypophthalmichthys molitrix* with the co-culture of grass carp *Ctenopharyngodon idella*

Bin Xia[1,2], Zhenlong Sun[1], Qin-Feng Gao[1,*], Shuanglin Dong[1], Fang Wang[1]

[1]Key Laboratory of Mariculture, Ministry of Education, Ocean University of China, Qingdao, Shandong 266003, PR China
[2]Marine Science and Engineering College, Qingdao Agricultural University, Qingdao, Shandong 266109, PR China

ABSTRACT: Fish farming activities have resulted in increasing nutrient pollution and subsequent deterioration of water quality in aquatic environments worldwide. Silver carp *Hypophthalmichthys molitrix* can efficiently remove excessive nutrient pollution by filtering the suspended particulate organic matter. To evaluate the feasibility and capacity of using silver carp as biofilters to remove the wastes released from the farming of grass carp *Ctenopharyngodon idella*, 3 mesocosms comprising grass carp and silver carp were developed. Carbon (C), nitrogen (N) and phosphorus (P) budgets of silver carp were measured every month from May to October in 2011. Owing to the changes in exogenous environmental conditions and autogenous physiological status such as water temperature, dissolved oxygen level and feeding behavior, the metabolic acquisition and expenditure of silver carp exhibited obvious temporal fluctuation. For a standardized silver carp with 30 cm body length, the average scope for growth of C, N and P were 54.83, 8.73 and 0.85 mg h^{-1}, respectively. Total nutrient assimilation capacities throughout the experimental period for C, N and P were 236.86, 37.70 and 3.67 g, respectively. Our findings show that silver carp with the co-culture of grass carp provides an economic and environmental win–win resolution to enhance aquaculture production and reduce organic pollution in water.

KEY WORDS: Silver carp · Grass carp · Biofiltration · Nutrient budget · Pollution · Fish farming

INTRODUCTION

The depletion of aquatic resources has stimulated the development of aquaculture worldwide. Meanwhile, the environmental risk of fish farming effluents is well recognized, especially in intensive aquaculture with high farming density and substantial supply of artificial feed (Muir 1982, Kestemont 1995, Gao et al. 2005, Xia et al. 2013c). Ackefors & Enell (1994) estimated that 78 kg nitrogen (N) and 9.5 kg phosphorus (P) per ton of fish production, on average, were released to the water column when the feed conversion coefficient is 1.5. Approximately 72% N and 70% P in feed were not retained by fish, which could become nutrient pollution in forms of uneaten pellets, feces and excretion. Untreated wastewater that is continuously discharged into aquatic environment could cause remarkable elevation of organic matters (Cao et al. 2007). Dissolved and particulate organic nutrients might provide an energy source for suspension feeders or an uptake source through membranes of multicellular and particularly unicellular organisms. However, the elevated inorganic nutrient loadings from decomposition of wasted feed and fish excreta could be cell-toxic in the case of N compounds, and stimulate toxic algae blooms, resulting in mass mortality of cultured species and subsequent economic loss.

Measures for reducing the negative impacts of aquaculture on the environment have become a major concern for the sustainable development of aquaculture (Gowen et al. 1990, Lin & Yi 2003, Pillay 2007). The high nutrient content of the suspended feed residues and other particulate organic matters in fish

*Corresponding author: qfgao@ouc.edu.cn

farming effluents may be a potential food source for suspension-feeding organisms (Gophen et al. 2003). In an integrated ecosystem combining fish and suspension-feeding organisms, the 'biofilters' can utilize the organic wastes as food sources, and the nutrient content in the organic matters is thus ingested and accumulated in the tissues of the organism instead of being dissolved in the water column or deposited on the sediment, which provides an economic and environmental win–win resolution scheme (Xie & Liu 2001, Gao et al. 2006, Yan et al. 2009).

Grass carp *Ctenopharyngodon idella* is the most commercially important freshwater fish species for aquaculture in China (MOAC 2013). Its production has increased rapidly over recent decades (Xia et al. 2013a, 2013b). However, the high stocking density with an artificial supply of formulated feed has resulted in a large amount of effluents. Silver carp *Hypophthalmichthys molitrix*, a typical suspension-feeding fish (Dong et al. 1992, Kadir et al. 2006), has drawn much attention worldwide because of its potential application for removing farming wastes without secondary pollution (Smith 1989, Starling 1993, Fukushima et al. 1999, Tucker 2006, Ke et al. 2007). It plays a positive role in improving the fish-farming environment (Burke et al. 1986, Starling & Rocha 1990). Chen et al. (1991), for example, reported that feeding by silver carp took 3.01% and 5.28% of the total N and P, respectively, out of the water in the Wuhan East Lake ecosystem. In spite of the ecological function of silver carp, metabolic acquisition and expenditure of nutrients including carbon (C), N and P by this species have not been examined so far. In the present study, co-culture systems combining grass carp and silver carp were developed, and temporal changes in C, N and P budgets of silver carp were examined so as to evaluate the feasibility and capacity of using silver carp as a biofilter to reduce nutrient pollution in grass carp ponds.

MATERIALS AND METHODS

Study area and experimental set-up

The experimental pond for fish culture was located in Jinan City, Shandong Province, China (36° 69' N, 116° 86' E). The freshwater used in the pond was pumped from a nearby well. The pond area was approximately 5 ha, with water depth ranging from 1.6 to 1.7 m. Three replicate enclosures of the same size (L × W × D = 8 × 8 × 2.5), which divided the culture pond into different parts, were constructed within the

pond. The enclosures were made of waterproof polyvinyl plastics with supporting timber piles around the boundaries of the enclosures. The separation of the pond area by enclosures avoided the water exchange between different experimental treatments. Juvenile grass carp and silver carp with body mass of ~100 g and mass proportion of 3:1 were released to the enclosures in March. The fish stocking density was 1 individual per square meter. Artificial feed was supplied to the grass carp with the ration of 5% total grass carp biomass daily. The main ingredients of the feed consisted of 28% protein, 6% fat and 33% carbohydrate. The experiment lasted for 6 months from May to October 2011.

Physiological measurements and standardizations

The filtration was determined *in situ* via an indirect biodeposition method in a flow-through system on the bank of the pond. Three silver carp from each enclosure were randomly collected to determine biofiltration capacity at monthly intervals. Each individual silver carp was kept in a separate aquarium (~25 l) and supplied with continuously flow-through water pumped from a larger aquarium (~75 l), which was used as a reservoir (Fig. 1). Three other flow-through systems without animals were used as controls. The water in the reservoir was pumped from the corresponding enclosure in the pond. The particles in the reservoir were kept in suspension by aeration and stirring. A preliminary experiment that was conducted to determine the appropriate flow rate used in the experiment showed that the reduction in particle

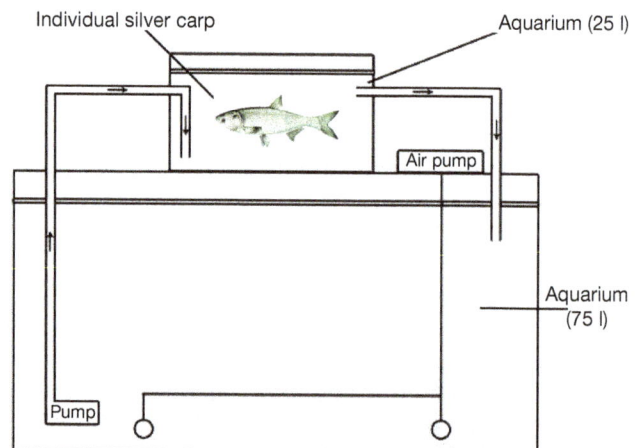

Fig. 1. Flow-through system used to determine the filtration of silver carp *Hypophthalmichthys molitrix in situ* via an indirect biodeposition method

concentration at a flow rate of $5 \, l \, min^{-1}$ was less than 20%. The 20% reduction in the particle concentration between inflow and outflow water did not obviously affect the feeding of silver carp. Measures were taken to minimize environmental stress on silver carp such as acclimation and shading.

Silver carp were cultured for 1 to 2 h, depending on the amount of feces collected. The collected water samples were filtered using glass fiber filters (Whatman GF/F) that were pre-combusted at 450°C for 6 h to remove any possible contamination of organic matter. The vacuum filtration was conducted under the suction of less than one-third atmospheric pressure to avoid the cell damage of phytoplankton. The filter papers were then dried in an oven at 80°C for 24 h, weighed to the nearest 0.1 mg, ashed in a muffle furnace at 450°C and reweighed to determine total particulate matter (TPM; $mg \, l^{-1}$), particulate organic matter (POM; $mg \, l^{-1}$), particulate inorganic matter (PIM; $mg \, l^{-1}$) and organic content ($f = POM/TPM$).

Fish feces were cautiously collected with a pipette, avoiding resuspension of the fecal pellets. Total feces egestion (ER; $mg \, h^{-1}$), organic matter egestion (OER; $mg \, h^{-1}$) and inorganic matter egestion (IER; $mg \, h^{-1}$) were determined by the same method described for the water samples.

Assuming that absorption of inorganic matter through the digestive system is negligible (Cranford & Grant 1990, Turker et al. 2003), clearance rate (CR; $l \, h^{-1}$), which is defined as the volume of water cleared per unit time, was then estimated as CR = IER/PIM. Filtration rate of TPM (FR; $mg \, h^{-1}$), which is defined as the food mass filtered by silver carp per unit time, was computed as FR = CR × TPM, and filtration rate of POM (OFR; $mg \, h^{-1}$) as OFR = CR × POM. Hence, the food absorption rate (AR; $mg \, h^{-1}$) could be estimated as AR = OFR − OER, and absorption efficiency (AE) as AE = AR/OFR.

Oxygen consumption and nutrient excretion of silver carp were determined after the filtration measurements. To determine oxygen consumption (V_{O_2}; $mg \, h^{-1}$), each silver carp was placed in a sealed 50 l polyethylene chamber (experimental chamber); 3 other empty chambers without silver carp served as control chambers. The sealed chambers were bathed in water to avoid the effect of temperature changes on the respiratory activities of the silver carp. After 0.5 to 1 h, depending on the animal size, the DO levels of the experimental and control chambers were measured using the iodometry method. V_{O_2} was calculated using the following equation:

$$V_{O_2} = (DO_C - DO_E) \times v/t \qquad (1)$$

where DO_C and DO_E are the DO concentrations of the control and experimental chambers, respectively, v is the volume of the chamber and t is the experimental time.

For the tests of N and P excretion rates, the silver carp were maintained in separate polyethylene boxes for 0.5 to 1 h, while 2 additional boxes without silver carp were used as controls. Results of previous studies showed that ammonium and phosphate are the predominant excretory products in teleosts (Ballestrazzi et al. 1998, Leung et al. 1999, Yang et al. 2002). Therefore, excretion rates of ammonium (V_N; $\mu g \, h^{-1}$) and phosphate (V_P; $\mu g \, h^{-1}$) were determined using the Nessler's reagent and phosphor-molybdate colorimetric methods, respectively. V_N and V_P were calculated as follows:

$$V_N \text{ (or } V_P) = (C_E - C_C) \times v/t \qquad (2)$$

where C_E and C_C are the nutrient concentrations of the experimental and control boxes respectively, v is the volume of the boxes and t is the experimental time.

To facilitate comparisons of physiological rates (filtration, oxygen consumption and nutrient excretion) at different monthly intervals, physiological rates were size-standardized to the mean silver carp body length of 30 cm according to the following equation (Strychar & MacDonald 1999):

$$Y_S = (X_S/X_O)^b \times Y_O \qquad (3)$$

where Y represents a physiological parameter and X the body length (cm) of silver carp, subscripts S and O represent the standard and observed values, respectively, and b is the power coefficient obtained from the respective monthly allometric equations relating physiological parameters with body length, i.e. $Y = aX^b$, where a is the regression coefficient. Analysis of covariance (ANCOVA) indicated that the slopes of the monthly allometric equations were unequal (i.e. no all-time pooled slope might be regressed). Therefore, b was the coefficient in the above allometric model derived from monthly data (Packard & Boardman 1987).

Nutrient analysis

Water samples (200–300 ml, depending on the concentration of particulate matter) were filtered through pre-combusted and weighed glass-fiber filters (Whatman GF/C) and dried at 80°C for 24 h. Particulate organic carbon (POC; $\mu g \, mg^{-1}$) and particulate organic nitrogen (PON; $\mu g \, mg^{-1}$) of seston were measured with a CHNS/O Analyzer (Vario EI), as were carbon and nitrogen contents of fecal pellets. Particulate organic phosphorus (POP; μg mg^{-1}) and total phosphorous content of fecal pellets were determined following the wet digestion method (Gao et al. 2008).

After collection, the silver carp samples were dried at 80°C for 2 d until constant weight. Nutrient (C, N and P) contents in the dried silver carp were determined using the method similar to that used to analyse fish feces. Production was then estimated by the net gain of nutrients in fish tissue.

Nutrient assimilation

The assimilation rates of C, N and P were measured as scope for growth (SFG; mg h^{-1}) for each element, which was defined as the difference between acquisition and expenditure. SFG of C, N or P was calculated as: SFG (mg h^{-1}) = AR – respiration or excretion rate.

Oxygen consumption was converted to C excretion based on a mean respiratory quotient of 0.85: 1 mg $O_2 \equiv 0.32$ mg C (Martin 1993). Assuming that the average SFG values from monthly measurements were constant, the assimilation of nutrients by silver carp was estimated as: Assimilation (mg) = SFG (mg h^{-1}) × 24 h × 30 d.

Statistical procedures

To obtain functional relationships of physiological processes to environmental factors such as food con-

ditions, temperature, DO level and transparency, a set of regression equations was fitted to experimental data following standard least-squares procedures. Regression analysis was applied by simple linear or non-linear procedures, depending on the most appropriate function to be fitted in each case (Zar 2009). Residuals were analyzed to check the normality and constant variance of predicted dependents. All statistical procedures were performed with SPSS for Windows Release 16.0.

RESULTS

Environmental conditions and nutrient contents

Temporal variations in environmental parameters, including temperature (T; °C), dissolved oxygen (DO; mg l^{-1}), transparency (SD; cm), food conditions (in terms of TPM [mg l^{-1}], POM [mg l^{-1}], PIM [mg l^{-1}] and organic content [f; %]) and nutrient contents including POC, PON and POP of the suspended particulate matter are listed in Table 1. Temperature underwent obvious temporal variations, with the highest value (~28.3°C) in July, decreasing afterward to the lowest (~23.4°C) in October. Because of high temperature and algae blooms in summer, DO levels and transparency remained at extremely low levels, especially in July and August. POM concentrations in July exhibited the highest values because of algal blooms. Owing to the feeding-driven re-suspension of particulate matter by silver carp, PIM concentrations in May and June were significantly higher than those in July and August, which led to the relatively

Table 1. Temporal changes in environmental conditions (T, temperature; DO, dissolved oxygen; SD, transparency), food conditions (TPM, total particulate matter; POM, particulate organic matter; PIM, particulate inorganic matter; f, organic content) and nutrient content of suspended particulate matter (POC, particulate organic carbon; PON, particulate organic nitrogen; POP, particulate organic phosphorus). With the exceptions of T, DO and SD, data are presented as means ± 1 SD (n = 8–9). Different superscript letters within a row indicate a significant difference between monthly sampling dates in 2011 (p < 0.05)

Parameter	May 20	Jun 20	Jul 20	Aug 20	Sep 20	Oct 15
T (°C)	23.5	26.5	28.3	27.9	25.2	23.4
DO (mg l^{-1})	8.9	7.4	6.5	5.7	7.9	8.3
SD (cm)	41.1	35.2	32.7	33.1	40.7	42.2
TPM (mg l^{-1})	39.30 ± 2.11[a,b]	41.11 ± 2.63[b]	37.89 ± 2.39[a,b,c]	36.54 ± 1.65[a,c,d]	34.47 ± 2.31[c,d]	32.96 ± 1.44[d]
POM (mg l^{-1})	14.72 ± 1.14[a]	15.54 ± 1.75[a]	22.32 ± 1.56[b]	20.87 ± 1.58[b]	13.37 ± 1.01[a,c]	12.07 ± 0.39[c]
PIM (mg l^{-1})	24.58 ± 2.06[a]	25.57 ± 2.75[a]	15.58 ± 3.12[b]	15.67 ± 4.14[b]	21.10 ± 2.38[a]	20.89 ± 1.41[a]
f (%)	0.37 ± 0.03[a]	0.38 ± 0.04[a]	0.59 ± 0.07[b]	0.58 ± 0.07[b]	0.39 ± 0.04[a]	0.37 ± 0.02[a]
POC (μg mg^{-1})	145.63 ± 13.32[a,b]	151.39 ± 12.71[a,b]	161.56 ± 7.98[a]	163.14 ± 9.89[a]	147.08 ± 13.22[a,b]	138.89 ± 10.77[b]
PON (μg mg^{-1})	20.62 ± 3.42[ab]	25.81 ± 2.12[c]	24.62 ± 3.40[b,c]	19.45 ± 1.91[a]	22.31 ± 2.01[a,b,c]	18.19 ± 0.08[a]
POP (μg mg^{-1})	1.75 ± 0.08[a]	2.18 ± 0.03[b]	3.08 ± 0.11[c]	1.89 ± 0.07[a]	2.30 ± 0.15[b]	1.09 ± 0.07[d]

Table 2. Allometric relationship between clearance rate (CR; $1 h^{-1}$) and body length (BL; cm), and standardized clearance rates (SCR; $1 h^{-1}$) of 30 cm silver carp *Hypophthalmichthys molitrix*. For SCR, data are presented as means ± 1 SD. Different superscript letters indicate a significant difference between monthly sampling dates in 2011 ($p < 0.05$)

Sampling date	Equation	n	r^2	p	SCR
May 20	$CR = 0.94 \times BL^{0.74}$	8	0.65	<0.01	11.77 ± 0.59^a
Jun 20	$CR = 1.14 \times BL^{0.75}$	5	0.40	<0.01	14.24 ± 0.92^b
Jul 20	$CR = 2.73 \times BL^{0.52}$	9	0.65	<0.01	15.69 ± 0.71^b
Aug 20	$CR = 0.0059 \times BL^{2.34}$	9	0.59	<0.01	15.48 ± 1.02^b
Sep 20	$CR = 0.67 \times BL^{1.00}$	7	0.45	<0.01	19.81 ± 1.16^c
Oct 15	$CR = 0.39 \times BL^{1.16}$	8	0.72	<0.01	19.60 ± 1.65^c

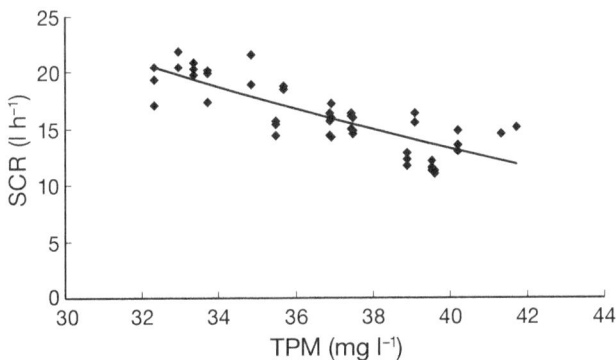

Fig. 3. Relationship between absorption efficiency (AE) of silver carp *Hypophthalmichthys molitrix* and total particulate matter (TPM)

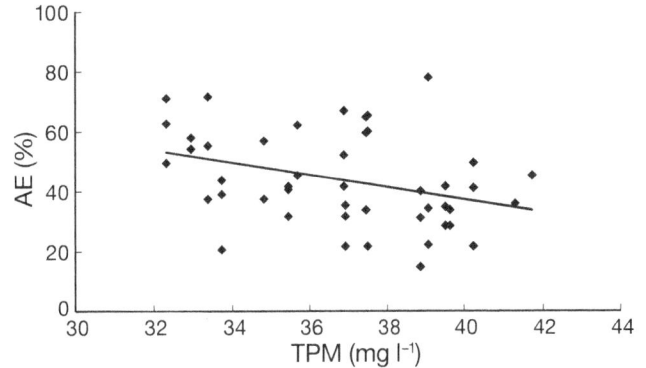

Fig. 2. Relationship between standardized clearance rate (SCR) of 30 cm silver carp *Hypophthalmichthys molitrix* and total particulate matter (TPM)

equation: SCR = 137.54 − 33.70 × ln TPM ($r^2 = 0.66$, $F_{1,44} = 64.51$, $p < 0.01$). Standard filtration rate (SFR) and standard filtration rate of POM (SOFR), were independent of TPM and organic content (f).

As shown in Fig. 3, absorption efficiency (AE) was negatively related to TPM according to the linear regression model: AE = 1.21 − 0.02 × TPM ($r^2 = 0.13$, $F_{1,44} = 2.27$, $p < 0.05$). Absorption rate (AR) was not significantly related to either food quantity (TPM) or food quality (f), in spite of the decreasing AE with increasing TPM.

Oxygen consumption rate (V_{O_2}) was significantly related to the body length of silver carp following the allometric models. These models and the corresponding standardized oxygen consumption rate (SV_{O_2}; mg h^{-1}), as well as the standardized carbon respiration rate (SV_C; mg h^{-1}) of a 30 cm silver carp for each monthly interval, are listed in Table 3. The allometric relationships of nitrogen excretion rate (V_N; μg h^{-1}) and phosphorus excretion rates (V_P; μg h^{-1}) to body length, in addition to the corresponding standardized V_N (SV_N; μg h^{-1}) and V_P (SV_P; μg h^{-1}), are presented in Tables 4 & 5, respectively.

lower f. Seasonal changes in POC and PON showed a similar pattern to that of POM, with relatively higher values in summer (June to August) and lower values in spring (May) and autumn (September and October).

Physiological rate

In each of the 6 sampling months, the CRs of the experimental silver carp were significantly related to individual body length. The regressive allometric equations and the standardized CR (SCR) of a 30 cm individual for each sampling month are listed in Table 2.

As shown in Fig. 2, SCR was negatively related to food quantity in terms of TPM following the

Table 3. Allometric relationship between oxygen consumption rate (V_{O_2}; mg h^{-1}) and body length (BL; cm), and corresponding standardized oxygen consumption rate (SV_{O_2}; mg h^{-1}) and carbon respiration rate (SV_C; mg h^{-1}) of a 30 cm silver carp *Hypophthalmichthys molitrix* for each experimental month. For SV_{O_2} and SV_C, data are presented as means ± 1 SD. Different superscript letters within each column indicate a significant difference between monthly sampling dates in 2011 ($p < 0.05$)

Sampling date	Equation	n	r^2	p	SV_{O_2} (mg h^{-1})	SV_C (mg h^{-1})
May 20	$V_{O_2} = 0.75 \times BL^{1.01}$	9	0.34	<0.01	22.39 ± 1.35^a	7.17 ± 0.43^a
Jun 20	$V_{O_2} = 4.04 \times BL^{0.57}$	9	0.58	<0.01	27.34 ± 1.15^b	8.75 ± 0.37^b
Jul 20	$V_{O_2} = 0.07 \times BL^{1.68}$	8	0.55	<0.01	18.53 ± 0.92^c	5.93 ± 0.30^c
Aug 20	$V_{O_2} = 0.34 \times BL^{1.16}$	9	0.71	<0.01	17.15 ± 1.01^c	5.49 ± 0.32^c
Sep 20	$V_{O_2} = 0.04 \times BL^{1.87}$	8	0.78	<0.01	22.06 ± 0.83^a	7.06 ± 0.27^a
Oct 15	$V_{O_2} = 0.16 \times BL^{1.45}$	9	0.62	<0.01	20.47 ± 0.96^a	6.55 ± 0.31^a

Table 4. Allometric relationship between nitrogen excretion rate (V_N; µg h^{-1}) and body length (BL; cm), and corresponding standardized nitrogen excretion rate (SV_N; µg h^{-1}) of a 30 cm silver carp *Hypophthalmichthys molitrix* for each experimental month. For SV_N, data are presented as means ± 1 SD. Different superscript letters indicate a significant difference between monthly sampling dates in 2011 (p < 0.05)

Sampling date	Equation	n	r^2	p	SV_N (µg h^{-1})
May 20	$V_N = 14.79 \times BL^{0.89}$	9	0.64	<0.01	297.28 ± 30.74[a]
Jun 20	$V_N = 103.17 \times BL^{0.65}$	9	0.76	<0.01	920.65 ± 15.58[b]
Jul 20	$V_N = 1.57 \times BL^{1.75}$	8	0.60	<0.01	571.04 ± 26.19[c]
Aug 20	$V_N = 10.20 \times BL^{1.24}$	8	0.73	<0.01	656.91 ± 17.73[d]
Sep 20	$V_N = 0.78 \times BL^{1.83}$	7	0.91	<0.01	368.36 ± 8.07[e]
Oct 15	$V_N = 1.59 \times BL^{1.44}$	8	0.58	<0.01	201.38 ± 9.66[f]

Table 5. Allometric relationship between phosphorus excretion rate (V_P; µg h^{-1}) and body length (BL; cm), and corresponding standardized phosphorus excretion rate (SV_P; µg h^{-1}) of a 30 cm silver carp *Hypophthalmichthys molitrix* for each experimental month. For SV_P, data are presented as means ± 1 SD. Different superscript letters indicate a significant difference between monthly sampling dates in 2011 (p < 0.05)

Sampling date	Equation	n	r^2	p	SV_P (µg h^{-1})
May 20	$V_P = 2.40 \times BL^{0.74}$	9	0.64	<0.01	28.98 ± 2.50[a]
Jun 20	$V_P = 0.89 \times BL^{1.46}$	9	0.43	<0.01	122.17 ± 9.59[b]
Jul 20	$V_P = 0.26 \times BL^{1.70}$	8	0.61	<0.01	81.57 ± 3.62[c]
Aug 20	$V_P = 0.025 \times BL^{2.22}$	9	0.54	<0.01	45.82 ± 3.12[d]
Sep 20	$V_P = 0.0015 \times BL^{2.90}$	7	0.42	<0.01	26.42 ± 4.60[a]
Oct 15	$V_P = 0.0025 \times BL^{2.42}$	8	0.73	<0.01	9.03 ± 2.33[e]

N and P excretion rates were temporally synchronic, with higher metabolic rates in summer (June, July and August) and lower in other months. Oxygen consumption rate appears lower values in July and August, and the highest values in June. As shown in Fig. 4, SV_{O_2}, SV_N and SV_P were significantly correlated with temperature according to the following equations: $SV_{O_2} = -600.3 + 49.1T - 0.96T^2$ ($r^2 = 0.54$, $F_{2,50} = 101.14$, p < 0.01); $SV_N = -32665 + 2497T - 46.67T^2$ ($r^2 = 0.72$, $F_{2,47} = 1439.1$, p < 0.01); and $SV_P = 0.00001T^{8.38}$ ($r^2 = 0.55$, $F_{1,48} = 594.2$, p < 0.01).

Nutrient assimilation

Net budgets of the nutrients, including C, N and P, which were represented by scope for growth (SFG), and the total assimilation of the 3 nutrients by silver carp are summarized in Table 6. SFGs in July and August appeared lower values relative to other months (June and September). The total nutrient assimilation capacities of individual silver carp for C, N and P during the culturing period were 236.86, 37.70 and 3.67 g, respectively.

DISCUSSION

Previous studies reported that the feeding process of fish was affected by numerous environmental factors, i.e. food availability (food quantity, in terms of TPM; food quality, in terms of POM and *f*), temperature, DO level, etc. (Stoner 2004, Lall & Tibbetts 2009, Vanella et al. 2012, Smith & Sanderson 2013). The relationship of SCR to TPM showed that food availability was the dominant factor affecting the feeding process of silver carp. The negative relationship of SCR to food quantity in terms of TPM suggests the presence of pre-ingestive regulation in silver carp in response to changes in food availability. With decreasing TPM, silver carp might actively regulate feeding behavior to maintain feeding quantity, such as opening the mouth gape, resulting in more water passing the oral cavity, increasing swimming speed, closing the gill arches to let water pass unfiltered, etc. Like the active regulation of clearance rate, filtration rate could be kept independent of food conditions. Turker et al. (2003) found that the filtration rate of silver carp increased to the maximum value and remained constant as suspended POC in the water column increased. In the present study, absorption rate was not related to TPM or *f*, and together with the downregulation of AE with increasing TPM, this indicates the presence of a post-ingestive regulative function in the feeding processes.

The metabolic rate of fish generally depends on numerous factors such as feeding habit, DO levels and temperature (Kutty 1972, Schurmann & Steffensen 1997, Wood 2001, Mallekh & Lagardère 2002). In the present study, the oxygen consumption rate and corresponding carbon respiration rate demonstrated an obvious temporal pattern. As indicated by the regressive model shown in Fig. 4A, the oxygen consumption rate of *Hypophthalmichthys molitrix* in the present study was significantly affected by water temperature. Previous studies showed that fish fed more vigorously and specific dynamic action (SDA) was accelerated as temperature increased within the thermal tolerance range, which would result in a high oxygen consumption rate (Aguiar et al. 2002, Lermen et al. 2004, Pang et al. 2010, Zhao et al. 2011a). In the present study, the oxygen consumption

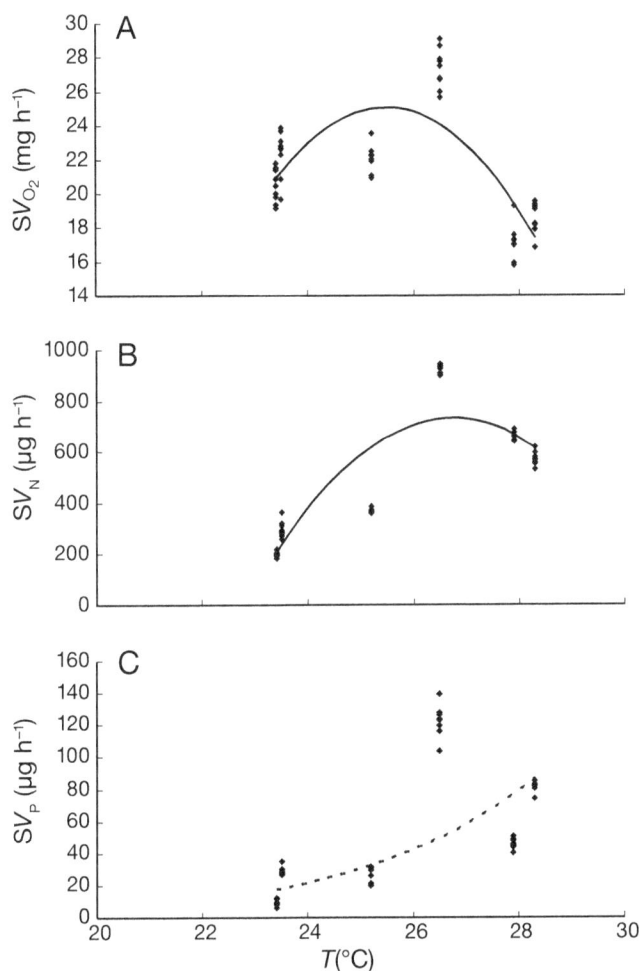

Fig. 4. (A) Standardized oxygen consumption rate (SV_{O_2}) versus water temperature (T). (B) Standardized nitrogen excretion rate (SV_N) versus water temperature. (C) Standardized phosphorus excretion rate (SV_P) versus water temperature

Table 6. Net budget rate (scope for growth [SFG]) and total assimilation of carbon, nitrogen and phosphorus by a silver carp *Hypophthalmichthys molitrix* individual at monthly intervals during the experimental period in 2011

Sampling date	SFG (mg h⁻¹ ind.⁻¹)			Assimilation (g ind.⁻¹)		
	C	N	P	C	N	P
May 20	31.48	4.39	0.41	22.67	3.16	0.29
Jun 20	55.32	9.53	0.98	39.83	6.86	0.71
Jul 20	40.35	7.26	0.80	29.05	5.23	0.57
Aug 20	45.59	6.32	0.77	32.82	4.55	0.55
Sep 20	85.25	13.82	1.36	61.38	9.95	0.98
Oct 15	70.99	11.04	0.237	51.11	7.95	0.56
	Average SFG			Total assimilation		
	54.83	8.73	0.85	236.86	37.70	3.67

rates of *H. molitrix* decreased despite the increasing temperatures in July and August, which might be due to the physiological inhibition caused by the relatively lower DO level. This result is similar to that of related studies on other teleost fish species (Rantin & Johansen 1984, Fernandes et al. 1995). In addition, Crocker & Cech (1997) reported that hypoxia significantly depressed the oxygen consumption rates of white sturgeon *Acipenser transmontanus* at all temperatures and body sizes. Zhao et al. (2011b) showed that under the conditions of elevated temperature and hypoxia, silver carp showed an anti-filtering response, i.e. decreased oxygen consumption and associated lower feeding rates in terms of SCR and SFR.

N and P excretion rates of silver carp exhibited positive correlations with temperature in the present study. Excretion rates of both N and P increased in summer, which might be induced by the following two mechanisms. First, the increased metabolic demands in response to the elevated temperature were partially met by means of deamination of amino acids and catabolism of phospholipids (Forsberg & Summerfelt 1992). Jobling (1981) reported that excretion rates of young plaice (*Pleuronectes platessa* L.) increased with increasing temperature. Second, glycogen reserves and synthesis were unavailable under conditions of low oxygen levels, while more protein and lipid were exploited as internal energy sources. A study by Kutty (1972) showed the increased anaerobic energy utilization and protein metabolism of *Tilapia* at low oxygen levels, which might be of advantage for preventing acidosis.

SFGs of C, N and P showed temporal changes during the experimental period, and decreased in July and August accompanied with the elevated temperature and lower DO levels. Silver carp showed the positive values of nutrient assimilation, indicating that they could efficiently accumulate nutrients throughout the experimental period. Xia et al. (2013c) reported that food sources of silver carp included POM, residue of fish feed and fish feces in co-culture ponds of grass carp and silver carp. Accordingly, the direct nutrient removal by silver carp from farming wastes including feed residue and fish feces were 91.19, 14.51 and 1.41 g per individual for C, N and P, respectively. In contrast, uptake of POM by silver carp further implied the potential of the indirect removal of dissolved nutrients, because the growth of microalgae, which was the principal constituent of POM, accumulated a considerable amount of dissolved nutrients from the water column. As such, the co-culture of silver carp with grass carp can effectively relieve the nutrient pollution due to the release of farming wastes in the forms of feed residue and fish excreta. Moreover, additional culture of silver carp improved the utilization of food resources in the aquaculture waters and increased the industrial benefit. Such polyculture systems containing various species belonging to different feeding guilds hence provide an economic and environmental win–win resolution scheme to reduce organic pollution and enhance aquaculture production.

Acknowledgements. The work described in this paper was funded by a grant from the National Basic Research Program of China ('973' Program).

LITERATURE CITED

Ackefors H, Enell M (1994) The release of nutrients and organic matter from aquaculture systems in Nordic countries. J Appl Ichthyol 10:225–241

Aguiar LH, Kalinlin AK, Rantin FT (2002) The effects of temperature on the cardio-respiratory function of the neotropical fish *Piaractus mesopotamicus.* J Therm Biol 27:299–308

Ballestrazzi R, Lanari D, Agaro ED (1998) Performance, nutrient retention efficiency, total ammonia and reactive phosphorus excretion of growing European sea-bass (*Dicentrarchus labrax,* L.) as affected by diet processing and feeding level. Aquaculture 161:55–65

Burke JS, Bayne DR, Rea H (1986) Impacts of silver and bighead carp on plankton communities of channel catfish ponds. Aquaculture 55:59–68

Cao L, Wang W, Yang Y, Yang C, Yuan Z, Xiong S, Diana J (2007) Environmental impact of aquaculture and countermeasures to aquaculture pollution in China. Environ Sci Pollut Res Int 14:452–462

Chen SL, Liu XF, Hua L (1991) The role of silver carp and bighead in the cycling of nitrogen and phosphorus in the East Lake ecosystem. Acta Hydrobiol Sin 15:8–26

Cranford PJ, Grant G (1990) Particle clearance absorption of phytoplankton and detritus by the sea scallop *Placopecton magellanicus* (Gmelin). J Exp Mar Biol Ecol 137: 105–121

Crocker CE, Cech JJ Jr (1997) Effects of environmental hypoxia on oxygen consumption rate and swimming activity in juvenile white sturgeon, *Acipenser transmontanus,* in relation to temperature and life intervals. Environ Biol Fishes 50:383–389

Dong SL, Li DS, Bing XW, Shi QF, Wang F (1992) Suction volume and filtering efficiency of silver carp (*Hypophthalmichthys molitrix* Val.) and bighead carp (*Aristichthys nobilis* Rich.). J Fish Biol 41:833–840

Fernandes MN, Barrionuevo WR, Rantin FT (1995) Effects of thermal stress on respiratory responses to hypoxia of South American prochilodontidae fish, *Prochilodus scrofa.* J Fish Biol 46:123–133

Forsberg JA, Summerfelt RC (1992) Effect of temperature on diel ammonia excretion of fingerling walleye. Aquaculture 102:115–126

Fukushima M, Takamura N, Sun L, Nakagawa M, Matsushige K, Xie P (1999) Changes in the plankton community following introduction of filter-feeding planktivo-

rous fish. Freshw Biol 42:719–735

Gao QF, Cheung KL, Cheung SG, Shin PKS (2005) Effects of nutrient enrichment derived from fish farming activities on macroinvertebrate assemblages in a subtropical. Mar Pollut Bull 51:994–1002

Gao QF, Shin PKS, Lin GH, Chen SP, Cheung SG (2006) Stable isotopic and fatty acid evidence for uptake of organic wastes by green-lipped mussels *Perna viridis* in a polyculture fish farm system. Mar Ecol Prog Ser 317:273–283

Gao QF, Xu WZ, Liu XS, Cheung SG, Shin PKS (2008) Seasonal changes in C, N and P budgets of green-lipped mussels *Perna viridis* and removal of nutrients from fish farming in Hong Kong. Mar Ecol Prog Ser 353:137–146

Gophen M, Tsipris Y, Meron M, Bar-Ilan I (2003) The management of Lake Agmon wetlands (Hula Valley, Israel). Hydrobiologia 506–509:803–809

Gowen RJ, Ezzi I, Rosenthal H, Maekinen T (1990) Environmental impact of aquaculture activities. European Aquaculture Society Special Publication, Ghent, p 257–283

Jobling M (1981) Some effects of temperature, feeding and body weight on nitrogenous excretion in young plaice *Pleuronectes platessa* L. J Fish Biol 18:87–96

Kadir A, Kundu RS, Milstein A, Wahab MA (2006) Effects of silver carp and small indigenous species on pond ecology and carp polycultures in Bangladesh. Aquaculture 261: 1065–1076

Ke ZX, Xie P, Guo LG, Liu YQ, Yang H (2007) *In situ* study on the control of toxic *Microcystis* blooms using phytoplanktivorous fish in the subtropical Lake Taihu of China: a large fish pen experiment. Aquaculture 265:127–138

Kestemont P (1995) Different systems of carp production and their impacts on the environment. Aquaculture 129: 347–372

Kutty MN (1972) Respiratory quotient and ammonia excretion in *Tilapia mossambica.* Mar Biol 16:126–133

Lall SP, Tibbetts SM (2009) Nutrition, feeding, and behavior of fish. Vet Clin North Am Exot Anim Pract 12:361–372

Lermen CL, Lappe R, Crestani M, Vieira VP and others (2004) Effect of different temperature regimes on metabolic and blood parameters of silver catfish *Rhamdia quelen.* Aquaculture 239:497–507

Leung KMY, Chu JCW, Wu RSS (1999) Effects of body weight, water temperature and ration size on ammonia excretion by the areolated grouper (*Epinephelus areolatus*) and mangrove snapper (*Lutjanus argentimaculatus*). Aquaculture 170:215–227

Lin CK, Yi Y (2003) Minimizing environmental impacts of freshwater aquaculture and reuse of pond effluents and mud. Aquaculture 226:57–68

Mallekh R, Lagardère JP (2002) Effect of temperature and dissolved oxygen concentration on the metabolic rate of the turbot and the relationship between metabolic scope and feeding demand. J Fish Biol 60:1105–1115

Martin KLM (1993) Aerial release of CO_2 and respiratory exchange ratio in intertidal fishes out of water. Environ Biol Fishes 37:189–196

MOAC (Ministry of Agriculture, China) (2013) China fisheries yearbook (2012) China Agriculture Publisher, Beijing

Muir JF (1982) Recirculated water systems in aquaculture. In: Muir JF, Roberts RJ (eds) Recent advances in aquaculture. Croom Helm, London, p 357–446

Packard GC, Boardman TJ (1987) The misuse of ratios to scale physiological data that vary allometrically with body size. In: Feder ME, Bennett AF, Burggern WW,

Huey RB (eds) New directions in ecological physiology. Cambridge University Press, Cambridge, p 216–239

Pang X, Cao ZD, Peng JL, Fu SJ (2010) The effects of feeding on the swimming performance and metabolic response of juvenile southern catfish, *Silurus meridionalis*, acclimated at different temperatures. Comp Biochem Physiol A 155:253–258

Pillay T (2007) Aquaculture and the environment, 2nd edn. Blackwell, Rome.

Rantin FT, Johansen K (1984) Responses of the teleost *Hoplias malabarieus* to hypoxia. Environ Biol Fishes 11:221–228

Schurmann H, Steffensen JF (1997) Effects of temperature, hypoxia and activity on the metabolism of juvenile Atlantic cod. J Fish Biol 55:1166–1180

Smith DW (1989) The feeding selectivity of silver carp, *Hypophthalmichthys molitrix* Val. J Fish Biol 34:819–828

Smith JC, Sanderson SL (2013) Particle retention in suspension-feeding fish after removal of filtration structures. Zoology 116:348–355

Starling FLRM (1993) Control of eutrophication by silver carp, *Hypophthalmichthys molitrix*, in the tropical Paranoa Reservoir (Brasilia, Brazil): a mesocosm experiment. Hydrobiologia 257:143–152

Starling FLRM, Rocha AJA (1990) Experimental study of the impacts of planktivorous fishes on plankton community and eutrophication of a tropical Brazilian reservoir. Hydrobiologia 200–201:581–591

Stoner AW (2004) Effects of environmental variables on fish feeding ecology: implications for the performance of baited fishing gear and stock assessment. J Fish Biol 65:1445–1471

Strychar KB, MacDonald BA (1999) Impacts of suspended particle on feeding and absorption rates in cultured eastern oysters (*Crassostrea virginica*, Gmelin). J Shellfish Res 18:437–444

Tucker CS (2006) Low-density silver carp *Hypophthalmichthys molitrix* (Valenciennes) polyculture does not prevent cyanobacterial off-flavours in channel catfish *Ictalurus punctatus* (Rafinesque). Aquacult Res 37:209–214

Turker H, Eversole AG, Brune DE (2003) Comparative Nile tilapia and silver carp filtration rates of Partitioned Aquaculture System phytoplankton. Aquaculture 220:449–457

Vanella FA, Boy CC, Fernandez DA (2012) Temperature effects on growing, feeding, and swimming energetic in the Patagonian blennie *Eleginops maclovinus* (Pisces: Perciformes). Polar Biol 35:1861–1868

Wood CM (2001) Influence of feeding, exercise, and temperature on nitrogen metabolism and excretion. Fish Physiol 20:201–238

Xia B, Gao QF, Li HM, Dong SL, Wang F (2013a) Turnover and fractionation of nitrogen stable isotope in tissues of grass carp *Ctenopharyngodon idella*. Aquacult Environ Interact 3:177–186

Xia B, Gao QF, Dong SL, Wang F (2013b) Carbon stable isotope turnover and fractionation in grass carp *Ctenopharyngodon idella* tissues. Aquat Biol 19:207–216

Xia B, Gao QF, Dong SL, Shin PKS, Wang F (2013c) Uptake of farming wastes by silver carp *Hypophthalmichthys molitrix* in polyculture ponds of grass carp *Ctenopharyngodon idella*: evidence from C and N stable isotopic analysis. Aquaculture 404–405:8–14

Xie P, Liu JC (2001) Practical success of biomanipulation using filter-feeding fish to control cyanobacteria blooms: a synthesis of decades of research and application in a subtropical hypereutrophic lake. ScientificWorldJournal 1:337–356

Yan LL, Zhang GF, Liu QG, Li JL (2009) Optimization of culturing the freshwater pearl mussels, *Hyriopsis cumingii* with filter feeding Chinese carp (bighead carp and silver carp) by orthogonal array design. Aquaculture 292:60–66

Yang SD, Liou CH, Liu FG (2002) Effects of dietary protein level on growth performance, carcass composition and ammonia excretion in juvenile silver perch (*Bidyanus bidyanus*). Aquaculture 213:363–372

Zar JH (2009) Biostatistical analysis, 5th edn. Prentice Hall, Upper Saddle River, NJ

Zhao ZG, Dong SL, Wang F, Tian XL, Gao QF (2011a) Respiratory response of grass carp (*Ctenopharyngodon idella*) to temperature changes. Aquaculture 322–323:128–133

Zhao ZG, Dong SL, Wang F, Tian XL, Gao QF (2011b) The measurements of filtering parameters under breathing and feeding of filter-feeding silver carp (*Hypophthalmichthys molitrix* Val.). Aquaculture 319:178–183

9

Methane distribution, sources, and sinks in an aquaculture bay (Sanggou Bay, China)

Jing Hou[1,2], Guiling Zhang[1,2,*], Mingshuang Sun[1], Wangwang Ye[1], Da Song[1]

[1]Key Laboratory of Marine Chemistry Theory and Technology, Ministry of Education, Ocean University of China, Qingdao 266100, PR China

[2]Qingdao Collaborative Innovation Center of Marine Science and Technology, Ocean University of China, Qingdao 266100, PR China

ABSTRACT: From 2012 to 2015, we investigated methane (CH_4) distribution, air–sea fluxes, and sediment–water fluxes in an aquaculture bay (Sanggou Bay, China), and estimated the input of CH_4 from potential land sources including rivers and groundwater. Surface water CH_4 in the bay ranged from 3.0 to 302 nM, while bottom CH_4 was usually higher due to sediment release. Water column CH_4 in summer and autumn was 3 to 10 times that in spring and winter due to seasonal variation in water temperature and land source inputs. Surface CH_4 was higher in kelp and scallop polyculture zones than in other culture zones and outside the bay, suggesting the influence of aquaculture activities. CH_4 concentrations were 123 to 2190 nM in rivers around the bay, and 1.6 to 405 nM in groundwater along the shoreline; both showed great spatial and temporal variations. Sediment–water CH_4 fluxes ranged from 0.73 to 8.26 µmol m^{-2} d^{-1}, with those in bivalve culture zones higher than in polyculture zones. Sea–air CH_4 fluxes ranged from 2.1 to 123.2 µmol m^{-2} d^{-1} (mean 48.2 µmol m^{-2} d^{-1}) and showed seasonal variations. CH_4 budget in Sanggou Bay showed that groundwater input (4.2×10^5 mol yr^{-1}) was the largest source of CH_4, followed by sediment release (2.6×10^5 mol yr^{-1}) and riverine input (1.4×10^5 mol yr^{-1}), while sea-to-air release (2.5×10^6 mol yr^{-1}) and export from the bay to the Yellow Sea (8.8×10^5 mol yr^{-1}) were the dominant CH_4 sinks. Net water column production-oxidation was estimated preliminarily to produce 1.7×10^5 mol CH_4 yr^{-1}. However, there was a great imbalance of sources and sinks, with an apparent missing source of 2.4×10^6 mol yr^{-1} that was mostly due to an underestimate of *in situ* water column production and CH_4 release from the sediments.

KEY WORDS: CH_4 · Sanggou Bay · Production · Sediment–water exchanges · Air–sea fluxes · Aquaculture

INTRODUCTION

Methane (CH_4), the most abundant hydrocarbon in the atmosphere, plays an important role in regulating the Earth's radiation balance and atmospheric chemistry in the troposphere (Cicerone & Oremland 1988, Lashof & Ahuja 1990). Although the current atmospheric mixing ratio of CH_4 (~1.8 ppm) is much less than that of CO_2 (~390 ppm), it is actually responsible for about 20% of the greenhouse effect (IPCC 2013). The atmospheric CH_4 mixing ratio has increased by a factor of 2.5, from 722 ppb in 1750 to 1803 ppb in 2011, as a result of human activities since the Industrial Revolution (IPCC 2013).

Oceans are a natural source of atmospheric CH_4, and there are large spatial and temporal variations of oceanic CH_4 emissions. Typically, oligotrophic waters are only slightly supersaturated (by about 5%) in CH_4 with respect to atmospheric equilibrium (Bates et al. 1996, Bange et al. 1998, Karl et al. 2008), resulting in low sea-to-air CH_4 fluxes. High sea-to-air CH_4 emissions can occur in biologically productive regions such as estuaries and coastal and upwelling areas, which contribute to about 75% of the oceanic

*Corresponding author: guilingzhang@ouc.edu.cn

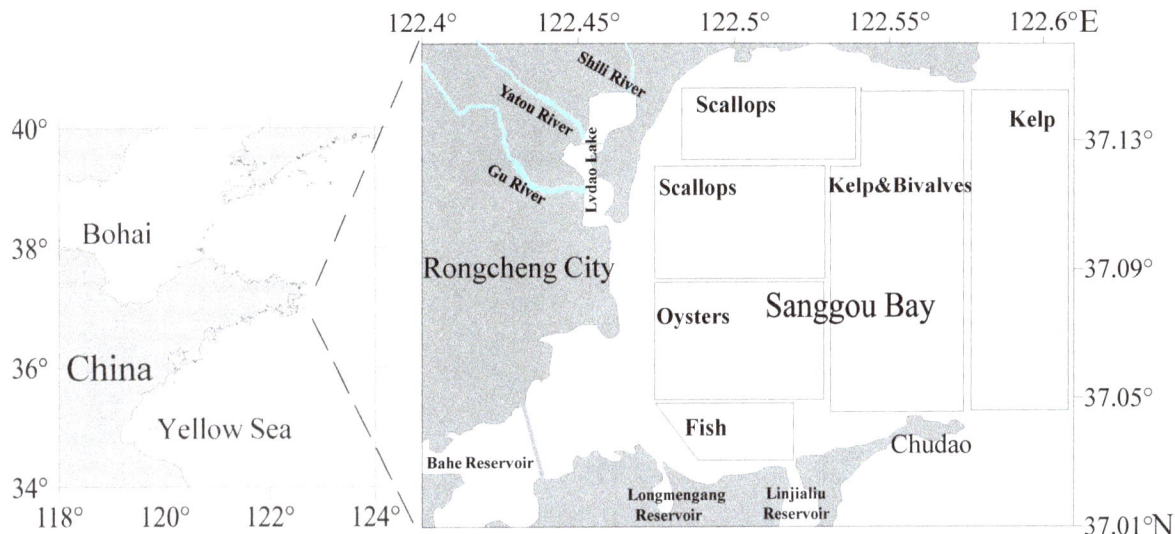

Fig. 1. Location of Sanggou Bay, China, and the main aquaculture practices

CH$_4$ emissions (Bange et al. 1994, EPA 2010). However, CH$_4$ emissions from coastal areas still have great uncertainties, due to poor coverage of CH$_4$ measurements and great spatial variations. Potential sources of CH$_4$ in coastal waters include riverine inputs, *in situ* water column production, and sediment release (Martens & Klump 1980, Kelley et al. 1990, Hornafius et al. 1999, Mau et al. 2007, Canet et al. 2010). It is generally assumed that CH$_4$ production and emission from coastal waters will be enhanced with the increase in human perturbations such as increased nutrient loading and intensive marine aquaculture. With the continuous decline in fishery harvests, aquaculture has become the world's fastest growing sector of food production, increasing nearly 60-fold during the last 5 decades, to meet the increasing demand for seafood (FAO 2007). However, the rapid increase in aquaculture production may also cause some environmental concerns, such as the release of greenhouse gases (Ferrón et al. 2007, Green et al. 2012). The discharge of effluent with high concentrations of organic matter and nutrients from coastal aquaculture systems to adjacent marine waters can lead to organic pollution and provide favorable conditions for the production of CH$_4$. However, most studies on coastal methane emissions have focused on estuaries and coastal waters, and few studies have been conducted on coastal aquaculture systems. The aims of this study were to determine the temporal and spatial distributions of CH$_4$ in an intensive coastal aquaculture bay (Sanggou Bay) in China, to identify various CH$_4$ sources and sinks, and to evaluate CH$_4$ emissions from this bay and the possible impact of aquaculture activities.

MATERIALS AND METHODS

Study area

Sanggou Bay (SGB) is a semi-circular bay on the north-eastern coast of China (Fig. 1). The bay is crescent-shaped, facing the Yellow Sea in the east, and has an average water depth of approximately 7.5 m and an area of approximately 144 km^2 (Zhang et al. 2009). Water renewal between the bay and the Yellow Sea is driven by a semi-diurnal tide with the largest tide range of 3.5 m. When the tide is rising, the tidal water enters the bay from the north, rotates counterclockwise, and flows out from the south through the west coast of the bay. The ebb tide has the reverse process, and the velocity of the residual flow is slow (Chinese Gulf Compilation Committee 1991). The main rivers emptying into the bay include the Sanggan, Ba, Shili, and Gu Rivers, with the total annual water discharge ranging between 1.7×10^8 and 2.3×10^8 m^3 yr^{-1} (Jiang et al. 2015). The largest river, i.e. the Gu, provides almost 70% of the total water discharge.

SGB has been used for aquaculture since the mid-1980s and is among the largest aquaculture sites in China (Guo et al. 1999, Zhang et al. 2009, Jiang et al. 2015). About 2/3 of the bay area is used for farming of bivalve shellfish, seaweed, and fish, with 4 major types of culture model, i.e. the monoculture of kelps, the monoculture of scallops, the monoculture of oysters, and the polyculture of kelps and bivalves (Fig. 1) (Shi et al. 2013). The main cultivation method is long-line culture, and cultivated species include kelp *Saccharina japonica*, scallops *Chlamys farreri*, and oys-

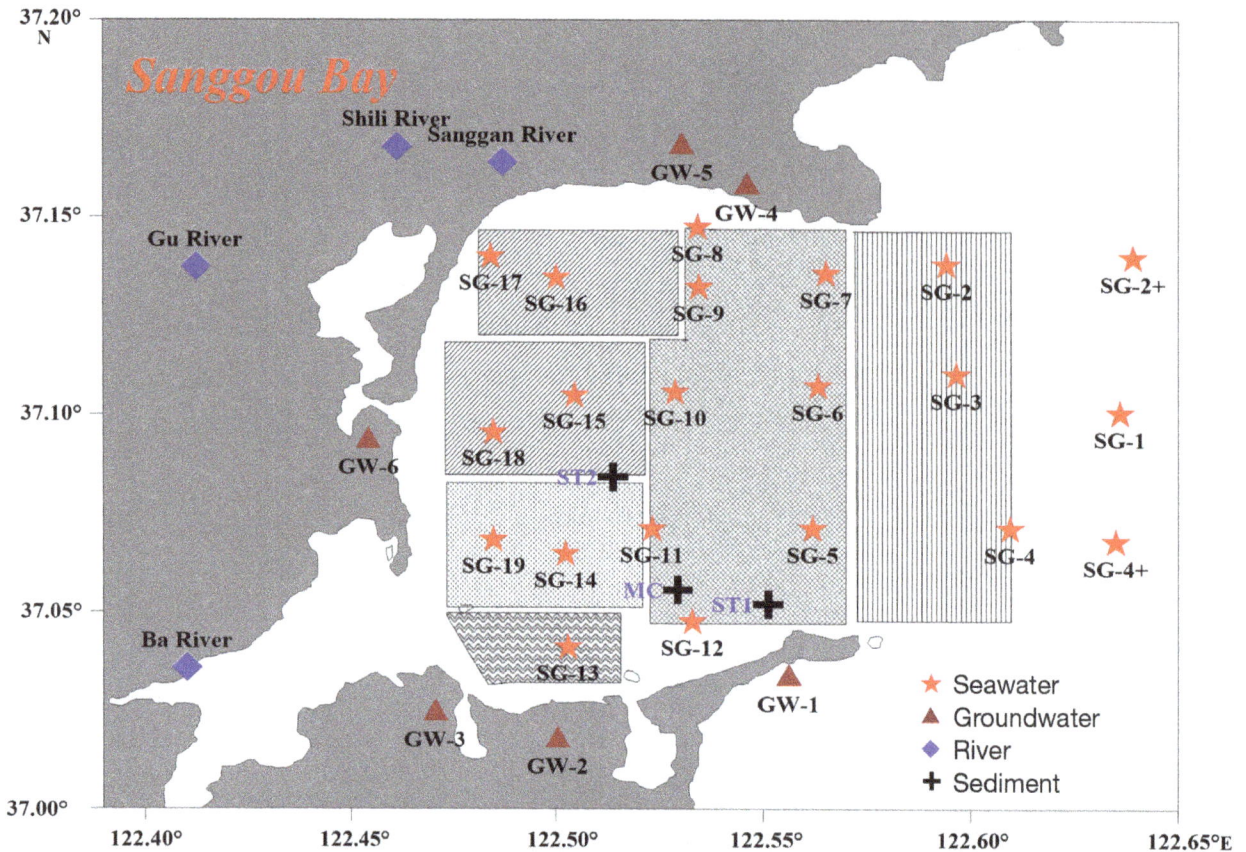

Fig. 2. Sampling locations in Sanggou Bay. Polygons delineate aquaculture types, see Fig. 1

ters *Crassostrea gigas* (Zhang et al. 2009). Kelp is tied to ropes and scallops are contained either in lantern nets or by ear-hanging (Zeng et al. 2015). Macroalgae (i.e. kelp) are grown outside the bay and only between November and May (Fang et al. 1996). During the seeding and harvesting period, kelp competes with phytoplankton for the assimilation of dissolved inorganic nitrogen. The aquaculture of bivalve shellfishes occurs from early spring to November. Shellfishes filtrate and ingest particulate matter and digest phytoplankton and particulate organic matter (POM), especially oysters when they have spawned in August, demanding more energy and stored substances (Mao et al. 2006). Most fish culture is clustered together in the southern part of the bay where water is calm and cages are within easy access from the shore (Fig. 1).

Water sampling and analysis

Eight cruises were carried out in SGB during June and September 2012, April, July, and October 2013, January and May 2014, and May 2015. The sampling locations are shown in Fig. 2. Duplicate samples of surface and bottom seawater were collected using 10 l Niskin bottles, and then filled into 116 ml glass bottles. After overflow of approximately 1.5- to 2-fold of bottle volume, 1 ml of saturated solution of $HgCl_2$ was added to inhibit microbial activity. The sample bottle was then immediately sealed with a butyl rubber stopper and an aluminum cap and stored upside down in a dark box (Zhang et al. 2008). All water samples were analyzed after return to the shore laboratory within 60 d of collection (Zhang et al. 2004). Salinity and seawater temperature were measured with a multi-parameter probe (WTW 350i), and wind speeds were measured with an anemometer at about 10 m above the sea surface.

To evaluate the CH_4 input from potential terrestrial sources, water samples were collected from 4 rivers (Sanggan, Ba, Shili, and Gu) and 6 groundwater wells (GW1–GW6, Fig. 2) along the shoreline of SGB in June and September 2012, April, July, and October 2013, and January 2014. River water and groundwater were collected using a 5 l plastic sampler, and the samples were processed as for seawater described above.

Dissolved CH_4 in seawater was measured using a gas-stripping method described by Zhang et al. (2004). After purging with high-purity N_2, samples were passed through a drying tube with calcium chloride to remove water vapor. CH_4 was then separated on a 3 m × 3 mm i.d. stainless steel column packed with 80/100 mesh Porapak Q and measured with a gas chromatograph (Shimadzu, model GC-14B) equipped with a flame ionization detector (FID) (Zhang et al. 2004). FID responses were calibrated using known volumes of CH_4 standards (2.05, 4.22, and 50.4 ppmV; Research Institute of China National Standard Materials). The FID response signal and CH_4 concentration had a linear relationship, so a multi-point calibration method was used to determine CH_4 concentration based on chromatographic peak areas. The precision of this method was about 3% (Zhang et al. 2004).

Sediment sampling and incubation experiments

CH_4 emission from the sediments was measured by the closed chamber incubation method previously described by Sun et al. (2015), which was modified from Barnes & Owens (1999). Sediment samples were collected by a box corer at different sampling stations (Fig. 1), and only samples with undisturbed sediment surfaces were used. At each station, 15 sediment cores were collected using plexiglass tubes (i.d. = 5 cm, height = 30 cm) and sealed using air-tight rubber bungs. After ambient bottom water was added carefully with no gas headspace, the core was capped with a plexiglass top with 2 sampling ports. All cores were placed in a water-filled tank held at ambient room temperature, and the overlying water was stirred by magnetic stirrers rotated at 60 rpm. Ten glass bottles filled with ambient bottom water were placed in the same tank as water column controls. Cores and bottled waters were incubated in the dark for ~24 to 48 h. Overlying water samples (56.5 ml) from 3 cores were collected each time at intervals of 4 to 8 h to measure CH_4 concentration. At the same time, 2 bottled water samples were also treated with 0.5 ml $HgCl_2$ as a water column control. The CH_4 concentrations of all samples were measured by the gas-stripping method described above. Sediment–water CH_4 flux was estimated from the slope of the CH_4 increase in the overlying water versus time. The discrepancy in the CH_4 emission rate that resulted from differences between incubation and *in situ* temperatures was calibrated by the Arrhenius empirical equation as described by Aller et al. (1985) and Song et al. (2016).

Water incubation experiments

Time series incubation experiments were conducted to determine net CH_4 production-oxidation rates and understand the potential production mechanism in April, July, and October 2013 and January, May, and September 2014. To test for the effects of methylated compounds on CH_4 production, surface water samples were incubated with or without the addition of dimethylsulfoniopropionate (DMSP; final concentration 50 µM), or trimethylamine (final concentration 1 µM), or with added 2-bromoethane sulfonic acid (BES; final concentration 10 mM) to inhibit methanogenesis, and the CH_4 concentrations were monitored for more than 9 d in October 2013. Seawater for incubation experiments was transferred from the 10 l polyvinylchloride sampling bottles into clean polycarbonate carboys before the start of each experiment. DMSP, trimethylamine, or BES were added to the final concentration, and subsamples were transferred into 56.5 ml glass serum bottles that were capped with gas-tight Teflon-lined silicone stoppers and crimp-sealed with aluminum caps. Subsamples with no added reagents were used as controls. The incubations were conducted at approximately *in situ* water temperatures (3°C for January, 10°C for April, 15°C for May, 26°C for July, 23°C for September, 18°C for October) under an approximately 12:12 h light:dark cycle. Duplicate water samples were poisoned by addition of 0.5 ml of saturated $HgCl_2$ solution at 1 to 2 d intervals and analyzed for CH_4 as described above. Two additional samples for dissolved oxygen (DO) were collected and measured using the Winkler titration method (Bryan et al. 1976). Net CH_4 production-oxidation rates were estimated from the initial slope of the increase of CH_4 over time.

Saturation and flux calculation

The saturation (R, %) and sea-to-air fluxes of CH_4 (F, µmol m^{-2} d^{-1}) were calculated using the following equations:

$$R = C_{obs}/C_{eq} \times 100\% \qquad (1)$$

$$F = k_w(C_{obs} - C_{eq}) \qquad (2)$$

where C_{obs} is the observed concentration of dissolved CH_4, and C_{eq} is the air-equilibrated seawater CH_4 concentration calculated from the *in situ* temperature and salinity using the equation of Wiesenburg & Guinasso (1979). Atmospheric CH_4 was not measured during these cruises. Therefore, mean atmospheric

CH_4 mixing ratios of 1.896, 1.901, and 1.929 ppm by volume (ppmv) at 3 observation stations near the coastal seas of China (NOAA Stns LLN, TAP, and SDZ) for 2012, 2013, and 2014, from the NOAA/ESRL Global Monitoring Division *in situ* program (www. esrl.noaa.gov/gmd), were used for calculations. We found that the variation of assumed atmospheric CH_4 concentrations in the range of 1.85 to 1.95 ppmv make differences less than ±2% in the computed air–sea CH_4 fluxes. Hence use of the annual mean atmospheric CH_4 concentration from monitoring networks for the sea–air flux calculation will not introduce significant errors. k_w is the gas transfer coefficient in cm h^{-1}, which is a function of wind speed and Schmidt number (Sc). Various empirical equations were employed to estimate k, among which the equations from Liss & Merlivat (1986) and Wanninkhof (1992) were used most frequently and represent the estimation in a lower and higher level, respectively. Nightingale et al. (2000) proposed a gas exchange relationship that shows a dependence on wind speed, and the corresponding value lies near the median of extensive methods and models. Wanninkhof (2014) recently updated the most frequently used method of Wanninkhof (1992), and this update reflects advances that have occurred over the last 2 decades in quantifying the gas transfer coefficient. Hence the methods from Nightingale et al. (2000) and Wanninkhof (2014) (hereafter N2000 and W2014) were chosen to calculate air–sea fluxes in this study.

RESULTS

Water column CH_4 and other parameters

Table 1 shows the temperature, salinity, and CH_4 concentration in surface and bottom waters of SGB during the 8 cruises. Water temperature ranged from 3.1 to 23.2°C, with the extremes in January and September. Salinity varied slightly, from 29.3 to 31.6 psu, with the lowest in September. DO in the water column ranged from 4.6 to 12.3 mg l^{-1} with an average of 10.2 ± 2.6 for June 2012, 6.3 ± 0.8 for September 2012, 9.8 ± 0.8 for April 2013, 9.1 ± 1.3 for July 2013, 10.5 ± 1.9 for January 2014, and 6.2 ± 0.4 mg l^{-1} for May 2015. DO in the surface water is usually comparable or slightly higher than at the bottom. Suspended particulate matter (SPM) in surface and bottom waters, respectively, was 18.6 ± 2.8 (mean ± SD) and 26.9 ± 13.9 mg l^{-1} for September 2012, 13.6 ± 9.1 and 14.9 ± 11.6 for April 2013, 11.3 ± 6.9 and 37.6 ± 26.8 for July 2013, 14.1 ± 5.7 and 27.6 ± 15.0 for October 2013, and 15.9 ± 14.1 and 13.5 ± 12.6 mg l^{-1} for January 2014. Obvious high bottom SPM was observed during the period from July to October. CH_4 concentrations in the water column ranged between 3.0 and 356 nM, and showed clear seasonal variation, with higher levels occurring in summer and autumn and lower levels in winter and early spring. CH_4 concentrations in autumn were comparable to those in summer and about 7- to 8-fold higher than those in winter and spring. Bottom CH_4 concentrations were higher than those at the surface during all cruises except in September 2012, during which surface CH_4 was 50% higher than in bottom water together with the lowest salinity among all cruises.

Geographical distributions of CH_4 in SGB

Fig. 3 shows the geographical distributions of temperature, salinity, and CH_4 in surface waters of SGB measured during this study. Two cruises were carried out in spring (April 2013 and May 2015), during which the kelp thrived and the long kelp enhanced frictional effects in the upper layers and influenced the water exchange between the bay and the Yellow

Table 1. Water temperature (°C), salinity (psu), and CH_4 concentrations (nM) in surface and bottom waters of Sanggou Bay, China, during 8 cruises. Mean ± SD

Date (yyyy-mm)	Station	Temperature Surface	Temperature Bottom	Salinity Surface	Salinity Bottom	Surface CH_4 Range	Surface CH_4 Average	Bottom CH_4 Range	Bottom CH_4 Average
2014-01	21	3.5 ± 1.1	3.1 ± 1.1	31.3 ± 0.1	31.3 ± 0.1	4.3–10.1	6.2 ± 1.4	4.3–19.4	8.7 ± 4.95
2013-04	20	7.7 ± 1.2	7.8 ± 1.2	31.6 ± 0.2	31.6 ± 0.2	5.2–10.1	7.2 ± 1.3	5.6–10.6	8.0 ± 1.5
2014-05	10	13.8 ± 2.2	14.1 ± 1.1	31.3 ± 0.4	31.3 ± 0.3	3.0–24.9	9.3 ± 7.9	3.0–29.5	12.5 ± 8.4
2015-05	21	14.2 ± 2.4	12.6 ± 1.8	31.5 ± 0.2	31.4 ± 0.1	5.4–43.4	22.6 ± 10.1	5.3–46.7	25.0 ± 11.0
2012-06	20	18.6 ± 3.2	17.3 ± 2.1	30.6 ± 0.3	30.4 ± 0.2	12.4–91.3	38.3 ± 21.9	5.8–98.4	43.9 ± 27.9
2013-07	27	21.3 ± 1.9	18.4 ± 0.8	29.8 ± 1.2	30.4 ± 0.1	10.8–82.1	53.0 ± 17.3	7.5–114	75.9 ± 23.4
2012-09	13	23.2 ± 0.9	22.3 ± 1.0	29.3 ± 1.0	29.8 ± 0.7	23.5–128	53.8 ± 33.5	21.7–136	35.5 ± 32.8
2013-10	21	18.6 ± 1.4	18.6 ± 0.9	30.0 ± 0.2	29.9 ± 0.1	21.5–302	63.8 ± 59.5	34.3–356	99.6 ± 81.4

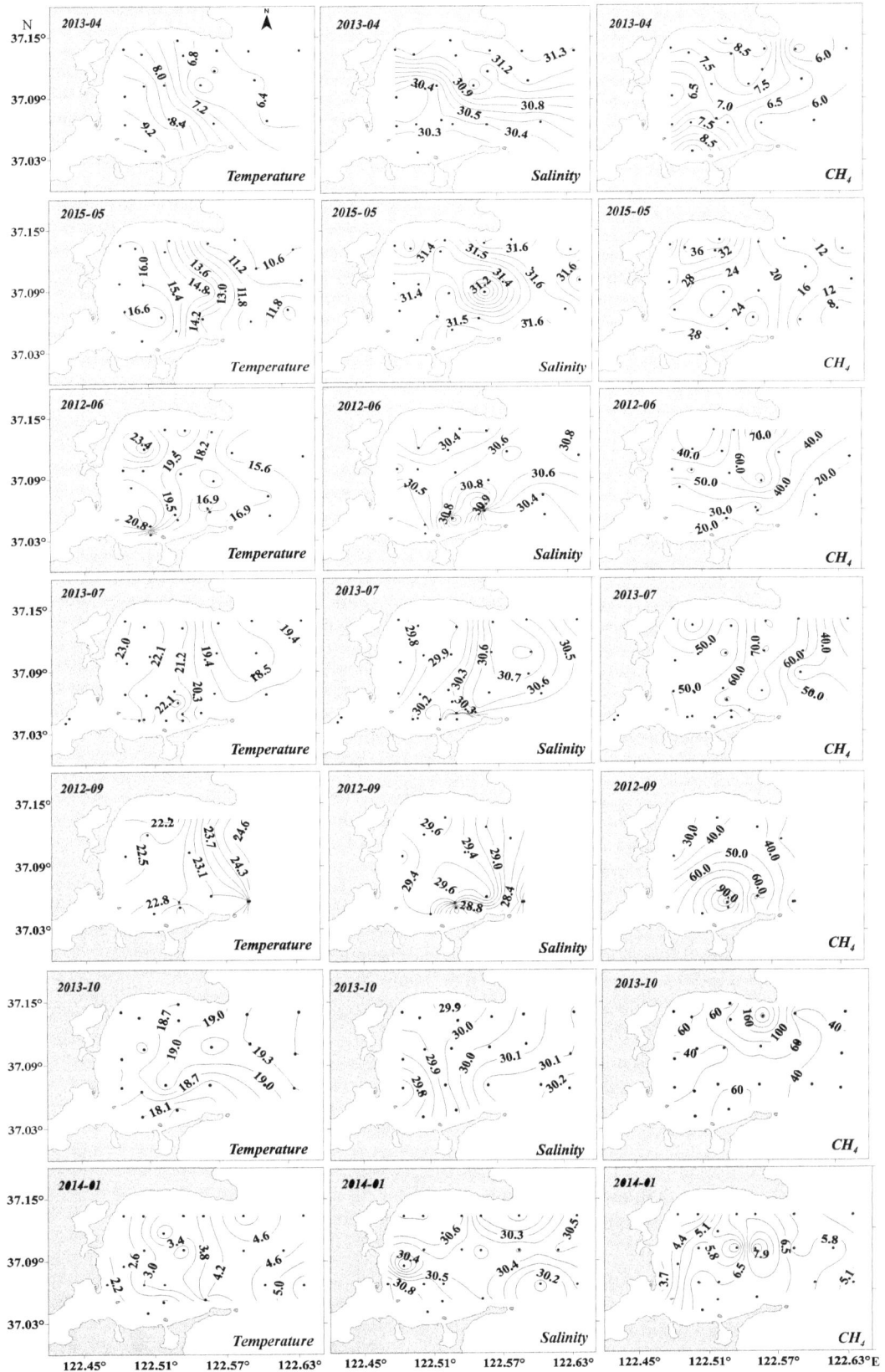

Fig. 3. Geographical distributions of water temperature (°C), salinity, and CH_4 concentration (nM) in surface waters of Sanggou Bay (SGB) during 2012 to 2015 (dates are yyyy-mm)

Sea (Zeng et al. 2015). Surface water temperature was higher in May than in April and presented a similar trend, which decreased gradually from nearshore to offshore, and showed an obvious gradient. In contrast, surface salinity increased from nearshore to offshore in May, while it decreased gradually from the northeast to the southwest in April. Dissolved CH_4 in April decreased gradually from the southwest and the northeast to the central bay, with concentrations ranging between 6 and 8 nM. In May, CH_4 concentrations in the inner bay were higher than those in outer bay, with highest dissolved CH_4 occurring in the northern part of the bay.

During summer cruises (June 2012 and July 2013), kelps had already been harvested, and water exchange with the Yellow Sea was not influenced by suspended kelp. Water temperature had a similar trend as in spring, i.e. decreasing from nearshore to offshore. Surface salinity varied in a narrow range (30.4–31.0) in June and increased slightly from the coast to the center of the bay, while in July, salinity increased gradually from the inner to the outer bay with low salinity (<30) in nearly half of the bay. CH_4 concentrations were much lower in June (mean ± SD: 38.3 ± 21.9 nM) than in July (53.0 ± 17.3 nM), and decreased gradually from the inner to the outer bay. In July, dissolved CH_4 concentrations increased from about 50 nM in the inner bay to >70 nM at the mouth of the bay, then decreased to <40 nM in the outer bay.

During autumn (September 2012 and October 2013), water temperature increased gradually from the inner to the outer bay, but the gradient was less pronounced. Salinity was relatively low (<30) compared to other seasons due to heavy rainfall and freshwater input. In September, low salinity (<29) together with high CH_4 (>90 nM) was observed in the southern part of the bay. However, CH_4 concentrations presented an opposite trend in October, with the highest value (>140 nM) measured in the northeastern part of the bay, while salinity increased gradually from inner to outer bay.

During winter (January 2014), the Bohai South Coast Current enters the bay from the north and flows out from the south through the west coast of the bay (Sun et al. 2007). Surface seawater temperature decreased gradually from the outer to the inner bay, while surface salinity (30.2–30.6) showed little variation over the whole bay. Dissolved CH_4 concentrations in January were the lowest during the whole year, with highest CH_4 (8 nM) occurring near the center of the bay. It then decreased rapidly seaward to <4 nM at the bay mouth.

Riverine and groundwater input

Rivers and groundwater are potential sources for dissolved CH_4 in SGB. CH_4 concentrations in the main rivers around the bay are shown in Fig. 4, which ranged from 123 to 2190 nM and were 1 to 2 orders of magnitude higher than those (3–356 nM) observed in the water column of the bay. Riverine CH_4 also presented obvious spatial and seasonal variations. For example, lowest CH_4 values usually occurred in winter, and the highest values occurred in summer and early fall for the smaller rivers (i.e. Sanggan, Shili, and Ba Rivers), while higher CH_4 values occurred in winter and summer, and lower CH_4 occurred in spring and late fall in the Gu River.

Considering that the runoff from the Sanggan, Shili, and Ba Rivers is limited and we lack discrete flow rate data for each river, we attributed 70% of the total runoff to the Gu River and 30% to the other rivers (Jiang et al. 2015). Riverine CH_4 flux to the SGB was estimated to be 1.4×10^5 mol yr^{-1}, using the average CH_4 concentrations in the Gu River (665 nM) and other small rivers (857 nM), and mean annual runoff (2.0×10^8 m^3).

Dissolved CH_4 concentrations in groundwater ranged from 1.6 to 405 nM and showed large spatial and temporal variations (Fig. 5). Groundwater near the mouth of the bay (GW4) had the highest CH_4 concentrations (26.2–255 nM, mean 98.6 nM), and Stn GW1 had medium values (6.7–109 nM, mean 40.7 nM), while those in the other areas usually had low CH_4 (<10 nM). For each region, lowest groundwater CH_4 usually occurred in winter (January) and early spring (April), while the highest values all occurred in late summer (September). In general, CH_4 concentrations in groundwater were much lower than those in rivers, but the submarine groundwater discharge to the SGB was ~50 times larger than river

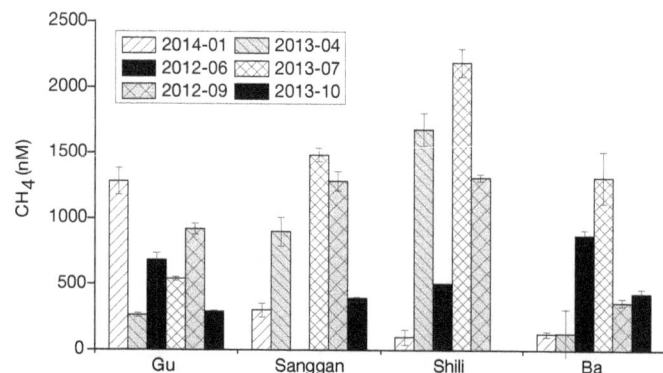

Fig. 4. CH_4 concentrations (mean ± SD) in different rivers flowing into Sanggou Bay in different seasons and years (yyyy-mm)

runoff, and estimated to be $(2.59-3.07) \times 10^7$ m^3 d^{-1} based on the non-conservative inventory of ^{226}Ra and ^{228}Ra in the water column (Wang et al. 2014). Hence, we estimated the CH$_4$ flux to the bay via the submarine groundwater by multiplying mean CH$_4$ concentration (40.2 nM) in the end-member well samples by the radium-derived mean submarine groundwater discharge (2.83×10^7 m^3 d^{-1}). It yielded a flux of 4.2×10^5 mol yr^{-1}, which was 3 times that of riverine CH$_4$ flux.

Sediment–water CH$_4$ fluxes

CH$_4$ fluxes across the sediment–water interface were measured at Stns MC, ST1, and ST2 (Fig. 6), among which Stns MC and ST1 were located in the kelp and bivalve polyculture zone, and Stn ST2 was located in the oyster monoculture zone. Sediment–water CH$_4$ fluxes from Stn ST1 ranged from 0.73 to 1.65 µmol m^{-2} d^{-1} with a mean ($\pm$SD) of 1.19 $\pm$

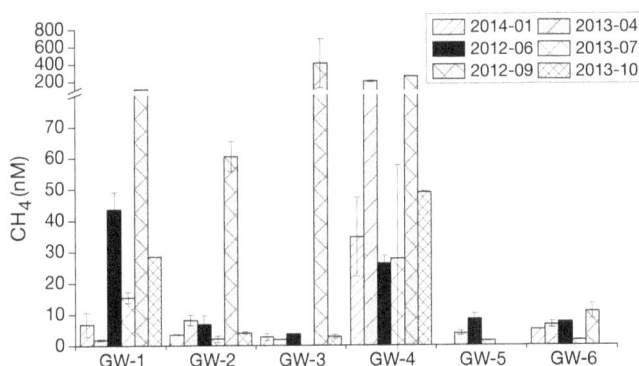

Fig. 5. CH$_4$ concentrations in groundwater measured at 6 stations (GW1–GW6) along the shoreline of Sanggou Bay in different seasons and years (yyyy-mm)

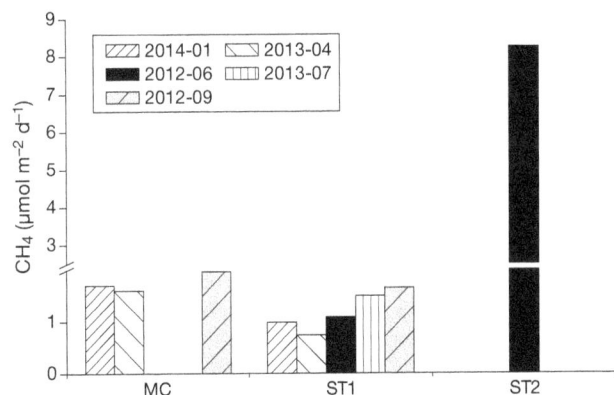

Fig. 6. Sediment–water CH$_4$ fluxes (µmol m^{-2} d^{-1}) at Stns MC, ST1, and ST2 in Sanggou Bay (see Fig. 2) in different seasons and years (yyyy-mm)

0.38 µmol m^{-2} d^{-1}, which showed obvious seasonal variation and correlated well with bottom water temperature ($F = 0.039T + 0.66$, n = 5, r^2 = 0.67). Sediment–water CH$_4$ fluxes from Stn MC ranged from 1.60 to 1.97 µmol m^{-2} d^{-1} with a mean of 1.76 $\pm$ 0.19 µmol m^{-2} d^{-1}. Sediment–water CH$_4$ fluxes from Stn ST1 (1.19 µmol m^{-2} d^{-1}) was lower than that from Stn ST2 (8.26 µmol m^{-2} d^{-1}) in June 2012. Although the sediment–water CH$_4$ flux from the oyster culture zone was measured only during 1 cruise, considering the seasonal variation trend in the polyculture zone, it is reasonable to deduct this flux as an annual average value. Based on the total surface area (about 144 km^2) (Zhang et al. 2009), average sediment–water CH$_4$ fluxes of 8.26 µmol m^{-2} d^{-1} (bivalve culture zone) and 1.48 µmol m^{-2} d^{-1} (polyculture zone), and assuming that the area ratio of the 2 culture zones is 1:1, annual CH$_4$ emission from sediments of SGB was estimated to be about 2.6×10^5 mol.

Water column methane production-oxidation

Net water column CH$_4$ production-oxidation rates (CH$_4$ formation-CH$_4$ oxidation) were estimated to be 0.25, 0.41, 0.19, 0.39, and 0.31 nM d^{-1} for Stn MC in April and July 2013 and January, May, and September 2014, which showed significant seasonal variation and correlated well with temperature ($R = 0.01T + 0.19$, n = 5, r^2 = 0.70). Net water column CH$_4$ production-oxidation rates were 0.28 and 0.67 nM d^{-1} for Stns ST2 and SG-3 in September 2014. If we take the mean value of 0.42 nM d^{-1} at these stations as the net CH$_4$ production rate in the water column, together with the area of 144 km^2 and a mean water depth of 7.5 m, total net CH$_4$ production-oxidation in the water column was estimated to be 1.7×10^5 mol yr^{-1}.

Surface CH$_4$ saturation and sea-to-air fluxes

CH$_4$ saturation in the surface waters of SGB ranged from 202 to 2734 % (Fig. 7), with great spatial and temporal variation. Average saturation was higher in autumn and summer than in spring and winter. In general, the surface waters of SGB were all over-saturated with CH$_4$, except for a few stations during winter. Thus, SGB is a net source of atmospheric CH$_4$.

Sea-to-air CH$_4$ fluxes calculated with the W2014 equation ranged from 2.1 to 123 µmol m^{-2} d^{-1} with a mean of 48.2 µmol m^{-2} d^{-1}, which was comparable to the results from the N2000 equation (2.3–126 µmol

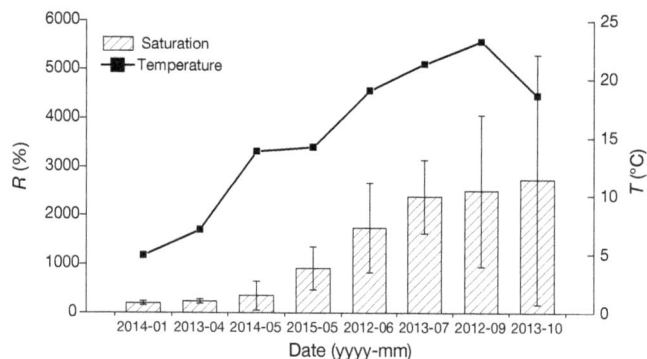

Fig. 7. CH$_4$ saturation (R, %; mean ± SD) and water temperature (T, °C) in surface waters of Sanggou Bay

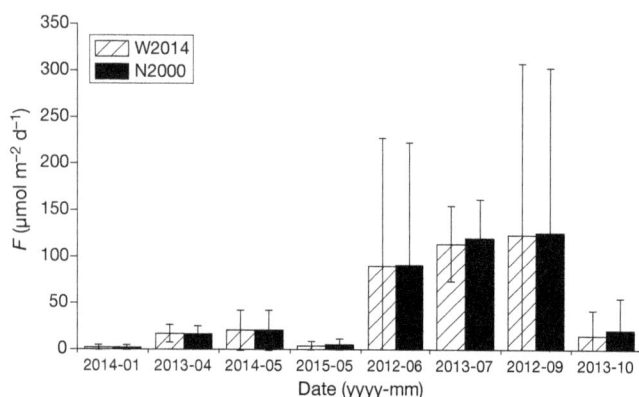

Fig. 8. Seasonal variation of air–sea CH$_4$ fluxes (F, µmol m^{-2} d^{-1}; mean ± SD) in Sanggou Bay, estimated using 2 different equations (W2014 and N2000); see 'Materials and methods' for details

m^{-2} d^{-1} with a mean of 50.3 µmol m^{-2} d^{-1}; Fig. 8). CH$_4$ fluxes showed clear seasonal variation, with those in early autumn (September) comparable to those in summer (July and June), and more than 8-fold higher than those in spring (April) and autumn (October); the lowest values occurred in winter. In addition, we estimated the CH$_4$ emission from SGB to be 2.5 × 10^6 mol yr^{-1} based on the annual mean atmospheric CH$_4$ flux (48.2 µmol m^{-2} d^{-1}) from this study and the area of SGB (144 km^2).

CH$_4$ exchange with the Yellow Sea

Considering that dissolved CH$_4$ in SGB was much higher than in the adjacent Yellow Sea, water exchange with the Yellow Sea should cause net loss of CH$_4$ from the bay. Aquaculture activities may influence water exchange between SGB and the Yellow Sea; however, suspended kelp culture mainly changed

the spatial pattern of the tidal flux but not the tidal prism (Zeng et al. 2015). Jiang et al. (2015) estimated the annual water exchange volume to be about 8.3 × 10^{10} m^3. CH$_4$ concentrations gradients between the inner and outer bay were 3.6 nM for September 2012, 1.5, 11.9, and 40.7 nM for April, July, and October 2013, and 0.9 nM for January 2014, respectively. During the kelp seeding to harvesting period (November to May), the CH$_4$ concentration gradient (mean: 1.2 nM) was lower than during the non-aquaculture period (mean: 18.7 nM). Hence, based on the annual mean observed CH$_4$ concentration gradient between the inner and outer bay (10.0 nM), and the annual water exchange volume of the bay (8.3 × 10^{10} m^3), we estimated the CH$_4$ flux exported out of the bay to be 8.8 × 10^5 mol yr^{-1}.

DISCUSSION

Comparison with previous research

Previous studies have reported CH$_4$ concentrations for some bays, estuaries, and coastal and shelf areas of northern China (Table 2). Average CH$_4$ concentrations in SGB during the corresponding seasons were lower than those reported for the adjacent Rushan Bay (mean ± SD: 59.9 ± 7.8 nM, May 2007; Wang et al. 2008), Dalian Bay (59.4 ± 82.0 nM, January 2010; 56.0 ± 69.4 nM, November 2009; Wang et al. 2011), and Jiaozhou Bay (137 ± 224 nM for August 2003 and 34.8 ± 75.5 for December 2003; Zhang et al. 2007), but obviously higher than the Yellow River estuary (3.9–14.3 nM; Gu et al. 2011) and Changjiang estuary (7.95 ± 5.24 nM; Zhang et al. 2008). CH$_4$ concentrations of SGB were 2- to 10-fold of those in the adjacent marginal shelf sea, i.e. the Yellow Sea (3.4–12.0 nM; Zhang et al. 2004, Yang et al. 2010, Ye et al. 2016).

The estimated atmospheric CH$_4$ fluxes from the SGB in this study were close to those from Jiaozhou Bay (Zhang et al. 2007) and Dalian Bay (Wang et al. 2011), and slightly higher than those from some estuaries, e.g. the Changjiang estuary (Zhang et al. 2008). However, CH$_4$ fluxes from SGB were far higher than those from shelf areas, e.g. 2-fold higher than those from the adjacent North Yellow Sea during Spring (April), and 5 times higher than during winter (January) and summer (July for SGB, August for North Yellow Sea), respectively, but comparable to those from the adjacent North Yellow Sea during October (Yang et al. 2010). Hence SGB is a hot spot of CH$_4$ emissions to the atmosphere.

Table 2. Compilation of surface concentrations and sea-to-air fluxes of CH_4 in different coastal areas of China (see Fig. 2 for station locations). Data are mean ± SD

Sea area	Date	Sampling station	CH_4 (nM)	Flux (μmol m^{-2} d^{-1})	Reference
Jiaozhou Bay	Aug 2003	16	137 ± 224	132 ± 220	Zhang et al. (2007)
	Dec 2003	14	34.8 ± 75.5	36.9 ± 87.3	Zhang et al. (2007)
Rushan Bay	May 2007	8	59.90 ± 7.75	–	Wang et al. (2008)
Dalian Bay	Jan 2010	17	59.41 ± 81.97	133.07 ± 193.37[a], 286.96 ± 416.98[b]	Wang et al. (2011)
	Nov 2009	17	56.01 ± 69.39	52.88 ± 68.42[a], 113.46 ± 146.82[b]	Wang et al. (2011)
Yellow River estuary	Jun 2009	28	3.9–14.3	7.2[a], 14.2[b]	Gu et al. (2011)
Changjiang estuary	Dec 2004	10	7.95 ± 5.24	21.1 ± 9.6[a], 41.1 ± 18.7[b]	Zhang et al. (2008)
East China Sea	Apr 2001	29	3.24 ± 0.59	1.63 ± 1.67[a], 2.77 ± 2.71[b]	Zhang et al. (2004)
	Aug 2013	65	6.26 ± 4.96	6.5 ± 7.4[a], 11.5 ± 111.9[b]	Ye et al. (2016)
Bohai Sea	Aug 2008	28	5.87 ± 2.02	3.1 ± 1.6[a], 8.1 ± 4.2[b]	Li et al. (2010)
North Yellow Sea	Jan 2007	78	3.40 ± 0.58	0.2 ± 1.0[a], 0.4 ± 1.7[b]	Yang et al. (2010)
	Aug 2007	76	6.43 ± 2.52	4.2 ± 4.7[a], 6.9 ± 7.3[b]	Yang et al. (2010)
	Apr 2007	59	5.70 ± 3.50	11.8 ± 10.2[a], 21.1 ± 16.4[b]	Yang et al. (2010)
	Oct 2006	80	12.02 ± 7.51	8.5 ± 12.7[a], 14.6 ± 22.3[b]	Yang et al. (2010)
Yellow Sea	Mar-Apr 2001	14	3.43 ± 0.23	0.81 ± 0.50[a], 1.33 ± 0.76[b]	Zhang et al. (2004)
Yellow and East China Seas	Aug 2011	38	8.21 ± 6.02	16.98 ± 21.40[c], 17.65 ± 21.72[d]	Sun et al. (2015)
	Oct 2011	55	5.03 ± 1.68	9.88 ± 9.97[c], 9.87 ± 9.61[d]	Sun et al. (2015)
	Dec 2011	59	4.07 ± 0.63	6.82 ± 6.86[c], 6.64 ± 6.44[d]	Sun et al. (2015)

[a]k_w was estimated by the LM86 equation (Liss & Herlivat 1986)
[b]k_w was estimated by the W92 equation (Wanninkhoft 1992)
[c]k_w was estimated by the W2014 equation (see 'Materials and methods' for details)
[d]k_w was estimated by the N2000 equation (see 'Materials and methods' for details)

Factors influencing spatial and temporal distribution of CH_4 in SGB

The concentration, saturation, water column production, and sediment–water fluxes of CH_4 in SGB all had obvious seasonal variation and were closely related to water temperature. Mean water column CH_4 concentrations and saturations during different cruises correlated positively with mean water temperature (CH_4 conc. = $2.33T - 4.70$, r^2 = 0.93, n = 11, p < 0.0001; R(CH_4) = $124.8T - 380.4$, r^2 = 0.8 n = 7, p < 0.008). This is consistent with the positive correlations observed for net water column CH_4 production-oxidation rates at Stn MC ($R = 0.01T + 0.19$, n = 5, r^2 = 0.82, p < 0.09) and sediment–water CH_4 fluxes at Stn ST1 ($F = 0.039T + 0.66$, n = 5, r^2 = 0.82, p < 0.09), which showed that CH_4 production rates in both water column and sediments increase with rising temperature. Temperature mainly controls the organic matter decomposition and the activity of methanogenesis, and the rising temperature may increase the relative abundance and diversity of methanogenic communities (Metje & Frenzel 2005, Høj et al. 2008). Yvon-Durocher et al. (2014) also reported seasonal variation in CH_4 emissions from diverse ecosystems using meta-analysis, and showed that CH_4 emissions increased significantly with seasonal increases in temperature. Hence our results suggest that water temperature plays a significant role in regulating the seasonal variation of CH_4 in SGB.

Although the salinity of SGB only showed a slight fluctuation throughout the year, mean water column CH_4 concentrations during different cruises correlated negatively with mean salinity (S) (CH_4 conc. = $-0.33S + 5.43$, r^2 = 0.94, n = 11, p < 0.001), suggesting that terrestrial input (i.e. rivers and groundwater) also play a role in the seasonal variation of CH_4 in the bay. In general, rivers and groundwater are primary routes for delivery of dissolved and particulate carbon and nutrients from land to coastal areas, and they are usually supersaturated with CH_4 (Taniguchi et al. 2002, Striegl et al. 2012). Observed CH_4 (123.3 to 2189.7 nM) in rivers around SGB are within the CH_4 range (5–5000 nM) reported for rivers worldwide (de Angelis & Lilley 1987, Upstill-Goddard et al. 2000) and much higher than those in the water column of the bay. Especially during the wet seasons (summer and early autumn), high discharges of river water and groundwater with rich CH_4 enter the bay, affecting its spatial distribution. For example, observed surface CH_4 in the bay was 50 % higher than that in bottom water in September 2012 together with the lowest salinity (Table 1), while CH_4 concentrations were usually higher at the bottom than at the surface during other cruises.

CH_4 distribution in SGB may also be influenced by aquaculture activities. The bay is extensively used for culture of macroalgae and shellfish. Previous studies showed that CH_4 can be produced in anaerobic microenvironments of SPM and digestive tracts of zooplankton and fish in oxygenic surface water (Marty 1993, Karl & Tilbrook 1994). Obvious high bottom SPM was observed during the shellfish culture period from July to October. Dense populations of bivalve shellfish (i.e. scallop and oyster) in shallow water can produce a large amount of feces and pseudo-feces, hence the digestive tract of shellfish and their waste provide favorable environments for potential water column methanogenesis. Suspended shellfish culture also accelerates biodeposition and results in sediments with rich organic matter and high microbial activity (Green et al. 2012), which in turn enhances rates of anaerobic decomposition of organic matter and lead to high CH_4 production in the sediment (Nizzoli et al. 2006, Jiang et al. 2015). Due to the sediment release, bottom CH_4 concentrations in SGB were usually higher than those at the surface, especially in the culture zones with bivalve shellfish in summer (Fig. 9). For example, we observed CH_4 concentrations at Stn ST2 in the oyster culture zone to be 4.0 nM at the surface (0 m), 3.8 nM at 2 m, 5.9 nM at 4 m, and 12.0 nM at the bottom (6 m) in May 2014. We also observed significant high sediment–water CH_4 fluxes in the oyster culture zone (8.26 μmol m^{-2} d^{-1}) compared to the polyculture zone (1.09 μmol m^{-2} d^{-1}) in June 2012. Green et al. (2012) observed greater CH_4 emissions from the sediment in areas that had the highest cover of oysters compared to areas with medium cover. They attributed this to more CH_4 produced by the reduction of CO_2 and the stimulation by the 'priming effect,' whereby the addition of fresh labile organic matter (such as from oyster biodeposits) temporarily stimulates microbial decomposition, including that of older, buried, recalcitrant organic matter (Green et al. 2012).

Recent research has shown that under certain nutrient-limited conditions, a variety of methyl-rich organic phosphorus or sulfur compounds are likely to be utilized by microorganism and serve as precursors of CH_4 production in aerobic surface waters (Damm et al. 2008, Karl et al. 2008, Zindler et al. 2013). Integrated multi-trophic aquaculture in SGB can enhance the recycling of organic matter and nutrients and provide favorable conditions for aerobic CH_4 production. Bivalve shellfish ingest algae, POM, bac-

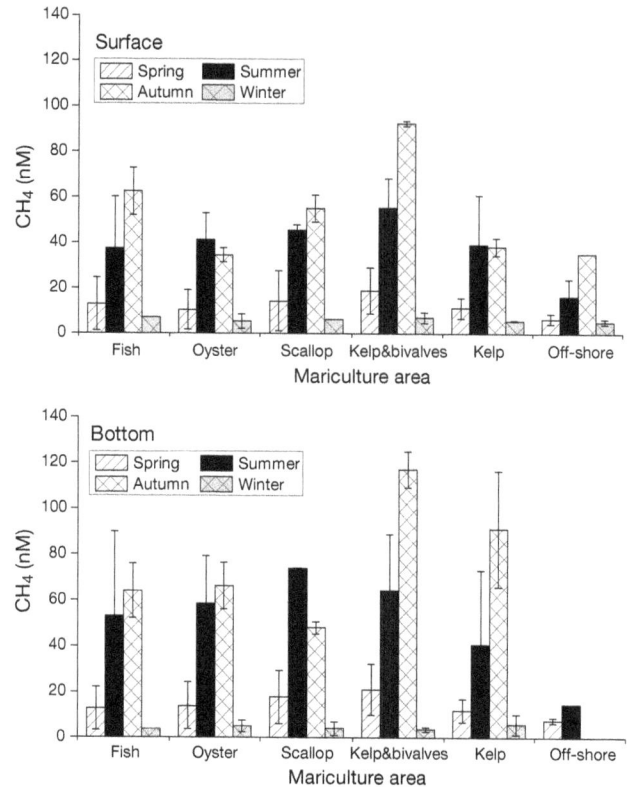

Fig. 9. Comparison of seasonal mean (±SD) CH_4 concentrations in surface and bottom waters in different culture areas of Sanggou Bay

teria, and other micro-organisms, while kelp absorbs organic and inorganic wastes from the shellfish (Fang et al. 1996). Commonly cultivated seaweeds (*Gracilaria lemaneiformis* and *Laminaria japonica*) in SGB have a high nutrient uptake efficiency (Mao et al. 2009, Xu et al. 2011) and may lead to nitrogen and phosphorus limitation in spring as well as phosphorus limitation in summer (Zhang et al. 2010), which in turn implies a potential formation of CH_4 from methyl-rich organic phosphorus or sulfur compounds (Damm et al. 2008, Karl et al. 2008, Zindler et al. 2013). DMSP can be produced by macroalgae, and grazing by bivalves appears to facilitate the release of DMSP from kelp (Smit et al. 2007). Hence the water column in polyculture zones may contain higher levels of DMSP and enhance the production of CH_4. Seasonal variation in mean CH_4 in different mariculture areas of SGB support this hypothesis and show that higher surface CH_4 concentrations usually occur in the polyculture areas of kelp and bivalve shellfish (Fig. 9). We also observed a highly significant and rapid increase in CH_4 during incubations with DMSP spikes in October 2013 (Fig. 10). During the incubations, CH_4 concentration increased sharply by more than 60-fold and reached 460 nM on Day 3.5, then decreased

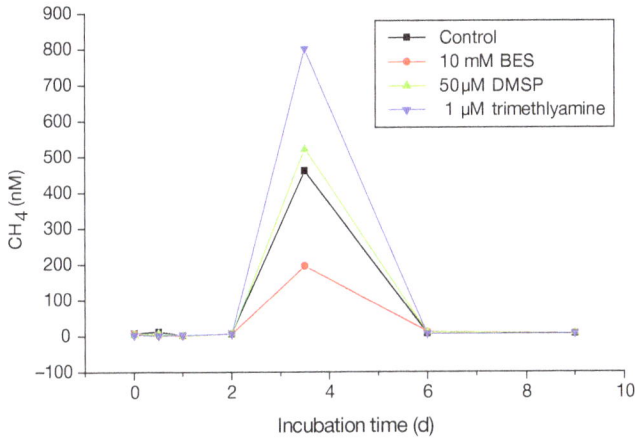

Fig. 10. CH_4 time series incubation experiments at Station MC (see Fig. 2) during October 2013. BES: 2-bromoethane sulfonic acid, DMSP: dimethylsulfoniopropionate

sharply to the initial concentration around 7 nM. The addition of 10 µM BES reduced the CH_4 production by more than half, while the addition of 50 µM DMSP and 1 µM trimethylamine significantly enhanced CH_4 production. During all of these incubations, DO in the bottles ranged from 5 to 8 mg l^{-1}, suggesting that CH_4 might be produced under aerobic conditions with the degradation of methylated compounds such as DMSP and trimethylamine in the water column in SGB.

Preliminary CH_4 budget and its implication

In order to understand the contributions of different sources and sinks to dissolved CH_4 in SGB, a preliminary CH_4 budget was constructed, although there are still great uncertainties in the estimate of each term. Considering that the sea–air flux values from W2014 and N2000 were quite similar, we took the results estimated by W2014 for the budget estimation. From the budget (Fig. 11), we can see that the groundwater input (4.2×10^5 mol yr^{-1}) was the largest quantified CH_4 source, followed by sediment

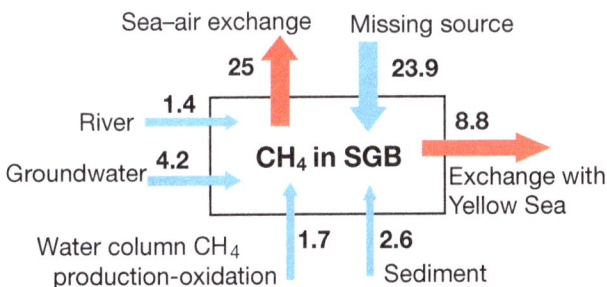

Fig. 11. Preliminary CH_4 budget (10^5 mol yr^{-1}) estimation for Sanggou Bay (SGB)

release (2.6×10^5 mol yr^{-1}) and riverine input (1.4×10^5 mol yr^{-1}), while sea-to-air release (2.5×10^6 mol yr^{-1}) and export from the bay to the Yellow Sea (8.8×10^5 mol yr^{-1}) were the dominant sinks for CH_4 in SGB. Net water column production-oxidation was estimated preliminarily to produce 1.7×10^5 mol CH_4 yr^{-1}. However, there is still a large imbalance between the sources and sinks of methane in the water column of SGB, with an apparent missing source of 2.4×10^6 mol yr^{-1} needed to balance the budget, although this value might be overestimated due to the propagation of the errors in the other terms in the budget.

Because of the large spatial and temporal variations in CH_4 concentrations in the groundwater samples, our ability to provide an accurate estimate for the CH_4 flux via submarine groundwater discharge is rather limited due to the small number of groundwater end-member samples used in the calculation (n = 6) and lack of seasonal variation in groundwater fluxes. Hence the source item of groundwater may have large uncertainties.

Previous studies have demonstrated that sediments are a significant CH_4 source for bays and coastal waters (Sansone et al. 1998, Ferrón et al. 2010). Ferrón et al. (2010) found that benthic CH_4 fluxes from the shelf of the Gulf of Cádiz ranged from 0.5 to 24.1 µmol m^{-2} d^{-1} with an average of 5 ± 6 µmol m^{-2} d^{-1} using benthic chambers. Sansone et al. (1998) reported that CH_4 benthic fluxes from Tomales Bay ranged from 0.4 to 16 µmol m^{-2} d^{-1} with an average of 5.5 and 2.5 µmol m^{-2} d^{-1} for summer and winter, respectively. These results are comparable to or slightly higher than our results (0.7–8.3 µmol m^{-2} d^{-1}) for SGB. However, given the high rates of labile organic matter loading in SGB, this strongly indicates that the fluxes assessed during the sediment incubations in this study might be underestimated. The most influential factors for this underestimation are likely to be an insufficient number of sampling stations and the *ex situ* sediment incubation method we used. Sediment–water CH_4 fluxes from SGB showed large spatial variation, although we only measured benthic fluxes at 3 stations and did not measure benthic fluxes from eelgrass beds in the southern region of SGB near Chudao. Previous studies showed that CH_4 benthic fluxes from eelgrass beds may be about 10-fold higher than those from unvegetated areas in Tomales Bay (Sansone et al. 1998). Due to lack of a benthic chamber, the emission of CH_4 from sediments was measured by a modified closed chamber incubation method (Barnes & Owens 1999). This method changes the environment (i.e. pressure and temperature) and cannot simulate real *in situ* conditions such as resuspension, deposit feeding, burrowing, and irrigation, which often significantly change the geochemical characteristics of sediments and overlying water and increase the benthic fluxes (Sansone et al. 1998, Upstill-Goddard et al. 2000). Hence, an underestimation of the CH_4 released from the sediments may account for part of the missing source.

Underestimating *in situ* water column production is also likely to contribute to the missing source. As discussed above, CH_4 might be produced in aerobic water columns of the bay with the degradation of methylated compounds such as DMSP produced from the macroalgae, and grazing by bivalves appeared to facilitate the release of DMSP from kelp (Smit et al. 2007). However, the time series incubation method employed in this study only focuses on the microbial activity in the water column itself and neglects the interaction of algae, shellfish, and microbes, which may result in great underestimation of potential contributions from *in situ* production in the water column. If the *in situ* production in the water column is indeed the only missing source for CH_4 budget in the bay, the net water column production-oxidation rate is estimated to be about 7 nM d^{-1}. This is reasonable based on our incubation results in October 2013, which showed a net water column CH_4 increase of ~40 nM d^{-1} with a DMSP spike. However, our incubation results also showed that CH_4 produced might be oxidized rapidly. Hence it is difficult to evaluate how much CH_4 is accumulated in the water column over time. Considering the complicated interactions between macroalgae, shellfish, and microbes, mesocosm experiments should be carried out in the future to further understand the *in situ* water column CH_4 production and consumption and to understand the CH_4 budget in multi-trophic aquaculture systems like that in SGB.

CONCLUSIONS

CH_4 concentrations in SGB showed obvious seasonal and spatial variation. CH_4 concentrations were 3 to 10 times higher in summer and autumn than in spring and winter. Bottom CH_4 concentrations were obviously higher than those in the surface water due to sediment release. Higher surface CH_4 concentrations occurred in the polyculture areas of kelp and bivalves. Seasonal variation in water temperature, terrestrial freshwater input, and aquaculture activities play significant roles in regulating the spatial and temporal variation of CH_4 in the bay. Ground-

water input (4.2×10^5 mol yr^{-1}) was the largest quantified source of CH_4, followed by sediment release (2.6×10^5 mol yr^{-1}), and riverine input (1.4×10^5 mol yr^{-1}), while sea-to-air release (2.5×10^6 mol yr^{-1}) and export from the bay to the Yellow Sea (8.8×10^5 mol yr^{-1}) were the dominant CH_4 sinks. Net water column production-oxidation was estimated preliminarily to produce 1.7×10^5 mol CH_4 yr^{-1}; however, this value may have been underestimated due to the neglect of interactions between algae, shellfish, and microbes. There was a great imbalance of sources and sinks, with an apparent missing source of 2.4×10^6 mol yr^{-1}, most of which might be attributed to underestimates of *in situ* water column production and CH_4 released from the sediments. Benthic chamber measurements and mesocosm experiments should be carried out in the future to further understand the CH_4 budget in multi-trophic aquaculture systems like that in SGB.

Acknowledgements. We thank Professors Jianguang Fang and Zengjie Jiang from the Yellow Sea Fisheries Research Institute, and colleagues from the Laboratory of Marine Biogeochemistry, Ocean University of China, for their assistances in field sample collections. This study was funded by the Ministry of Science and Technology of China through Grant no. 2011CB409802, supported by the National Science Foundation of China through Grant no. 41521064, and by the 111 Project (B13030).

LITERATURE CITED

ä Aller RC, Mackin JE, Ullman WJ, Wang CH and others (1985) Early chemical diagenesis, sediment-water solute exchange, and storage of reactive organic matter near the mouth of the Changjiang, East China Sea. Cont Shelf Res 4:227–251

ä Bange HW, Bartell U, Rapsomanikis S, Andreae MO (1994) Methane in the Baltic and North Seas and a reassessment of the marine emissions of methane. Global Biogeochem Cycles 8:465–480, doi:10.1029/94GB02181

ä Bange HW, Dahlke S, Ramesh R, Meyer-Reil LA, Rapsomanikis S, Andreae M (1998) Seasonal study of methane and nitrous oxide in the coastal waters of the Southern Baltic Sea. Estuar Coast Shelf Sci 47:807–817

ä Barnes J, Owens N (1999) Denitrification and nitrous oxide concentrations in the Humber estuary, UK, and adjacent coastal zones. Mar Pollut Bull 37:247–260

ä Bates TS, Kelly KC, Johnson JE, Gammon RH (1996) A reevaluation of the open ocean source of methane to the atmosphere. J Geophys Res 101:6953–6961

ä Bryan J, Rlley J, Williams PL (1976) A Winkler procedure for making precise measurements of oxygen concentration for productivity and related studies. J Exp Mar Biol Ecol 21:191–197

ä Canet C, Prol-Ledesma RM, Dando PR, Vázquez-Figueroa V and others (2010) Discovery of massive seafloor gas seepage along the Wagner Fault, northern Gulf of California. Sediment Geol 228:292–303

Chinese Gulf Compilation Committee (1991) Chinese Gulf,

Vol 3. Ocean Press, Beijing

ä Cicerone RJ, Oremland RS (1988) Biogeochemical aspects of atmospheric methane. Global Biogeochem Cycles 2: 299–327

ä Damm E, Kiene R, Schwarz J, Falck E, Dieckmann G (2008) Methane cycling in Arctic shelf water and its relationship with phytoplankton biomass and DMSP. Mar Chem 109: 45–59

ä de Angelis MA, Lilley MD (1987) Methane in surface waters of Oregon estuaries and rivers. Limnol Oceanogr 32: 716–722

EPA (Environmental Protection Agency) (2010) Methane and nitrous oxide emissions from natural sources. http:// nepis.epa.gov/

Fang J, Kuang S, Sun H, Li F, Zhang A, Wang X, Tang T (1996) Mariculture status and optimising measurements for the culture of scallop *Chlamys farreri* and kelp *Laminaria japonica* in Sanggou Bay. Mar Fish Res 17:95–102

FAO (Food and Agriculture Organization of the United Nations) (2007) Fisheries and Aquaculture Information and Statistics Service. http://www.fao.org

ä Ferrón S, Ortega T, Gómez-Parra A, Forja J (2007) Seasonal study of dissolved CH_4, CO_2 and N_2O in a shallow tidal system of the Bay of Cádiz (SW Spain). J Mar Syst 66: 244–257

ä Ferrón S, Ortega T, Forja JM (2010) Temporal and spatial variability of methane in the north-eastern shelf of the Gulf of Cádiz (SW Iberian Peninsula). J Sea Res 64: 213–223

ä Green DS, Boots B, Crowe TP (2012) Effects of non-indigenous oysters on microbial diversity and ecosystem functioning. PLOS ONE 7:e48410

Gu PP, Zhang GL, Li PP, Han Y, Zhao YC (2011) Effect of water–sediment regulation on dissolved methane in the lower Yellow River estuary and it's [sic] adjacent marine area. China Environ Sci 31:1821–1828

Guo X, Ford SE, Zhang F (1999) Molluscan aquaculture in China. J Shellfish Res 18:19–31

ä Høj L, Olsen RA, Torsvik VL (2008) Effects of temperature on the diversity and community structure of known methanogenic groups and other archaea in high Arctic peat. ISME J 2:37–48

Hornafius JS, Quigley D, Luyendyk BP (1999) The world's most spectacular marine hydrocarbon seeps (Coal Oil Point, Santa Barbara Channel, California): quantification of emissions. J Geophys Res 104:20703–20711

IPCC (Intergovernmental Panel on Climate Change) (2013) Climate change 2013: the physical science basis. Cambridge University Press, New York, NY

ä Jiang Z, Li J, Qiao X, Wang G and others (2015) The budget of dissolved inorganic carbon in the shellfish and seaweed integrated mariculture area of Sanggou Bay, Shandong, China. Aquaculture 446:167–174

ä Karl DM, Tilbrook BD (1994) Production and transport of methane in oceanic particulate organic matter. Nature 368:732–734

ä Karl DM, Beversdorf L, Björkman KM, Church MJ, Martinez A, Delong EF (2008) Aerobic production of methane in the sea. Nat Geosci 1:473–478

ä Kelley CA, Martens CS, Chanton JP (1990) Variations in sedimentary carbon remineralization rates in the White Oak River estuary, North Carolina. Limnol Oceanogr 35: 372–383

ä Lashof DA, Ahuja DR (1990) Relative contributions of greenhouse gas emissions to global warming. Nature 344: 529–531

Li P, Zhang G, Zhao Y, Liu S (2010) Study on distributions and flux of methane dissolved in the Bohai Sea in summer. Adv Mar Sci 28:478–487

Liss PS, Merlivat L (1986) Air-sea gas exchange rates: introduction and synthesis. In: Buat-Ménard P (ed) The role of air-sea exchange in geochemical cycling. NATO ASI Ser 185:113–127

ä Mao Y, Zhou Y, Yang H, Wang R (2006) Seasonal variation in metabolism of cultured Pacific oyster, *Crassostrea gigas*, in Sanggou Bay, China. Aquaculture 253:322–333

ä Mao Y, Yang H, Zhou Y, Ye N, Fang J (2009) Potential of the seaweed *Gracilaria lemaneiformis* for integrated multi-trophic aquaculture with scallop *Chlamys farreri* in North China. J Appl Phycol 21:649–656

▶ Martens CS, Klump JV (1980) Biogeochemical cycling in an organic-rich coastal marine basin—I. Methane sediment-water exchange processes. Geochim Cosmochim Acta 44:471–490

ä Marty DG (1993) Methanogenic bacteria in seawater. Limnol Oceanogr 38:452–456

ä Mau S, Valentine DL, Clark JF, Reed J, Camilli R, Washburn L (2007) Dissolved methane distributions and air sea flux in the plume of a massive seep field, Coal Oil Point, California. Geophys Res Lett 34:L22603

▶ Metje M, Frenzel P (2005) Effect of temperature on anaerobic ethanol oxidation and methanogenesis in acidic peat from a northern wetland. Appl Environ Microbiol 71: 8191–8200

ä Nightingale PD, Malin G, Law CS, Watson AJ and others (2000) In situ evaluation of air-sea gas exchange parameterizations using novel conservative and volatile tracers. Global Biogeochem Cycles 14:373–387

ä Nizzoli D, Welsh DT, Fano EA, Viaroli P (2006) Impact of clam and mussel farming on benthic metabolism and nitrogen cycling, with emphasis on nitrate reduction pathways. Mar Ecol Prog Ser 315:151–165

ä Sansone FJ, Rust TM, Smith SV (1998) Methane distribution and cycling in Tomales Bay, California. Estuaries 21: 66–77

ä Shi H, Zheng W, Zhang X, Zhu M, Ding D (2013) Ecological–economic assessment of monoculture and integrated multi-trophic aquaculture in Sanggou Bay of China. Aquaculture 410-411:172–178

▶ Smit AJ, Robertson-Andersson DV, Peall S, Bolton JJ (2007) Dimethylsulfoniopropionate (DMSP) accumulation in abalone *Haliotis midae* (Mollusca: Prosobranchia) after consumption of various diets, and consequences for aquaculture. Aquaculture 269:377–389

ä Song G, Liu S, Zhu Z, Zhai W, Zhu C, Zhang J (2016) Sediment oxygen consumption and benthic organic carbon mineralization on the continental shelves of the East China Sea and the Yellow Sea. Deep-Sea Res II 124: 53–63

▶ Striegl RG, Dornblaser M, McDonald C, Rover J, Stets E (2012) Carbon dioxide and methane emissions from the Yukon River system. Global Biogeochem Cycles 26: GB0E05

ä Sun MS, Zhang GL, Cao XP, Mao XY, Li J, Ye WW (2015) Methane distribution, flux, and budget in the East China Sea and Yellow Sea. Biogeosci Discuss 12:7017–7053

Sun P, Zhang Z, Hao L, Wang B and others (2007) Analysis of nutrient distributions and potential eutrophication in seawater of the Sanggou Bay. Adv Mar Sci 25:436–445

▶ Taniguchi M, Burnett WC, Cable JE, Turner JV (2002) Investigation of submarine groundwater discharge. Hydrol Processes 16:2115–2129

Upstill Goddard RC, Barnes J, Frost T, Punshon S, Owens NJ (2000) Methane in the southern North Sea: low salinity inputs, estuarine removal, and atmospheric flux. Global Biogeochem Cycles 14:1205–1217

Wang J, Qu K, Xu Y, Shan B (2008) The distribution and releasing rate of methane in sediment of culture area in Rushan Bay. Mar Fish Res 29:101–107

Wang Q, Guan D, Li M, Li D, Wang J, Yao Z, Zhao H (2011) Distribution and atmospheric fluxes of CO_2, CH_4 and N_2O in surface water of Dalian Bay. Mar Environ Sci 30: 398–403

► Wang X, Du J, Ji T, Wen T, Liu S, Zhang J (2014) An estimation of nutrient fluxes via submarine groundwater discharge into the Sanggou Bay—a typical multi-species culture ecosystem in China. Mar Chem 167:113–122

► Wanninkhof R (1992) Relationship between wind speed and gas exchange over the ocean. J Geophys Res 97: 7373–7382, doi:10.1029/92JC00188

► Wanninkhof R (2014) Relationship between wind speed and gas exchange over the ocean revisited. Limnol Oceanogr Methods 12:351–362

► Wiesenburg DA, Guinasso NL Jr (1979) Equilibrium solubilities of methane, carbon monoxide, and hydrogen in water and sea water. J Chem Eng Data 24:356–360

► Xu D, Gao Z, Zhang X, Qi Z, Meng C, Zhuang Z, Ye N (2011) Evaluation of the potential role of the macroalga *Laminaria japonica* for alleviating coastal eutrophication. Bioresour Technol 102:9912–9918

► Yang J, Zhang GL, Zheng LX, Zhang F, Zhao J (2010) Seasonal variation of fluxes and distributions of dissolved methane in the North Yellow Sea. Cont Shelf Res 30: 187–192

► Ye W, Zhang G, Zhu Z, Huang D, Han Y, Wang L, Sun M

(2016) Methane distribution and sea-to-air flux in the East China Sea during the summer of 2013: impact of hypoxia. Deep-Sea Res II 124:74–83

► Yvon-Durocher G, Allen AP, Bastviken D, Conrad R and others (2014) Methane fluxes show consistent temperature dependence across microbial to ecosystem scales. Nature 507:488–491

► Zeng D, Huang D, Qiao X, He Y, Zhang T (2015) Effect of suspended kelp culture on water exchange as estimated by in situ current measurement in Sanggou Bay, China. J Mar Syst 149:14–24

Zhang G, Zhang J, Kang Y, Liu S (2004) Distributions and fluxes of methane in the East China Sea and the Yellow Sea in spring. J Geophys Res 109:C07011

► Zhang GL, Zhang J, Xu J, Ren JL, Liu SM (2007) Distributions, land-source input and atmospheric fluxes of methane in Jiaozhou Bay. Water Air Soil Pollut Focus 7: 645–654

► Zhang G, Zhang J, Liu S, Ren J, Xu J, Zhang F (2008) Methane in the Changjiang (Yangtze River) estuary and its adjacent marine area: riverine input, sediment release and atmospheric fluxes. Biogeochemistry 91:71–84

► Zhang J, Hansen PK, Fang J, Wang W, Jiang Z (2009) Assessment of the local environmental impact of intensive marine shellfish and seaweed farming—application of the MOM system in the Sungo Bay, China. Aquaculture 287:304–310

► Zhang J, Jiang Z, Wang W (2010) Seasonal distribution and variation of nutrients and nutrients limitation in Sanggou bay. Prog Fish Sci 31:16–25

► Zindler C, Bracher A, Marandino CA, Taylor B, Torrecilla E, Kock A, Bange HW (2013) Sulphur compounds, methane, and phytoplankton: interactions along a north-south transit in the western Pacific Ocean. Biogeosciences 10: 3297–3311

10

Parameterisation and application of dynamic energy budget model to sea cucumber *Apostichopus japonicus*

Jeffrey S. Ren[1], Jeanie Stenton-Dozey[1], Jihong Zhang[2,3,*]

[1]National Institute of Water and Atmospheric Research, 10 Kyle Street, PO Box 8602, Christchurch 8440, New Zealand
[2]Yellow Sea Fisheries Research Institute, Chinese Academy of Fishery Sciences, 106 Nanjing Road, Qingdao 266071, PR China
[3]Function Laboratory for Marine Fisheries Science and Food Production Processes,
Qingdao National Laboratory for Marine Science and Technology, 1 Wenhai Road, Aoshanwei, Jimo, Qingdao 266200, PR China

ABSTRACT: The sea cucumber *Apostichopus japonicus* is an important aquaculture species in China. As global interest in sustainable aquaculture grows, the species has increasingly been used for co-culture in integrated multitrophic aquaculture (IMTA). To provide a basis for optimising stocking density in IMTA systems, we parameterised and validated a standard dynamic energy budget (DEB) model for the sea cucumber. The covariation method was used to estimate parameters of the model with the DEBtool package. The method is based on minimisation of the weighted sum of squared deviation for datasets and model predictions in one single-step procedure. Implementation of the package requires meaningful initial values of parameters, which were estimated using non-linear regression. Parameterisation of the model suggested that the accuracy of the lower (T_L) and upper (T_H) boundaries of tolerance temperatures are particularly important, as these would trigger the unique behaviour of the sea cucumber for hibernation and aestivation. After parameterisation, the model was validated with datasets from a shellfish aquaculture environment in which sea cucumbers were co-cultured with the scallop *Chlamys farreri* and Pacific oyster *Crassostrea gigas* at various combinations of density. The model was also applied to a land-based pond culture environment where the sea cucumber underwent a fast growth period in spring and non-growth periods during winter hibernation and summer aestivation. Application of the model to datasets showed that the model is capable of simulating the physiological behaviour of the sea cucumber and responds adequately to the wide range of environmental and culture conditions.

KEY WORDS: Sea cucumber · Dynamic energy budget model · DEB model · Parameterisation · Covariation method · Application

INTRODUCTION

The sea cucumber *Apostichopus japonicus*, a commercially valuable species, is commonly found in shallow intertidal habitats off the west Pacific Ocean. The vertical distribution ranges from intertidal areas down to depths of 20–30 m (Chen 1990). Culture of the species started in the early 1980s after success of the technique in commercial hatcheries and seed

production in China (Liao 1997). Since then, sea cucumber farming has become an important part of mariculture and the scale of the industry has increased dramatically (Chen 2003, Zhang et al. 2015).

Sea cucumbers are surface-deposit feeders. They are not only commercially important, but also play an important role in coastal ecosystems. They ingest and assimilate a large amount of organic matter deposited on the seafloor prior to extensive microbial pro-

*Corresponding author: zhangjh@ysfri.ac.cn

cessing. Sea cucumbers have often been called the earthworms of the sea, because they are responsible for the extensive shifting and mixing of the substrate and recycling of detrital matter (Lauerman et al. 1997, Miller et al. 2000, Yang et al. 2005). They grind sedimentary organic matter into finer particles, turning over the top layers of sediment in lagoons, reefs and other habitats, and allowing the penetration of oxygen (Lauerman et al. 1997). Sea cucumbers are also important in determining habitat structure for other species and can represent a substantial portion of the ecosystem biomass (Sibuet & Segonzac 1985).

The ability to turn biodeposition into valuable flesh product and being ecologically complementary with many aquaculture species make the sea cucumber an important species for co-cultures in integrated multitrophic aquaculture (IMTA) (e.g. Zhou et al. 2006, Slater et al. 2009, Zamora et al. 2014, Yuan et al. 2015). Over the past decade, experimental and commercial trials have been conducted to co-culture it with other species to improve water quality and mitigate environmental impacts (e.g. Li et al. 2001, Yang et al. 2001, Yu et al. 2014). These studies have shown an increase in system productivity and improvement of environmental conditions. Despite promising results, these experiments were based on the traditional technique of trial and error and experimentation, and the right proportion of stocking density for each co-cultured species could not be satisfactorily estimated. Maximum ecological and economic benefits of IMTA systems can only be achieved with the optimal density at which species are physiologically and ecologically complementary (Chopin et al. 2008, Barrington et al. 2009). The growth rate of the sea cucumber depends on the rates of biodeposition from other species within an IMTA system where the stocking density of co-cultured species can be optimised by means of mathematical models. Recently, a model was developed to optimise production in IMTA systems (Ren et al. 2012). A dynamic energy budget (DEB) model of the sea cucumber is one of the sub-models of the larger IMTA model.

Although DEB modelling has gained increasing popularity, most modelling development has been focused on shellfish to improve shellfish aquaculture (e.g. Ren & Schiel 2008, Rosland et al. 2009, Alunno-Bruscia et al. 2011), while little attention has been paid to deposit-feeders. The objective of this study is to develop a DEB model of the sea cucumber for subsequent application in IMTA systems. We followed a standard DEB model and the main focus of the study was on its parameterisation and validation.

MATERIALS AND METHODS

Model structure and parameterisation

A DEB model is based on a generic theory that biological organisms share common physiological behaviours (Kooijman 2010). The same functional structure could therefore be applied to different species with species-specific parameter values. Although some effort has been made to extend the model by coping with varying food quantity and quality (Saraiva et al. 2012), the standard form of the DEB model has been widely adopted to study energetics and growth of aquaculture species, particularly bivalves (e.g. Ren & Schiel 2008, Bourlès et al. 2009, Rosland et al. 2009). Our primary purpose was to develop and validate a simple model of the sea cucumber, which can be incorporated into a larger IMTA model for optimising production of IMTA. Therefore, a standard DEB model was ideal for this purpose. The model structure, including functional responses and energy allocation, has been described in detail (van der Meer 2006, Ren & Schiel 2008), and we only focus on parameterisation here. The same symbols were used as those in the standard DEB model.

Estimation of model parameters depends on the quantity and quality of the experimental data. However, comprehensive data are not usually available for most species. For some species with sufficient datasets, the estimated values of parameters are often inconsistent between datasets due to variation in experimental conditions (Saraiva et al. 2011). To account for this problem, van der Veer et al. (2006) suggested a standard procedure using all available datasets by means of weighted non-linear regression. Recently, a covariation method has been developed to estimate parameters of a standard DEB model. It is based on the simultaneous minimisation of the weighted sum of squared deviations between datasets and predictions in one single-step procedure (Lika et al. 2011). For the present modelling exercise, we adopt this method to estimate parameters of the sea cucumber model using the DEBtool package (www.bio.vu.nl/thb/deb/deblab/). Implementation of the method requires 2 types of datasets including real data and pseudo-data. The real data consist of zero-variate and univariate data, which are from actual observations of the sea cucumber at known temperatures and food conditions. The univariate data of length versus age are shown in Fig. 1. The data reflects the mean growth of the species from post-settlement at the natural environment. Lack of length at early stages in univariate data may

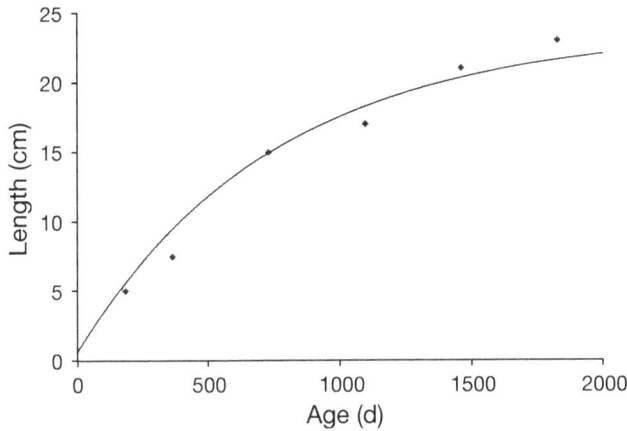

Fig. 1. Simulation of sea cucumber (*Apostichopus japonicus*) length using the primary parameters of the dynamic energy budget (DEB) model (line). Dots are observations and the line is the fitted growth curve

not considerably compromise the estimation of final parameter values, because zero-variate data consists of information on length at birth. The pseudo-data are a set of values of primary or compound parameters usually for a generalised animal obtained from a large collection of estimated parameters for a wide variety of species (Kooijman 2010, Saraiva et al. 2011). The concept of pseudo-data is used to avoid an unrealistic combination of parameters and to maintain the rules for covariation of the parameter values implied by physical laws (Saraiva et al. 2011). The pseudo-data include parameters for the Arrhenius temperature (T_A), volume-specific somatic maintenance coefficient $[p_M]$ and shape coefficient (δ_M). Because the initial values determine whether the results are meaningful, these values should be as close to 'real' as possible. For the present study, we estimated the parameter values from experimental data for *Apostichopus japonicus*, rather than generalised animal parameter values. A procedure proposed by van der Meer (2006) was followed to estimate initial values of these parameters. DEB theory assumes that all rates (e.g. assimilation, catabolic and maintenance rates) can be described by an Arrhenius relationship using a single Arrhenius temperature within a species-specific tolerance range. The estimation of parameters in the Arrhenius equation (T_A, T_{AL}, T_{AH}, T_L and T_H; see Table 2 for definitions) was based on data from physiological experiments within a range of temperatures (Li et al. 2002). The volume-specific somatic maintenance coefficient is the ratio of the costs of maintenance to structural volume synthesis. It was indirectly estimated from changes in energy content during aestivation over time (Zhou et al. 2006).

The shape coefficient is used for conversion between length and biovolume. It was estimated from the relationship between length and wet mass by assuming that the density of the biovolume equals 1 g cm^{-3}. The DEB assumes that the energy allocation follows the κ-rule, which assumes that a fixed proportion κ of the energy from the reserves goes to maintenance and growth. The remaining fraction $1-\kappa$ goes to development and reproduction. The initial value for the fraction of energy utilisation rate spent on maintenance plus growth (κ) can be determined from reproduction information, i.e. the maximum gonadal mass fraction relative to other tissue (van der Veer et al. 2006). Based on reproduction information (Chen 1990), the value was estimated following van der Veer et al. (2006). The use of an area-specific assimilation rate, $\{p_{Am}\}$, is no longer needed with the covariation method, because it is derived from some additional parameters. For comparison, we still estimate its value with this method. The real data and pseudo-data are listed in Table 1.

There is still a lack of data for the initial values of a few parameters including maturity at birth (E_{Hb}), maturity at metamorphosis (E_{Hj}), maturity at puberty (E_{Hp}) and Weibull aging acceleration (h_a). Following the methodology of Lika et al. (2011), default values were initially used and subsequently tuned until all parameters could be determined by observational data. It should be noted that the Gompertz stress coefficient (s_G) could be particularly important for the sea cucumber, because this species would undergo winter hibernation and summer aestivation, and may also eviscerate when an external stress occurs. This would cause drastic consequences on its energetics. The Gompertz stress coefficient controls the mean lifespan. Increasing the value decreases the mean lifespan, but increases survival at young age. According to Kooijman (2010), negative values can occur if damage-inducing compounds can be degraded. The sea cucumber would fall in this category and hence a negative value is assigned for s_G. The final values of the parameters are listed in Table 2.

Validation datasets

The model was validated using a few datasets collected from growth experiments of the sea cucumber. The first dataset was from a co-culture experiment to determine the growth performance of the sea cucumber with shellfish (Zhou et al. 2006). In this experiment, sea cucumbers were co-cultured with either scallops *Chlamys farreri* or oysters *Crassostrea*

Table 1. Data used in parameter estimation procedure. gWW: grams of wet weight

Symbol	Value	Unit	Definition	Source
Zero-variate				
a_b	8	d	Age at birth (20°C)	Qiu et al. (2015)
a_p	300	d	Age at puberty (20°C)	Yamana et al. (2010)
a_m	7×365	d	Lifespan	Sun (2013)
L_b	0.08	cm	Length at birth	Ito & Kitamura (1998), Qiu et al. (2015)
L_p	7.5	cm	Length at puberty	Hamano et al. (1989), Ito & Kitamura (1998)
L_i	29.7	cm	Maximum length	Yamana et al. (2010), Hamano et al. (1989)
W_b	$0.064L_b^3$	gWW	Wet weight at birth	Sun (2013)
W_p	$0.064L_p^3$	gWW	Wet weight at puberty	Sun (2013)
W_i	$0.064L_i^3$	gWW	Ultimate weight	Sun (2013)
Pseudo-data				
κ	0.85	–	Allocation fraction	Chen (1990)
κ_G	0.95	–	Growth efficiency	Liu et al. (1996)
$\dot{v}$	0.02	cm	Energy conductance	Kooijman (2010)
$[\dot{p}_M]$	4.8	J cm^{-3} d^{-1}	Volume-specific somatic maintenance	Liu et al. (1996)
δ_m	0.28	–	Shape coefficient	Sun (2013)
T_A	7300	K	Arrhenius temperature	Li et al. (2002)
$\{\dot{p}_{Am}\}$	154	J cm^{-3} d^{-1}	Maximum area-specific assimilation rate	Zhou et al. (2006)
Univariate				
Growth data		yr vs. cm	Time vs. length	Sun (2013), Qiu et al. (2015)

gigas in the shellfish farm of Sishili Bay (37° 32′ N, 120° 30′ E) off Shandong Peninsula, China. A few different combinations of co-culture practices were used.

For the present study, we chose 2 datasets from co-cultures of sea cucumbers with scallops and oysters, respectively. For co-culture of sea cucumber with scallops, 10 sea cucumbers and 350 scallops were placed in each lantern net. The initial average body weight of the sea cucumber was 56.9 g and the scal-

lop shell length was 2.3 cm. For co-culture of sea cucumbers with oysters, the animals were cultured at a density of 10 sea cucumbers with 120 oysters. The initial average body weight of sea cucumbers was 39 g and the oyster length 7.7 cm. The experiment lasted ~8 mo. The body weights of the sea cucumbers were measured monthly for the first 3 mo and at the end of the experiment (see details in Zhou et al. 2006). Temperature was not recorded during the experiment. Because the variation of water tempera-

Table 2. Final dynamic energy budget (DEB) model parameters for sea cucumber *Apostichopus japonicus* and other parameters

Symbol	Value	Unit	Definition	Source
κ	0.85	–	Catabolic flux to growth and maintenance	Chen (1990)
κ_R	0.8	–	Reproduction efficiency	This study
$\{\dot{p}_{Am}\}$	129.5	J cm^{-3} d^{-1}	Area-specific maximum assimilation rate	This study
$[\dot{p}_M]$	12.1	J cm^{-3} d^{-1}	Volume-specific maintenance rate	This study
$[E_m]$	4300	J cm^{-3}	Maximum storage density	This study
$[E_G]$	1700	J cm^{-3}	Volume-specific costs for structure	This study
T_A	5500	K	Arrhenius temperature	Li et al. (2002)
T_{AL}	70 000	K	Rate of decrease at lower boundary	Yang et al. (2005)
T_{AH}	35 500	K	Rate of decrease at upper boundary	Yang et al. (2005)
T_L	273	K	Lower boundary of tolerance range	Zhang et al. (2004), Yang et al. (2005)
T_H	290	K	Upper boundary of tolerance range	Liu et al. (1996), Yang et al. (2005)
δ_m	0.28	–	Shape coefficient	Sun (2013)
$\dot{v}$	0.03	cm d^{-1}	Energy conductance	This study
E_{Hb}	4.7	J	Maturity at birth	This study
E_{Hp}	7200	J	Maturity at puberty	This study
E_{Hj}	5.1	J	Maturity at metamorphosis	This study
h_a	1.4×10^{-7}	J	Weibull aging acceleration	This study
s_G	−0.2	–	Gompertz stress coefficient	This study

ture in the bay follows an annual pattern, the annual cycle data were used for the model simulation.

The second dataset was obtained from a land-based growth experiment near Qingdao off Shandong Peninsula, China (Yu & Song 1999). The experiment lasted 1 yr. Juveniles with an average body weight of ~2.5 g were transplanted from a hatchery to a culture pond. The animals were fed with artificial food made of algae meal, but feeding was suspended during the aestivation period in summer. Body weight and water temperature were measured monthly.

SIMULATIONS AND RESULTS

Temperature and food density are forcing variables used to run the model. There was significant difference in water temperature between the experiments (ANOVA, $p < 0.01$). It was considerably higher in the land-based experiment than in the sea-based ones from February through August. This pattern reversed during the rest of the year (Fig. 2).

The model was applied to the first dataset. The optimal fit of the model to the observation resulted in f-values of 0.51 and 0.62, respectively under co-culture practices with oysters and scallops (Fig. 3). (Because the functional response (f) describes the relationship between food uptake and density, f-value is an indicator of food availability). This suggested that food in the co-culture with scallops was more abundant than with oysters. The daily growth rates were 1.3 and 1.6% for co-culture with oysters and scallops, respectively. The model simulations appropriately reflected differences in environmental conditions, particularly food availability.

The model was also applied to the growth of the sea cucumber in the land-based culture (2nd dataset). The model with the fixed f-value of 0.59 would result in prediction being consistent with the observation (Fig. 4), which indicates that the model can reasonably reproduce the physiological behaviour of the sea cucumber. Little growth occurred during winter and summer months, but the growth was high during the spring months when the temperature was optimal for growth. When temperature was low in winter, animals considerably reduced feeding and hibernation may have occurred. Similarly, when water temperature rises from mid-June, aestivation was induced and animals stopped feeding until September when temperature decreases to the growth range. Both situations resulted in ceasing growth of structural volume and the energy requirement for mainte-

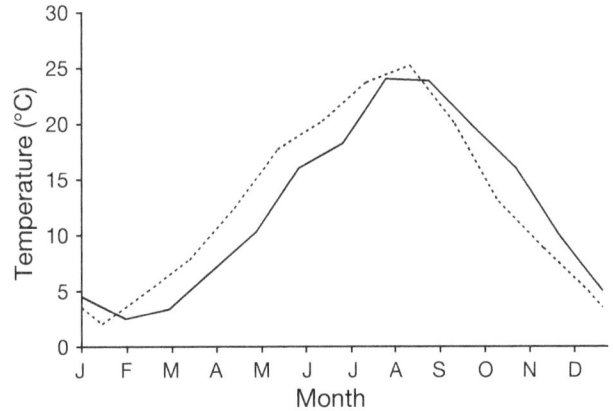

Fig. 2. Annual variation of temperatures in Sishili Bay (solid line) and culture pond (dotted line) off Shandong Peninsula, China

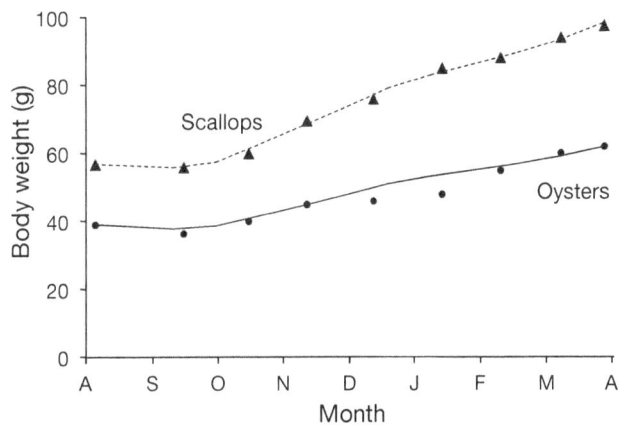

Fig. 3. Comparison between the observed (dots and triangles) and simulated (solid and dashed lines) growth of sea cucumber *Apostichopus japonicus* in the co-culture environment during August 2000–April 2001. Solid and dashed lines are the modelled body weight from co-cultures with oysters *Crassostrea gigas* and scallops *Chlamys farreri*, respectively

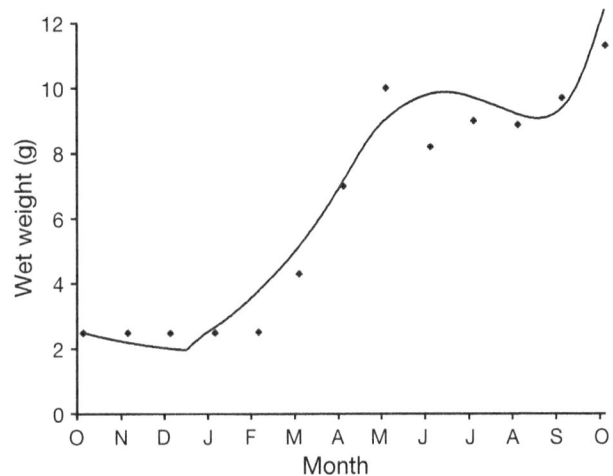

Fig. 4. Comparison between observed (dots) and simulated (line) growth of sea cucumber *Apostichopus japonicus* from a land-based culture system during October 1981–October 1982

nance is met from reserves, which caused the decline of body weight due to this net negative energy flow. Overall, food availability is similar to that in the scallop co-culture environment, but slightly higher than in the oyster co-culture system.

DISCUSSION

The present modelling exercise aimed to parameterise and validate a standard DEB model of the sea cucumber for subsequent application in IMTA ecosystems. The model parameters were mainly estimated using the covariation method. Application of the model to data proved successful, as the model is capable of reproducing the observed patterns of growth in different environmental and culture conditions, including co-culture with oysters and scallops, and land-based pond culture.

In parameterisation of the DEB model, there are a few noticeable differences of some parameter values estimated between the covariation method and weighted non-linear regression. The maximum area-specific assimilation, $\{\dot{p}_{Am}\}$, is lower, and volume-specific somatic maintenance, $[\dot{p}_M]$, is higher than the initial values. This suggests that the ability of the sea cucumber to capture food is low but the animal spends more energy for somatic maintenance than those from experimental data. Similarly, the reserve capacity $[E_m]$ from the covariation method is higher than the initial value. These differences may largely be due to variations in experimental conditions, which magnify the estimation of parameter values. In addition, parameter values with a poorer fit may have much more physical realism, because observations cannot include all possible aspects of energy budgets (Lika et al. 2011). For example, the maximum reserve $[E_m]$ cannot directly be measured and is usually estimated from starvation experiments, but such measurement depends on the initial condition of an organism and experimental conditions over the starvation period. In addition, the sea cucumber is extremely sensitive to changes in environmental conditions, particularly temperature, which can induce atypical behaviours including evisceration, hibernation and aestivation. Therefore, given sufficient information on the required data, the final parameter values estimated from the covariation method would appropriately reflect the behaviour of the species.

The use of the covariation method in parameter estimation is a considerable improvement over previous ones. It is based on minimisation of the weighted sum of squared deviation for datasets and model predictions in one single-step procedure (Lika et al. 2011). It also includes physiological constraints by introducing pseudo-data to reduce deviation of parameters related to variations in experimental datasets. Development of the covariation method would help increase the application of the DEB model. A standard DEB model provides a framework for functional structure which is applicable to different species with species-specific parameter values, while the covariation method offers a general technique for parameterisation of the DEB model. However, implementation of the method requires completeness of zero-variate data including age, length and weight at birth and puberty with temperature information, and lifespan, ultimate total length and weight. These data may not be always available in the literature for many species. It also requires a consistent choice of the data and a careful establishment of assumptions (Saraiva et al. 2011), which is subject to physiological and biological knowledge of the studied organisms. Therefore, this method is suitable for species with sufficient information on both zero-variate data and pseudo-data, in which it would estimate the optimal parameters for a DEB model to best predict observations. However, it may potentially compromise the accuracy of parameters for species without these comprehensive datasets.

Temperature is particularly imperative in the physiological processes of the sea cucumber due to the unique characteristic of hibernation and aestivation. Because it is difficult to estimate parameters related to the Arrhenius relationship using the covariation method, these parameters were estimated following a standard procedure by van der Meer (2006). Although the sea cucumber can survive in a wide range of temperatures between −1.5 and 30.5°C in a short period (Zhang et al. 2004, Yang et al. 2005), the optimal range for growth is 10–15°C (Yu & Song 1999). Temperatures beyond the optimal range would cause dramatic declines in physiological activities. Therefore, it is essential to obtain accurate information for parameters governing both lower and upper boundaries. Unfortunately, the published data have shown great variability (Liu et al. 1996, Li et al. 2002, Yang et al. 2005). A physiological measurement in the laboratory has recorded that the maximum feeding rate occurs at 20°C, but was still reasonably high at 25°C (Yang et al. 2005). However, a field experiment has indicated that the feeding rate of 2 yr old individuals decreased considerably from 19°C, and all individuals stopped feeding when the temperature reached 22°C (Liu et al. 1996). Yu & Song (1999) reported that the suitable temperature for growth is within 5–17°C.

This discrepancy may be related to experimental conditions and/or physiological differences between the juvenile and adult. Adults are less tolerant and therefore the triggering temperature for aestivation is lower than that for juveniles (Yu & Song 1999, Yang et al. 2005). Despite evidence of size-dependent tolerance temperature, available information is not sufficient to extend the DEB model by incorporating a functional response. Furthermore, the feeding rate seems a quicker process than that of respiration at the upper boundary in the sea cucumber. A small increase in temperature may result in individuals completely stopping feeding, but continuing respiration (e.g. Liu et al. 1996). The rate of decrease in oxygen consumption at the upper boundary is considerably lower than that of the feeding rate (Li et al. 2002). The difference in T_H values for feeding and respiration has been reported in *Crassostrea gigas* (Bourlès et al. 2009). Although there is some evidence in the rate of changes between feeding and maintenance in the sea cucumber, the available information is insufficient to be incorporated into the model in the present study. Further experiments are required to investigate whether this discrepancy is caused by the existence of different Arrhenius relationships or experimental artefacts.

When applying the model, it is noted that the development of eggs is not part of a standard DEB model and needs to be tested separately. As the extension of a standard DEB model was beyond the scope of this study, we did not include the topic of egg development. And the parameters for egg development may not be correctly estimated with the covariation method either. Additionally, existing datasets are insufficient to take energy content of eggs into consideration when doing the parameterisation, and so further experiments are required to gather information for such purpose.

Despite economic and ecological benefits of sea cucumber–shellfish co-culture (Yang et al. 2000, Yuan et al. 2015), optimal ratios of the sea cucumber and shellfish have been investigated through trial and error and experimentation (e.g. Zhou et al. 2006, Yu et al. 2014). This traditional method is not only costly but also inaccurate. For the present simulation dataset, the ratio of sea cucumbers to scallops and oysters was 1:35 and 1:12, respectively (Zhou et al. 2006). The size of scallops (3.0 cm shell length) was smaller than oysters (7.7 cm shell length). Although it was not possible to calculate food availability for sea cucumbers from both cultures, the model simulation has shown that the scaled functional responses were $f = 0.62$ and 0.51, respectively, for co-culture with scallops and oysters. This suggests that the experimental design has resulted in more biodeposits in co-culture with scallops than with oysters. Based on the growth rate, Zhou et al. (2006) concluded that the maximum density in each lantern net is 34 scallops. However, our modelling result indicated that 34 individuals would still result in approximately two-thirds of the maximum food density for sea cucumber growth. Similarly, the growth of the sea cucumber co-cultured with the scallop *Chlamys farreri* at a density of 1 individual m^{-2} was reported to be 60% faster than those at a density of 2 individuals m^{-2} (Yang et al. 2001), but the optimal ratio could not be determined with the experimental result. Because the optimal ratio depends on dynamics of environmental conditions and energetics of co-cultured species, full assessment and design of co-culture practices can only be determined using an IMTA model (Ren et al. 2012). The development of the sea cucumber DEB model has provided a tool for growth predictions of the sea cucumber in various environmental conditions and a basis for optimising stocking density of IMTA operations.

Acknowledgements. The study was supported by the New Zealand Foundation of Research, Science and Technology (Contract Number C01X0513), The National Science & Technology Pillar Program of China (2011BAD13B06) and the National Natural Science Foundation of China (41076111; 41276172).

LITERATURE CITED

Alunno-Bruscia M, Bourlès Y, Maurer D, Robert S and others (2011) A single bio-energetics growth and reproduction model for the oyster *Crassostrea gigas* in six Atlantic ecosystems. J Sea Res 66:340–348

Barrington K, Chopin T, Robinson S (2009) Integrated multitrophic aquaculture (IMTA) in marine temperate waters. In: Soto D (ed) Integrated mariculture: a global review. FAO Fish Aquacult Tech Pap 529. FAO, Rome, p 7–46

Bourlès Y, Alunno-Bruscia M, Pouvreau S, Tollu G and others (2009) Modelling growth and reproduction of the Pacific oyster *Crassostrea gigas*: advances in the oyster-DEB model through application to a coastal pond. J Sea Res 62:62–71

Chen J (1990) Sea cucumber culture in China. In: Bueno P, Lovatelli A (eds) Brief introduction to mariculture of five selected species in China. FAO Corporate Document Repository, Rome

Chen J (2003) Overview of sea cucumber farming and sea ranching practices in China. SPC Beche-de-mer Inf Bull 18:18–23

Chopin T, Robinson SMC, Troell M, Neori A, Fang J (2008) Multitrophic integration for sustainable marine aquaculture. In: Jørgensen SE, Fath BD (eds) The encyclopedia of ecology, ecological engineering, Vol 3. Elsevier, Oxford, p 2463–2475

Hamano T, Amio M, Hayashi K (1989) Population dynamics

of *Stichopus japonicas* Selenka (Holothuroidea, Echinodermata) in an intertidal zone and on the adjacent subtidal bottom with artificial reefs for Sargassum. Suisanzoshoku 37:179–186

Ito S, Kitamura H (1998) Technical development in seed production of the Japanese sea cucumber, *Stichopus japonicus*. SPC Beche-de-mer Inf Bull 10:24–28

Kooijman SALM (2010) Dynamic energy budget theory for metabolic organisation. Cambridge University Press, Cambridge

Lauerman LML, Smoak JM, Shaw TJ, Moore WS, Smith KL (1997) ^{234}Th and ^{210}Pb evidence for rapid ingestion of settling particles by mobile epibenthic megafauna in the abyssal NE Pacific. Limnol Oceanogr 42:589–595

Li SK, Wang LC, Gao YG (2001) Co-culture of sea cucumbers and abalones in intertidal ponds. Fisheries Modernisation 2:16–18

Li B, Yang H, Zhang T, Zhou Y, Zhang C (2002) Effect of temperature on respiration and excretion of sea cucumber *Apostichopus japonicus*. Oceanol Limnol Sin 33: 182–187

Liao YL (1997) Fauna Sinica, Phylum Echinodermata, Class Holothuroidea. Science Press, Beijing

Lika K, Kearney MR, Freitas V, van der Veer HW and others (2011) The 'covariation method' for estimating the parameters of the standard dynamic energy budget model I: philosophy and approach. J Sea Res 66:270–277

Liu Y, Li F, Song B, Sun H, Zhang X, Gu B (1996) Study on aestivating habit of sea cucumber *Apostichopus japonicas* Selenka: ecological characteristics of aestivation. J Fish Sci China 3:41–48

Miller RJ, Smith CR, DeMaster DJ, Fornes WL (2000) Feeding selectivity and rapid particle processing by deep-sea megafaunal deposit feeders: a ^{234}Th tracer approach. J Mar Res 58:653–673

Qiu T, Zhang T, Hamel JF, Mercier A (2015) Development, settlement and post-settlement growth. In: Yang H, Hamel JF, Mercier A (eds) The sea cucumber *Apostichopus japonicus*: history, biology and aquaculture. Academic Press, London, p 111–130

Ren JS, Schiel DR (2008) A dynamic energy budget model: parameterisation and application to the Pacific oyster *Crassostrea gigas* in New Zealand waters. J Exp Mar Biol Ecol 361:42–48

Ren JS, Stenton-Dozey J, Plew DR, Fang J, Gall M (2012) An ecosystem model for optimising production in integrated multitrophic aquaculture systems. Ecol Modell 246:34–46

Rosland R, Strand Ø, Alunno-Bruscia M, Bacher C, Strohmeier T (2009) Applying dynamic energy budget (DEB) theory to simulate growth and bio-energetics of blue mussels under low seston conditions. J Sea Res 62:49–61

Saraiva S, van der Meer J, Kooijman SALM, Sousa T (2011) DEB parameters estimation for *Mytilus edulis*. J Sea Res 66:289–296

Saraiva S, van der Meer J, Kooijman SALM, Witbaard R, Philippart CJH, Hippler D, Parker R (2012) Validation of a dynamic energy budget (DEB) model for the blue mussel *Mytilus edulis*. Mar Ecol Prog Ser 463:141–158

Sibuet M, Segonzac M (1985) Abondance et répartition de l'épifaune mégabenthique. In: Laubier L, Monniot C (eds) Peuplements profonds du Golfe de Gascogne (Campagnes BIOGAS). IFREMER, Brest, p 143–156

Slater MJ, Jeffs AG, Carton AG (2009) The use of the waste from green-lipped mussels as a food source for juvenile sea cucumber *Australostichopus mollis*. Aquaculture 292:219–224

Sun G (2013) Shandong local standards – sea cucumber *Apostichopus japonicus*. www.docin.com/p-798281216.html (in Chinese)

van der Meer J (2006) An introduction to dynamic energy budget (DEB) models with special emphasis on parameter estimation. J Sea Res 56:85–102

van der Veer HW, Cardoso JFMF, van der Meer J (2006) The estimation of DEB parameters for various North Atlantic bivalve species. J Sea Res 56:107–124

Yamana Y, Hamano T, Goshima S (2010) Natural growth of juveniles of the sea cucumber *Apostichopus japonicus*: studying juveniles in the intertidal habitat in Hirao Bay, eastern Yamaguchi Prefecture, Japan. Fish Sci 76: 585–593

Yang H, Wang J, Zhou Y, Zhang T, Wang P, He W (2000) Comparison of efficiencies of different culture systems in the shallow sea along Yantai. J Fish Sci China 24: 140–145

Yang H, Zhou Y, Wang J, Zhang T, Wang P, He Y, Zhang F (2001) A modelling estimation of carrying capacities for *Chlamys farreri*, *Laminaria japonica* and *Apostichopus japonicus* in Sishiliwan Bay, Yantai, China. J Fish Sci China 7:27–31

Yang H, Yuan X, Zhou Y, Mao Y, Zhang T, Liu Y (2005) Effects of body size and water temperature on food consumption and growth in the sea cucumber *Apostichopus japonicus* (Selenka) with special reference to aestivation. Aquacult Res 36:1085–1092

Yu D, Song B (1999) Variation of survival rates and growth characteristics of pond cultural juvenile *Apostichopus japonicus*. J Fish Sci China 6:109–110

Yu Z, Zhou Y, Yang H, Hu C (2014) Bottom culture of the sea cucumber *Apostichopus japonicas* Selenka (Echinodermata: Holothuroidea) in a fish farm, Southern China. Aquacult Res 45:1434–1441

Yuan X, Zhou Y, Mao Y (2015) *Apostichopus japonicas*: a key species in integrated polyculture systems. In: Yang H, Hamel JF, Mercier A (eds) The sea cucumber *Apostichopus japonicus*: history, biology and aquaculture. Academic Press, London, p 323–332

Zamora LN, Dollimore J, Jeffs AG (2014) Feasibility of co-culture of the Australasian sea cucumber (*Australostichopus mollis*) with the Pacific oyster (*Crassostrea gigas*) in northern New Zealand. NZ J Mar Freshw Res 48: 394–404

Zhang T, Zhou W, Song Z (2004) Key technique in cultivation of sea cucumber (*Apostichopus japonicus*) using artificial reef with high yield and high benefit. Hebei Fisheries 3:71–75

Zhang L, Song X, Hamel JF, Mercier A (2015) Aquaculture, stock enhancement and restocking. In: Yang H, Hamel JF, Mercier A (eds) The sea cucumber *Apostichopus japonicus*: history, biology and aquaculture. Academic Press, London, p 289–322

Zhou Y, Yang H, Liu S, Yuan X and others (2006) Feeding and growth on bivalve biodeposits by the deposit feeder *Stichopus japonicas* Selenka (Echinodermata: Holothuroidea) co-cultured in lantern nets. Aquaculture 256:510–520

11

Modelling long-term recruitment patterns of blue mussels *Mytilus galloprovincialis*: a biofouling pest of green-lipped mussel aquaculture in New Zealand

Javier Atalah[1,*], Hayden Rabel[2], Barrie M. Forrest[1]

[1]Cawthron Institute, Private Bag 2, Nelson 7010, New Zealand
[2]Statistics New Zealand, Private Bag 4741, Christchurch 8140, New Zealand

ABSTRACT: The green-lipped mussel *Perna canaliculus* forms the cornerstone of the New Zealand aquaculture industry. Like shellfish farming globally, *P. canaliculus* aquaculture is susceptible to the detrimental effects of biofouling. One of the greatest ongoing threats results from the recruitment of blue mussels *Mytilus galloprovincialis* onto cultured *P. canaliculus* and ropes used for collecting or on-growing juvenile stock. Using a dataset spanning 40 yr, we modelled spatio-temporal patterns in *M. galloprovincialis* recruitment in relation to a suite of potential explanatory variables and tested the ability of the best model to forecast recruitment. Our goal was to identify locations (sites and water depths) and times of peak *M. galloprovincialis* abundances, to enable the industry to implement management practices to minimise operational risks. Despite large inter-annual and spatial variability in recruitment patterns, our analyses revealed an upward trend in the abundance of *M. galloprovincialis* over the last 2 decades. Generally, seasonal patterns in abundance showed a large recruitment peak in October and a smaller one in April. There was a strong negative effect of depth, and weak effects of sea surface temperature and Southern Oscillation Index (in the previous month) on *M. galloprovincialis* abundances. The best model fitted the data well ($R^2 = 0.72$); however, it provided only moderate forecasting power ($R^2 = 0.16$), highlighting the challenges in forecasting recruitment. These results have been incorporated into a web application that enables aquaculture companies to interactively investigate historic trends in *M. galloprovincialis* and forecast abundances in the month ahead.

KEY WORDS: Aquaculture · *Perna canaliculus* · Biofouling · Invasive species · Integrated nested Laplace approximations · Over-settlement · Spatio-temporal models

INTRODUCTION

Biofouling, the accumulation of organisms on submerged surfaces, is a significant operational problem for marine shellfish aquaculture globally (Adams et al. 2011, Fitridge et al. 2012). Suspended shellfish culture systems are particularly susceptible to the negative impacts of biofouling, in part because they provide a refuge from benthic predation and an ideal habitat for the proliferation of a range of problematic taxa. Biofouling can impact aquaculture operations at most production stages, including interference with spat (juvenile) recruitment, reduced growth, crop losses, decreased marketability and increased operational costs.

Spat are either wild-caught or hatchery-reared (Pérez Camacho et al. 1995, Lauzon-Guay et al. 2005, Díaz et al. 2014) and are a relatively vulnerable life-stage. As spat supply is at the start of the production chain, biofouling of spat can have significant 'downstream' effects on industry profitability. Since the cost and logistic barriers to managing problematic biofouling species after they have established on aquaculture installations are considerable, there is

*Corresponding author: javier.atalah@cawthron.org.nz

increasing global interest in preventative strategies that seek to minimise colonisation by pest species, not only at the spat settlement/recruitment stage, but also at the line seeding and crop grow-out stages of industry operations (Fitridge et al. 2012, Fletcher et al. 2013, Sievers et al. 2014).

Implementation of preventative approaches, for example avoiding or minimising peaks in colonisation by problematic species during key stages of production, requires knowledge of spatio-temporal patterns of their occurrence, and of the factors that determine their recruitment. This information could then be used to facilitate the development of prevention-based management measures. The timing and magnitude of recruitment may be controlled by various factors, including larval supply (Cáceres-Martínez & Figueras 1998a,b, Hoffmann et al. 2012), the presence of conspecifics (McGrath et al. 1988), water depth (Fuentes & Molares 1994) and environmental variables such as water temperature (Avendano & Cantillanez 2013), food availability (Toupoint et al. 2012), wind patterns (Barria et al. 2012), wave-exposure and water flow (Hunt & Scheibling 1996), and solar irradiance (Fuentes-Santos et al. 2016). For example, Fletcher et al. (2013) considered seasonal recruitment patterns of the problematic ascidian *Didemnum vexillum* in relation to sea surface temperature (SST), to identify low-risk 'windows' for

undertaking critical mussel industry activities such as spat deployment to nursery areas.

In New Zealand, biofouling is emerging as one of the significant constraints to the development of the mussel-farming industry, which is based on cultivation of the indigenous green-lipped mussel *Perna canaliculus* (hereafter *Perna*). In the country's most significant aquaculture region at the north end of the South Island (Fig. 1A), the native New Zealand blue mussel *Mytilus galloprovincialis* (hereafter *Mytilus*) has been a persistent and wide-spread biofouling problem for several decades. This species is part of the *Mytilus edulis* complex (Westfall & Gardner 2010, Oyarzún et al. 2016). It represents a Southern Hemisphere lineage of the Mediterranean mussel *Mytilus galloprovincialis*, with evidence of extensive hybridisation and backcrossing between New Zealand blue mussels and the invasive Northern Hemisphere lineage of *M. galloprovincialis* (Gardner et al. 2016). The Northern Hemisphere lineage is considered an invader in many parts of the world with the ability to rapidly become established (Apte et al. 2000), although it is also an important aquaculture species in many countries (FAO 2004–2016). In New Zealand, *Mytilus* can pre-empt space on *Perna* spat-catching ropes, and recruit directly onto *Perna* spat (a process referred to by industry as 'over-settlement'), thereby significantly reducing spat supply. For example,

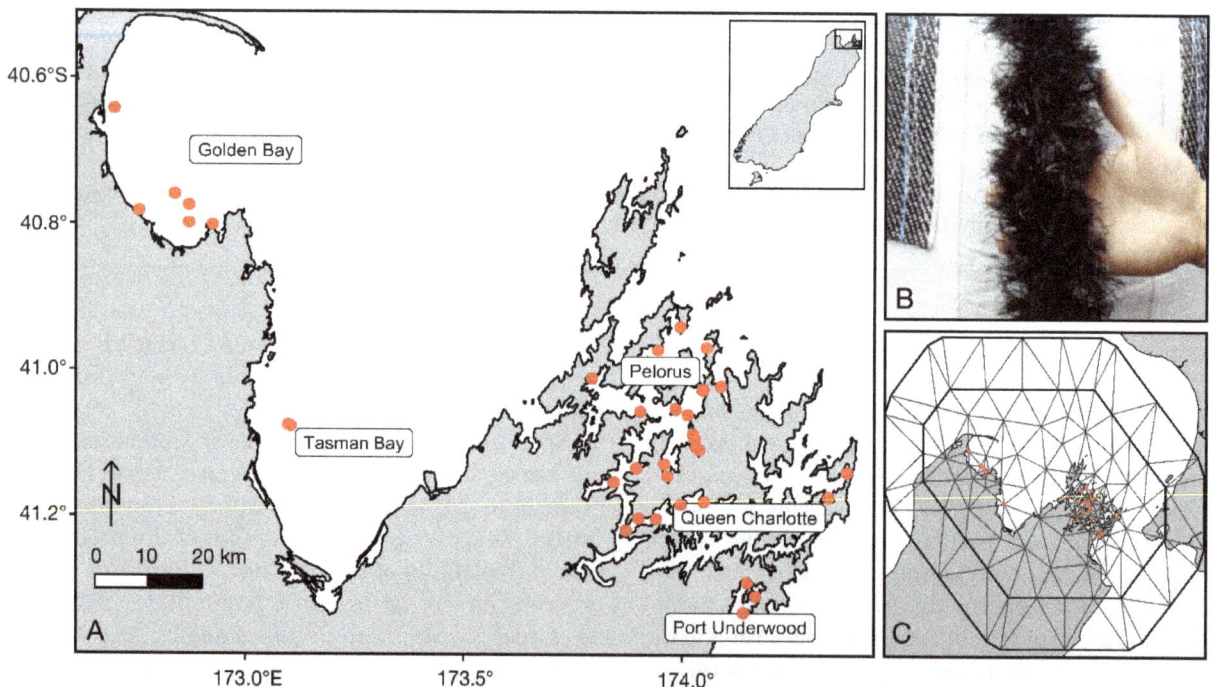

Fig. 1. (A) Study area, in relation to New Zealand's South Island (inset), showing the main sub-regions. (B) 'Xmas tree' rope used in collectors to sample *Mytilus* spat. (C) Mesh pattern used to calculate the effect of geographical location (Gaussian random field) and the location of the study sites (red dots)

Mytilus that is transferred onto nursery ropes during *Perna* seeding processes can significantly decrease *Perna* retention on the lines, with *Perna* seed losses in some cases exceeding 95% (Carton et al. 2007). *Mytilus* also has a range of additional impacts at subsequent stages of production, including causing significant losses during the final crop grow-out stage (Forrest & Atalah 2017).

In this paper, we used spatio-temporal Bayesian modelling to analyse a long-term regional dataset on patterns of *Mytilus* recruitment to settlement arrays deployed throughout the north of the South Island region, and relate patterns to a suite of potential explanatory variables. Specific goals were to: (1) identify times and locations of greatest *Mytilus* recruitment risk, (2) identify environmental drivers and other explanatory factors affecting recruitment patterns and (3) illustrate how such information can be used to develop mitigation measures that can be incorporated into husbandry practices. As part of the latter, we describe the development of an interactive data mining and forecasting tool that is of relevance to biofouling management in aquaculture internationally.

MATERIALS AND METHODS

Mytilus spat data

Our study used datasets obtained from the New Zealand Marine Farming Association which contained mussel spat monitoring data dating back to 1975. Over this period of ca. 40 yr, monitoring of *Mytilus* has been conducted alongside monitoring of *Perna* using spat collectors consisting of 0.5 m sections of mussel industry 'Xmas tree' rope (Fig. 1B). Monitoring is ongoing, and involves deploying the collectors for 1 wk periods at depths from 1 to 15 m, at multiple sites across the study area. These sites represent some key mussel aquaculture sub-regions, namely Pelorus Sound, Queen Charlotte Sound, Port Underwood, Tasman Bay and Golden Bay (Fig. 1A). After 1 wk, the spat collectors are retrieved and taken to a laboratory where *Mytilus* spat are counted using a binocular microscope. At this stage the recruits are ca. 0.5 mm in shell length.

Environmental data

Site-specific SST satellite data were obtained from the Group for High Resolution Sea Surface Tempera-

ture (https://www.ghrsst.org/). The data covered the time period from January 2002 until December 2015. Southern Oscillation Index (SOI, Trenberth 1984) data were obtained from the Department of Science, Information Technology and Innovation, Queensland, Australia (https://www.longpaddock.qld.gov.au). The SOI is a measure of the large-scale pressure differential between the western and eastern tropical Pacific coinciding with El Niño and La Niña episodes (Troup 1965). In general, smoothed time-series of the SOI correspond well with changes in ocean temperatures across the Pacific.

Statistical analysis

Initial data exploration revealed geographic and temporal gaps in the dataset, especially in the earlier years of the monitoring programme. As such, while we describe trends in *Mytilus* since 1975, the statistical analysis focused on a subset of data covering a continuous 14 yr period from 2002 to 2015 (Table A1 in the Appendix). The dataset from that period was relatively complete in terms of seasonal, geographical and water depth representation, and consisted of 19 968 observations of *Mytilus* spat count m^{-1} of rope (the response variable). The dataset also included a suite of potential continuous explanatory variables (year, month, depth, latitude/longitude, SOI, SST), which were used to model *Mytilus* spat abundances in space and time. The model was constructed using a Bayesian approach based on integrated nested Laplace approximations (INLA, Rue et al. 2009). This methodology is freely available in the statistics software R (R Core Team 2014) as part of the package R-INLA (www.r-inla.org). This package has become popular among applied research in a wide range of fields, including ecology, criminology and epidemiology (Blangiardo & Cameletti 2015). Bayesian methods provide a powerful and flexible approach for large or complex datasets, such as the *Mytilus* spat dataset described here.

The model was built by approximating the predicted distribution of spat m^{-1} given previously observed *Mytilus* occurrences and their associated explanatory variables as described above. As the raw data consisted of a large number of zero counts for *Mytilus* (55%) and high over-dispersion, we aggregated the total *Mytilus* spat m^{-1} to Year-Month-Site-Depth nodes (Fig. 1C) and included an 'offset' term to account for the number of samples that could have contributed to this total. This approach reduced the space and time dimensions of the model and allowed for more accurate approximations of the true probability distribution.

Our base model specified the covariates SOI and SST, lagged by 1 mo to imitate forecasting conditions, as linear fixed effects and water depth as a non-linear random effect. Long-term effects and inter-annual variability were included as spatio-temporal random effects, where: (1) the spatial effects were linked to the spatial random field at each mesh node (Fig. 1C), and (2) each month's spatial effect was estimated under the assumption that the spatial pattern changed each month and was correlated with the spatial effects of the previous and subsequent month. The structure of this model allowed us to infer *Mytilus* spat abundances at previously sampled sites as well as unobserved locations.

The response variable, total monthly *Mytilus* spat count m^{-1}, was assumed to follow a negative binomial distribution, with the parameters of this distribution (ϕ) linked to a structured additive predictor η through a link function $\mu = E(e^\eta)$, where log(E) is equal to the offset of η. The linear predictor η for the final model was defined as:

$$\eta = \beta_1 SOI_{t-1} + \beta_2 SST_{t-1} + \sum_{k=1}^{K} f_k (\text{depth}) w_k + f(s,t) \quad (1)$$

where β_1 and β_2 are the linear regression coefficients for the fixed effects, lag 1 (i.e. previous month) SOI and SST, respectively. The third term is the sum of random walk order 1 smooth functions defining the random effect of depth where coefficients vary depending on the observed depth values, w_k. The final term is the spatio-temporal random effect, with Matérn correlation structure defining the spatial effect, and additional autoregressive (order 1) correlation structures for consecutive months.

The strength of the association between observed counts of *Mytilus* spat versus fitted counts from the historical dataset was assessed using Pearson correlation. To test the power of the final model to forecast *Mytilus* spat abundances 1 mo ahead in relation to current environmental conditions at a given location and time, we used 1 mo out-of-sample cross-validation. This technique uses environmental conditions for a given month (i.e. SOI and SST) to forecast *Mytilus* abundance 1 mo ahead. This process was repeated for the previous 24 mo to capture any seasonal variation. Model performance was checked by plotting observed versus both fitted and forecasted values along with the 1:1 line. Additionally, model performance was summarised using 3 statistics; regression R^2, Nash-Sutcliffe efficiency (NSE) and

percent bias (PBIAS) for both fitted and forecasted values. The regression R^2 value is the coefficient of determination derived from a regression of the observations against the predictions. NSE indicates how closely the observations coincide with predictions (Nash & Sutcliffe 1970). NSE values range from $-\infty$ to 1. An NSE of 1 corresponds to a perfect match between predictions and the observations. An NSE of 0 indicates the model is only as accurate as the mean of the observed data, and values less than 0 indicate the model predictions are less accurate than using the mean of the observed data. Bias measures the average tendency of the predicted values to be larger or smaller than the observed values. Optimal bias is 0, whereas positive values indicate underestimation bias, and negative values indicate overestimation bias (Piñeiro et al. 2008). PBIAS is computed as the sum of the differences between the observations and predictions divided by the sum of the observations (Moriasi et al. 2007). A rule of thumb is that model predictions are satisfactory if NSE > 0.50 and if PBIAS < ±25%, and are good if 0.65 < NSE > 0.75 and if 25% < PBIAS < ±40% (Moriasi et al. 2007). Additionally, we used accepted conventions to interpret correlation strength according to R^2 values (Cohen 1988, Evans 1996).

RESULTS

Over the 40 yr period represented by the dataset, there was a large inter-annual variability in the recruitment of *Mytilus* across the study area (Fig. 2A). Low abundances of spat (mean < 103 ind. m^{-1}) were recorded until the early 1990s, with the exception of the first year of monitoring in 1975 (mean ± SE = 330 ± 105 ind. m^{-1}). However, after 1994, there was a general trend for a steady annual increase in the abundance of *Mytilus* spat, with peaks in mean abundance of 464 ± 145 and 797 ± 197 m^{-1} recorded in 1995 and 2005, respectively. The upward trend in the last 2 decades is described by the smoother time-series in Fig. 2A.

The final model consisted of the fixed effect SOI$_{t-1}$, SST$_{t-1}$, the non-linear effect of depth and the spatio-temporal effect (see 'Materials and methods'). Overall, the recruitment of *Mytilus* throughout the year had 2 distinctive peaks: a larger one in the Austral spring through to early summer (i.e. between October and December) and a smaller one in early autumn (i.e. April, Fig. 2B). However, at a finer resolution, the model predicted distinct geographic patterns in the distribution and abundance of *Mytilus*,

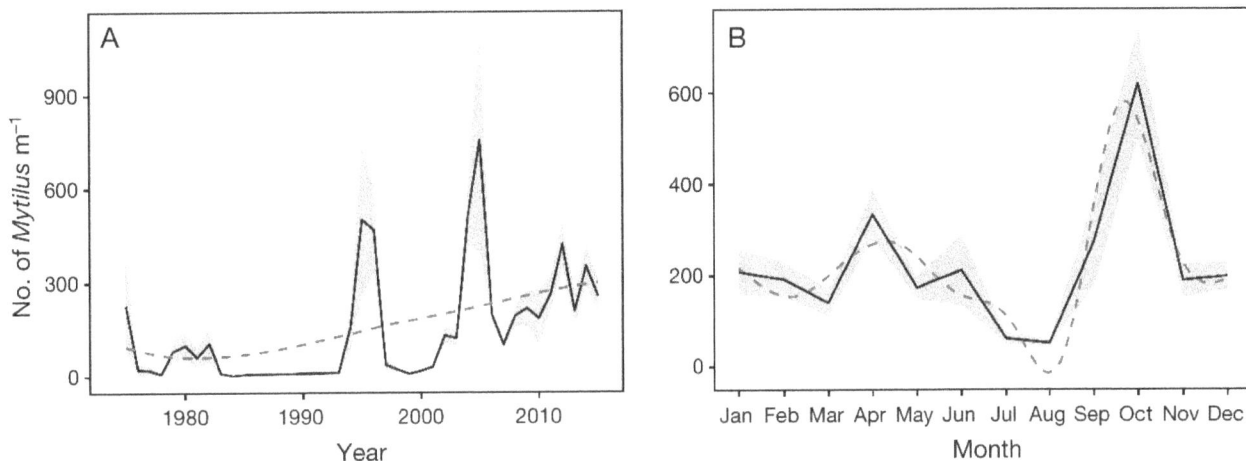

Fig. 2. (A) Long-term and (B) seasonal trend in mean *Mytilus* spat abundance m⁻¹ rope. The red dashed line is the smoothed time series using a cubic spline and 95% confidence intervals (grey shading)

which varied considerably across months and seasons (Fig. 3). At sites in the western sub-region (i.e. Golden and Tasman Bay) the autumn peak extended until May and was of comparable or even greater magnitude to the spring peak (Fig. 3). At sites in the easternmost sub-region (i.e. Queen Charlotte Sound) a strong early autumn peak was not evident, with the model revealing greatest spat abundances in June and October (Fig. 3). Generally, there were more pronounced peaks in mid-Pelorus Sound, compared to the other sub-regions (Fig. 3).

Increasing water depths were generally associated with lower expected *Mytilus* abundances (Fig. 4). There was a positive effect of water depth on spat abundances until 6 m, and a negative effect of depth between 6 and 15 m (Fig. 4). There was a marginal negative effect of SST (mean effect = −0.077, 95% CI = −0.312 to 0.157), indicating that lower *Mytilus* abundances were correlated with higher SST in the previous month. There was a small positive effect of SOI on spat abundances (mean effect = 0.009, 95% CI = −0.024 to 0.042). Although neither SST nor SOI were deemed significant based on 95% confidence intervals, they were retained in the final model, as cross-validation indicated that they increased the model's explanatory and forecasting power. Overall, the validation process indicated that the final model performance was good (Fig. 5A, $R^2 = 0.72$ and NSE = 0.72), and there was a minimal overestimation of 0.9% as indicated by the PBIAS statistic. On the other hand, under forecast conditions (1 mo ahead), the model performance was only moderate ($R^2 = 0.16$ and NSE = −0.33, Fig. 5) and over-predicted *Mytilus* spat abundance considerably, particularly at low abundances (PBIAS = 82%, Fig. 5B).

DISCUSSION

Biofouling poses a significant threat to shellfish aquaculture globally and may drastically affect spat (juvenile) supply, crop yield and marketability, and industry operations (Adams et al. 2011, Fitridge et al. 2012). Shellfish aquaculture that relies on wild-caught spat is particularly vulnerable to the negative impacts of biofouling. Given the lack of cost-effective management options to mitigate the impacts of established biofouling, a potentially feasible approach is the implementation of husbandry practices to minimise exposure to initial colonisation by problematic species (Sievers et al. 2014). Here we described high variability in spatio-temporal patterns in a regional dataset on *Mytilus* recruitment spanning 40 yr. A Bayesian spatio-temporal modelling approach was applied to a 14 yr subset of the data, enabling identification of locations and windows of time when *Perna* farms have been historically prone to colonisation by *Mytilus*. The model was simultaneously evaluated for its ability to forecast *Mytilus* spat abundances 1 mo ahead.

Spatio-temporal patterns

Large inter-annual variability in abundances, such as evident in the data analysed here, is an intrinsic characteristic of the recruitment of mussels (Hunt & Scheibling 1998, Porri et al. 2006) and other marine invertebrates (Rodríguez et al. 1993). Processes that cause such variation may include fluctuations in physical factors, resource availability and biological interactions (Underwood & Chapman 2000). In this

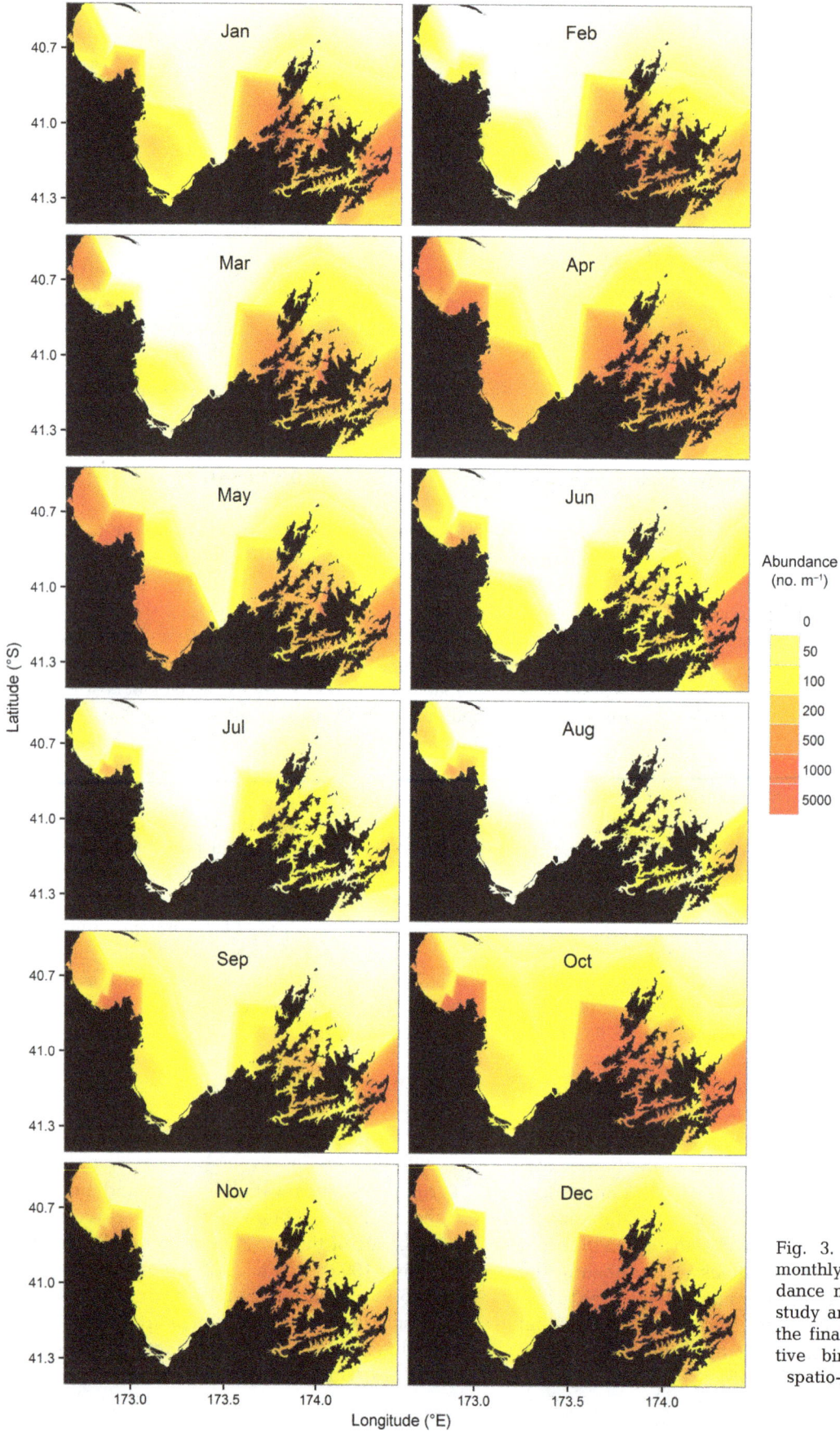

Fig. 3. Predicted average monthly *Mytilus* spat abundance m⁻¹ rope across the study area as estimated by the final model with negative binomial errors and spatio-temporal structure

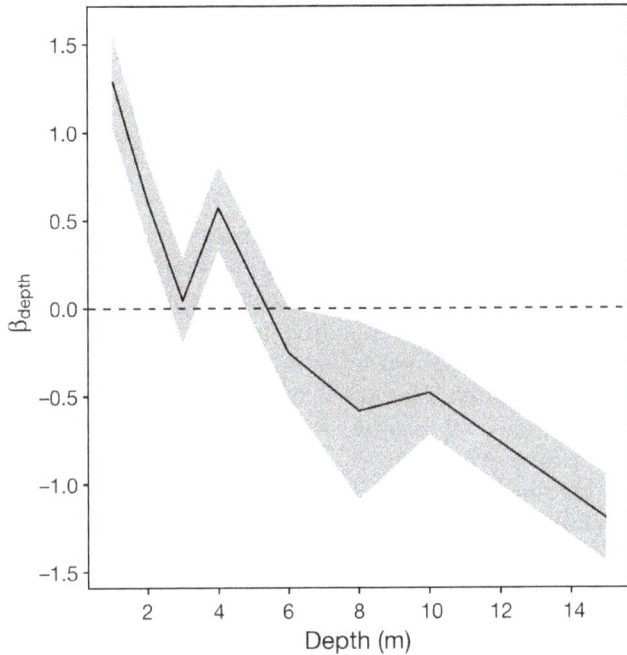

Fig. 4. Posterior mean of the smoothed depth regression effect (β_{depth}) on expected *Mytilus* spat with 95% confidence intervals (grey shading)

study, although *Mytilus* recruitment varied several-fold among years, there was an overall upward trend in abundances over the last 2 decades. Although the cause of the apparent long-term increase in recruitment cannot be inferred from this study, a possible

explanation relates to the considerable expansion of mussel cultivation since it began in the 1970s. Overall, there has been an increase in the mussel farming area, and hence habitat for *Mytilus*, from ca. 100 ha in the mid- to late-1970s to ca. 2500 ha present day (Handley 2015). In fact, in Pelorus Sound (where the majority of farms in our study region are located) there was a moderate positive correlation between the cumulative area of mussel farms and the annual median *Mytilus* abundance (r = 0.63, p < 0.001). An increased *Mytilus* population could lead to an increased larval inoculation pressure for ongoing establishment (Ruiz et al. 2000, Lockwood et al. 2009), such that increased abundances on mussel farms could become a self-perpetuating problem. A similar phenomenon was suggested for salmon farms in Norway, where the recruitment of *Ectopleura larynx* and other sessile biofouling species was dramatically enhanced within farms relative to control locations (Bloecher et al. 2015).

Although *Mytilus* recruitment was recorded throughout the year, there was a strong overall seasonal effect, with 2 peaks: a major and a minor peak in October and April, respectively. Seasonal fluctuations in *Mytilus* recruitment of comparable magnitude have been previously reported in New Zealand (Kennedy 1977, Meredyth-Young & Jenkins 1978) and other parts of the world (Dix & Ferguson 1984, Yildiz & Berber 2010, Yildiz et al. 2010). These fluctuations are most likely related to

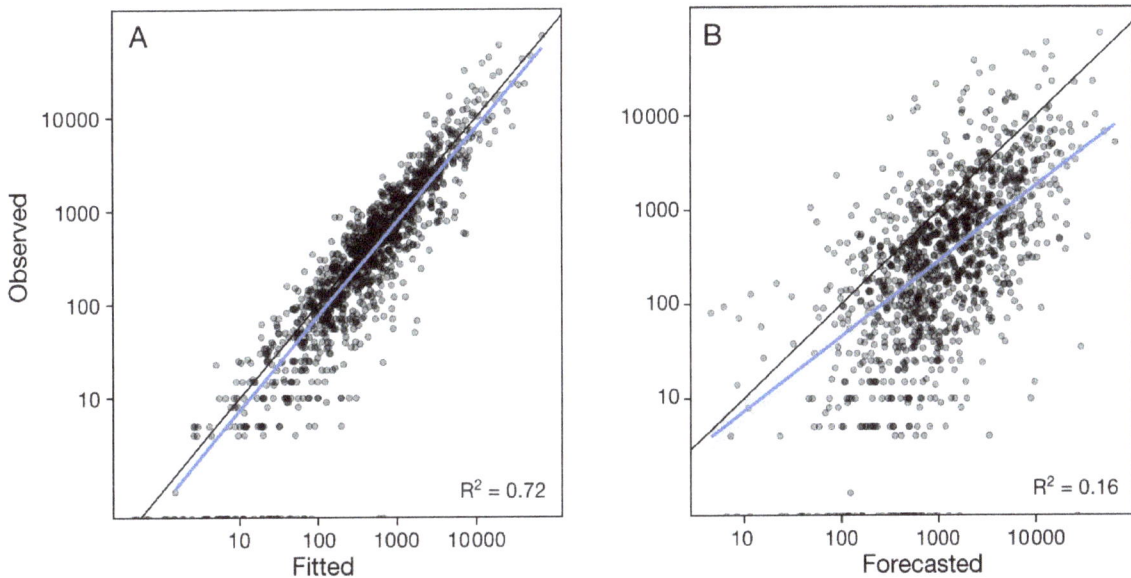

Fig. 5. (A) Fitted counts and (B) counts forecasted 1 mo ahead versus observed counts of *Mytilus* spat m^{-1}. Due to the overdispersed *Mytilus* distribution, both axes are log-scaled to aid in the visualisation of the relationships. The blue lines are ordinary least square regressions; grey shading represents 95% pointwise confidence intervals of the fitted values. The black lines are the one-to-one relationships

the reproductive seasonality observed in this species, with a peak spawning season in spring coinciding with increases in water temperature and chlorophyll a concentrations. Several spawning events can occur after the first one, generally until summer or late autumn (Cáceres-Martínez & Figueras 1998b). In autumn, a second peak in food availability (Gibbs et al. 1992, Gibbs & Vant 1997, Ogilvie et al. 2000, MacKenzie & Adamson 2004) may favour the accumulation of reserves for gametogenesis and a subsequent minor spawning event.

When the long-term and seasonal patterns in Mytilus abundance were considered by the main sub-regions, distinct geographic variation was evident (see Fig. 3). For example, Golden Bay sites showed a minor end of summer peak between April and May, rather than between March and April as observed in Pelorus Sound. The greatest Mytilus abundances overall occurred in mid-Pelorus Sound and Queen Charlotte Sound, during October and July. The mid-Pelorus Sound is the most intensely farmed area of the study region (Aquaculture New Zealand 2012). As discussed above, Mytilus populations established on these farms may act as a larval source and enhance subsequent recruitment in adjacent areas. Additionally, mid-Pelorus is a relatively enclosed area, with several inner bays having low water flow (<0.1 m s⁻¹) and slow flushing times (Gibbs et al. 1991), which may favour larval retention and settlement (McQuaid & Phillips 2000).

The overall pattern of higher Mytilus recruitment in shallower (<6 m) compared to deeper waters (>6 m) is consistent with the depth patterns reported in previous studies in Pelorus Sound (Meredyth-Young & Jenkins 1978) and elsewhere (Cáceres-Martínez & Figueras 1997, Holthuis et al. 2015). The occurrence of lower Mytilus spat abundances in deeper waters supports the existing practice of some mussel farmers, who submerge lines to avoid or minimise over-settlement by Mytilus. These depth differences are conceivably explained by the vertical distribution of Mytilus larvae in relation to environmental variables. Competent larvae of Mytilus are generally not homogeneously distributed across the water column and have a tendency to aggregate in surface waters above the thermocline/halocline (Dobretsov & Miron 2001). Mussel cultivation waters across our study region experience a thermocline at ca. 10 m or deeper, and in Pelorus Sound surface salinity at shallow depths <10 m can be reduced due to significant freshwater inputs (Gibbs et al. 1991, Gibbs 2001).

Utility and operational implications of the model

Relationships between Mytilus spat recruitment and environmental conditions, such as SST and SOI, were generally weak but did at least provide additional accuracy to the model. The marginal effect of SST (in the previous month) was not as great as might be anticipated given the likely importance of temperature in the reproductive seasonality of Mytilus noted above. More broadly, water temperature is generally recognized as a driving factor in the settlement and recruitment of marine invertebrates, as it can affect adult reproductive output, and the subsequent survival, growth and metamorphosis of larvae (Kinne et al. 1970, Coma et al. 2000, Bates 2005).

The significant spatial random component in the model likely reflects the effect of other unmeasured factors that influence the recruitment patterns observed. As well as the probable importance of chlorophyll a as suggested above, other environmental factors such as salinity, wind patterns and hydrographic characteristics may influence reproductive seasonality and recruitment. Additionally, a very recent study of settlement patterns in Mytilus highlighted solar irradiance as a key predictor variable, suggesting that solar irradiance during late winter indirectly drives the timing and intensity of settlement onset (Fuentes-Santos et al. 2016). Recognising that it can be difficult to attribute changes in biological responses to a single variable (Coma et al. 2000), it is clear that a broad range of variables would ideally be incorporated into our model. Predictive power may have increased by incorporating measures of mussel food sources, such as phytoplankton and detritus, or other environmental variables (e.g. salinity and wind). For example, the recruitment of many coastal species with long larval duration, such as Mytilus (2–4 wk) is driven by surface currents, which are largely controlled by wind patterns (Roughgarden et al. 1988). Such advances may be possible in future years, especially with the development of monitoring technologies that facilitate the collection of large and long-term environmental datasets (He et al. 2015). Even with comprehensive environmental data, however, recruitment variability is likely to be attributable, in part, to unrelated factors, including attributes that are intrinsic to the mussels themselves. For example, mussels have the capacity for secondary settlement by attaching and detaching from the substrate several times until they find a suitable substrate (Bayne 1964, Newell et al. 2010). In the case of Mytilus, the process of secondary settlement may occur for several weeks after initial larval settlement

(Bownes & McQuaid 2009). As such, the recruitment response patterns described in the present study, and indeed other studies of early post-larval mussel settlement and recruitment, may not reflect the final distribution patterns of adult mussels (Le Corre et al. 2013).

Despite the highly variable nature of the data, and the weak individual effects of potential environmental drivers, the model nonetheless provided relatively reliable estimates of *Mytilus* spat abundances (observed vs. fitted data, $R^2 = 0.72$). These findings can be used to guide farm management practices, such as the location and timing of deployment of *Perna* spat collection ropes to minimise *Mytilus* over-settlement. However, the moderate relationship between observed and forecasted data ($R^2 = 0.16$) suggests that the prediction of *Mytilus* recruitment 1 mo ahead should be viewed with caution. Clearly, there are phenomena contributing to recruitment patterns that have not been captured by the model and underlying data. We can nonetheless provide a reasonable model of the likely general abundance of *Mytilus* spat at given locations and times.

Considering that the regional economic loss of *Mytilus* on *Perna* aquaculture has been estimated at ca. US$16 million annually in the Marlborough Sounds region alone (Forrest & Atalah 2017), the ability to implement management practices to avoid high abundances of *Mytilus* may translate into considerable economic benefits. Predictive maps such as depicted in Fig. 3 provide a useful means to illustrate to marine farmers the risk periods and hotspots for *Mytilus* recruitment. For example, a *Mytilus* recruit count of >200 m^{-1}, evident in orange and red colours in Fig. 3, is the approximate threshold that the industry considers operationally significant. This information can be considered in relation to peak times of *Perna* settlement in the region (i.e. April, Atalah et al. 2016). Accordingly, the spatio-temporal model has recently been integrated into a web application (https://cawthron.shinyapps.io/BMOP/) described by Atalah et al. (2016). The web application currently displays spat count data for both *Mytilus* and *Perna*, for more than 50 monitoring sites across the study region. The application enables marine farmers to interactively explore historic settlement patterns of *Mytilus* and *Perna* spat at fine-scale resolution (i.e. at particular farm sites or bays), with a future step being to extend the forecasting model developed for *Mytilus* to also include *Perna*. This tool will therefore facilitate management decisions that aim to maximise *Perna* collection while minimising losses due to *Mytilus* (e.g. based on the ratio between *Perna* and

Mytilus spat abundance), and will therefore provide an integrated production and pest management tool for famers. This type of application is clearly of relevance to the management of any fouling pest in aquaculture for which spatio-temporal variation in occurrence provides opportunities for management (Fletcher et al. 2013, Sievers et al. 2014). Although the tool does not provide a panacea for the management of all high-risk biofouling, by addressing the most significant species it facilitates sustainable and efficient aquaculture production.

Acknowledgements. Many thanks to the Marine Farming Association for providing the spat dataset and for insightful discussions about *Mytilus* spat recruitment dynamics. We also thank Jim Jenkins for sample processing and Debbie Stone and Camille Poutrin for assistance with data management and preparation. Thanks to Weimin Jiang for obtaining sea surface temperature satellite data. This study was funded by the New Zealand Ministry of Business, Innovation and Employment under Contract CAWX1315 (The Cultured Shellfish Programme).

LITERATURE CITED

Adams CM, Shumway SE, Whitlatch RB, Getchis T (2011) Biofouling in marine molluscan shellfish aquaculture: a survey assessing the business and economic implications of mitigation. J World Aquacult Soc 42:242–252

Apte S, Holland BS, Godwin LS, Gardner JP (2000) Jumping ship: a stepping stone event mediating transfer of a nonindigenous species via a potentially unsuitable environment. Biol Invasions 2:75–79

Aquaculture New Zealand (2012) New Zealand aquaculture: a sector overview with key facts, statistics and trends. http://aquaculture.org.nz/wp-content/uploads/2012/05/NZ-Aquaculture-Facts-2012.pdf (accessed 7 January 2014)

Atalah J, Rabel H, Forrest B (2016) Blue mussel over-settlement predictive model and web application. Prepared for Marine Farming Association. Cawthron Report No. 2801. Cawthron Institute, Nelson

Avendano M, Cantillanez M (2013) Reproductive cycle, collection and early growth of *Aulacomya ater*, Molina 1782 (Bivalvia: Mytilidae) in northern Chile. Aquacult Res 44:1327–1338

Barria A, Gebauer P, Molinet C (2012) Spatial and temporal variability of mytilid larval supply in the Seno de Reloncavi, southern Chile. Rev Biol Mar Oceanogr 47:461–473

Bates WR (2005) Environmental factors affecting reproduction and development in ascidians and other protochordates. Can J Zool 83:51–61

Bayne BL (1964) Primary and secondary settlement in *Mytilus edulis* L. (Mollusca). J Anim Ecol 33:513–523

Blangiardo M, Cameletti M (2015) Spatial and spatio-temporal Bayesian models with R-INLA. John Wiley & Sons, West Sussex

Bloecher N, Floerl O, Sunde LM (2015) Amplified recruitment pressure of biofouling organisms in commercial salmon farms: potential causes and implications for farm management. Biofouling 31:163–172

Bownes SJ, McQuaid CD (2009) Mechanisms of habitat segregation between an invasive and an indigenous mussel: settlement, post-settlement mortality and recruitment. Mar Biol 156:991–1006

Cáceres-Martínez J, Figueras A (1997) Mussel (*Mytilus galloprovincialis* Lamarck) settlement in the Ria de Vigo (NW Spain) during a tidal cycle. J Shellfish Res 16:83–85

Cáceres-Martínez J, Figueras A (1998a) Distribution and abundance of mussel (*Mytilus galloprovincialis* Lmk) larvae and post-larvae in the Ria de Vigo (NW Spain). J Exp Mar Biol Ecol 229:277–287

Cáceres-Martínez J, Figueras A (1998b) Long-term survey on wild and cultured mussels (*Mytilus galloprovincialis* Lmk) reproductive cycles in the Ria de Vigo (NW Spain). Aquaculture 162:141–156

Carton A, Jeffs A, Foote G, Palmer H, Bilton J (2007) Evaluation of methods for assessing the retention of seed mussels (*Perna canaliculus*) prior to seeding for grow-out. Aquaculture 262:521–527

Cohen J (1988) Statistical power analysis for the behavioral sciences. Lawrence Erlbaum Associates, Hillsdale, NJ

Coma R, Ribes M, Gili JM, Zabala M (2000) Seasonality in coastal benthic ecosystems. Trends Ecol Evol 15:448–453

Díaz C, Figueroa Y, Sobenes C (2014) Seasonal effects of the seeding on the growth of Chilean mussel (*Mytilus edulis platensis*, d'Orbigny 1846) cultivated in central Chile. Aquaculture 428–429:215–222

Dix T, Ferguson A (1984) Cycles of reproduction and condition in Tasmanian blue mussels, *Mytilus edulis planulatus*. Mar Freshw Res 35:307–313

Dobretsov SV, Miron G (2001) Larval and post-larval vertical distribution of the mussel *Mytilus edulis* in the White Sea. Mar Ecol Prog Ser 218:179–187

Evans JD (1996) Straightforward statistics for the behavioral sciences. Brooks/Cole, Pacific Grove, CA

FAO (Food and Agriculture Organization of the United Nations) (2004–2016) *Mytilus galloprovincialis*. Cultured Aquatic Species Information Programme. www.fao.org/fishery/culturedspecies/Mytilus_galloprovincialis/en (accessed 7 December 2016)

Fitridge I, Dempster T, Guenther J, de Nys R (2012) The impact and control of biofouling in marine aquaculture: a review. Biofouling 28:649–669

Fletcher LM, Forrest BM, Atalah J, Bell JJ (2013) Reproductive seasonality of the invasive ascidian *Didemnum vexillum* in New Zealand and implications for shellfish aquaculture. Aquacult Environ Interact 3:197–211

Forrest BM, Atalah J (2017) Significant impact from blue mussel *Mytilus galloprovincialis* biofouling on aquaculture production of green-lipped mussels in New Zealand. Aquacult Environ Interact (in press)

Fuentes J, Molares J (1994) Settlement of the mussel *Mytilus galloprovincialis* on collectors suspended from rafts in the Ria de Arousa (NW of Spain)—annual pattern and spatial variability. Aquaculture 122:55–62

Fuentes-Santos I, Labarta U, Álvarez-Salgado XA, Fernández-Reiriz MJ (2016) Solar irradiance dictates settlement timing and intensity of marine mussels. Sci Rep 6:29405

Gardner JPA, Zbawicka M, Westfall KM, Wenne R (2016) Invasive blue mussels threaten regional scale genetic diversity in mainland and remote offshore locations: the need for baseline data and enhanced protection in the Southern Ocean. Glob Change Biol 22:3182–3195

Gibbs MM (2001) Sedimentation, suspension, and resuspension in Tasman Bay and Beatrix Bay, New Zealand, two contrasting coastal environments which thermally stratify in summer. NZ J Mar Freshw Res 35:951–970

Gibbs M, Vant W (1997) Seasonal changes in factors controlling phytoplankton growth in Beatrix Bay, New Zealand. NZ J Mar Freshw Res 31:237–248

Gibbs M, James M, Pickmere S, Woods P, Shakespeare B, Hickman R, Illingworth J (1991) Hydrodynamic and water column properties at six stations associated with mussel farming in Pelorus Sound, 1984–85. NZ J Mar Freshw Res 25:239–254

Gibbs M, Pickmere S, Woods P, Payne G, James M, Hickman R, Illingworth J (1992) Nutrient and chlorophyll *a* variability at six stations associated with mussel farming in Pelorus Sound, 1984–85. NZ J Mar Freshw Res 26:197–211

Handley S (2015) The history of benthic change in Pelorus Sound (Te Hoiere), Marlborough. Prepared for Marlborough District Council. NIWA client report no: NEL2015-001. NIWA, Nelson

He KS, Bradley BA, Cord AF, Rocchini D and others (2015) Will remote sensing shape the next generation of species distribution models? Remote Sens Ecol Conserv 1:4–18

Hoffmann V, Pfaff MC, Branch GM (2012) Spatio-temporal patterns of larval supply and settlement of intertidal invertebrates reflect a combination of passive transport and larval behavior. J Exp Mar Biol Ecol 418-419:83–90

Holthuis TD, Bergström P, Lindegarth M, Lindegarth S (2015) Monitoring recruitment patterns of mussels and fouling tunicates in mariculture. J Shellfish Res 34:1007–1018

Hunt HL, Scheibling RE (1996) Physical and biological factors influencing mussel (*Mytilus trossulus, M. edulis*) settlement on a wave-exposed rocky shore. Mar Ecol Prog Ser 142:135–145

Hunt HL, Scheibling RE (1998) Spatial and temporal variability of patterns of colonization by mussels (*Mytilus trossulus, M. edulis*) on a wave-exposed rocky shore. Mar Ecol Prog Ser 167:155–169

Kennedy VS (1977) Reproduction in *Mytilus edulis aoteanus* and *Aulacomya maoriana* (Mollusca: Bivalvia) from Taylors Mistake, New Zealand. NZ J Mar Freshw Res 11:255–267

Kinne O, Oppenheimer CH, Gessner F, Brett JR, Garside ET (1970) Temperature: general introduction. In: Kinne O (ed) Marine ecology: a comprehensive integrated treatise on life in oceans and coastal waters. Vol I: Environmental factors. Wiley Interscience, London, p 321–346

Lauzon-Guay JS, Dionne M, Barbeau MA, Hamilton DJ (2005) Effects of seed size and density on growth, tissue-to-shell ratio and survival of cultivated mussels (*Mytilus edulis*) in Prince Edward Island, Canada. Aquaculture 250:652–665

Le Corre N, Martel AL, Guichard F, Johnson LE (2013) Variation in recruitment: differentiating the roles of primary and secondary settlement of blue mussels *Mytilus* spp. Mar Ecol Prog Ser 481:133–146

Lockwood JL, Cassey P, Blackburn TM (2009) The more you introduce the more you get: the role of colonization pressure and propagule pressure in invasion ecology. Divers Distrib 15:904–910

MacKenzie L, Adamson J (2004) Water column stratification and the spatial and temporal distribution of phytoplankton biomass in Tasman Bay, New Zealand: implications for aquaculture. NZ J Mar Freshw Res 38:705–728

McGrath D, King PA, Gosling EM (1988) Evidence for the direct settlement of *Mytilus edulis* larvae on adult mussel beds. Mar Ecol Prog Ser 47:103–106

McQuaid CD, Phillips TE (2000) Limited wind-driven dispersal of intertidal mussel larvae: *in situ* evidence from the plankton and the spread of the invasive species *Mytilus galloprovincialis* in South Africa. Mar Ecol Prog Ser 201:211–220

Meredyth-Young JL, Jenkins RJ (1978) Depth of settlement of two mussel species on suspended collectors in Marlborough Sounds, New Zealand. NZ J Mar Freshw Res 12:83–86

Moriasi DN, Arnold JG, Van Liew MW, Bingner RL, Harmel RD, Veith TL (2007) Model evaluation guidelines for systematic quantification of accuracy in watershed simulations. Trans ASABE 50:885–900

Nash JE, Sutcliffe JV (1970) River flow forecasting through conceptual models. Part I. A discussion of principles. J Hydrol (Amst) 10:282–290

Newell CR, Short F, Hoven H, Healey L, Panchang V, Cheng G (2010) The dispersal dynamics of juvenile plantigrade mussels (*Mytilus edulis* L.) from eelgrass (*Zostera marina*) meadows in Maine, USA. J Exp Mar Biol Ecol 394:45–52

Ogilvie SC, Ross AH, Schiel DR (2000) Phytoplankton biomass associated with mussel farms in Beatrix Bay, New Zealand. Aquaculture 181:71–80

Oyarzún PA, Toro JE, Cañete JI, Gardner JPA (2016) Bioinvasion threatens the genetic integrity of native diversity and a natural hybrid zone: smooth-shelled blue mussels (*Mytilus* spp.) in the Strait of Magellan. Biol J Linn Soc 117:574–585

Pérez Camacho A, Labarta U, Beiras R (1995) Growth of mussels (*Mytilus edulis galloprovincialis*) on cultivation rafts: influence of seed source, cultivation site and phytoplankton availability. Aquaculture 138:349–362

Piñeiro G, Perelman S, Guerschman JP, Paruelo JM (2008) How to evaluate models: observed vs. predicted or predicted vs. observed? Ecol Model 216:316–322

Porri F, McQuaid CD, Radloff S (2006) Spatio-temporal variability of larval abundance and settlement of *Perna perna*: differential delivery of mussels. Mar Ecol Prog Ser 315:141–150

R Core Team (2014) R: a language and environment for statistical computing. R Foundation for Statistical Computing, Vienna. www.R-project.org/

Rodríguez SR, Ojeda FP, Inestrosa NC (1993) Settlement of benthic marine invertebrates. Mar Ecol Prog Ser 97: 193–207

Roughgarden J, Gaines S, Possingham H (1988) Recruitment dynamics in complex life cycles. Science 241:1460–1466

Rue H, Martino S, Chopin N (2009) Approximate Bayesian inference for latent Gaussian models by using integrated nested Laplace approximations. J R Stat Soc Ser B Stat Methodol 71:319–392

Ruiz GM, Fofonoff PW, Carlton JT, Wonhom MJ, Hines AH (2000) Invasion of coastal marine communities in North America: apparent patterns, processes and biases. Annu Rev Ecol Syst 31:481–531

Sievers M, Dempster T, Fitridge I, Keough MJ (2014) Monitoring biofouling communities could reduce impacts to mussel aquaculture by allowing synchronisation of husbandry techniques with peaks in settlement. Biofouling 30:203–212

Toupoint N, Gilmore-Solomon L, Bourque F, Myrand B, Pernet F, Olivier F, Tremblay R (2012) Match/mismatch between the *Mytilus edulis* larval supply and seston quality: effect on recruitment. Ecology 93:1922–1934

Trenberth KE (1984) Signal versus noise in the Southern Oscillation. Mon Weather Rev 112:326–332

Troup A (1965) The Southern Oscillation. QJR Meteorol Soc 91:490–506

Underwood AJ, Chapman MG (2000) Variation in abundances of intertidal populations: consequences of extremities of environment. Hydrobiologia 426:25–36

Westfall KM, Gardner JPA (2010) Genetic diversity of Southern hemisphere blue mussels (Bivalvia: Mytilidae) and the identification of non-indigenous taxa. Biol J Linn Soc 101:898–909

Yildiz H, Berber S (2010) Depth and seasonal effects on the settlement density of *Mytilus galloprovincialis* L. 1819 in the Dardanelles. J Anim Vet Adv 9:756–759

Yildiz H, Lök A, Acarli S, Serdar S, Köse A (2010) A preliminary survey on settlement and recruitment patterns of Mediterranean mussel (*Mytilus galloprovincialis*) in Dardanelles, Turkey. Kafkas Univ Vet Fak Derg 16(Suppl B): S319–S324

Appendix

Table A1. Summary table of the data (1994 to 2015 yearly averages) used for modelling. *Mytilus*: *Mytilus galloprovincialis* abundance, SST: sea surface temperature, SOI = southern oscillation index

Year	Effort (n)	*Mytilus* (ind. m^{-1})	SST$_{t-1}$ (°C)	SOI$_{t-1}$
1994	129	904.4	16.0	−10.5
1995	653	1403.5	15.4	−2.8
1996	627	1491.9	14.5	5.1
1997	455	144.3	15.9	−8.0
1998	409	82.7	15.9	−7.4
1999	311	35.5	17.2	8.2
2000	150	190.0	15.0	7.8
2001	884	213.5	16.6	1.3
2002	210	528.0	16.0	−4.8
2003	246	437.9	16.2	−4.7
2004	283	2409.6	15.7	−2.6
2005	362	2491.1	16.0	−7.6
2006	354	946.4	16.0	3.3
2007	549	509.9	15.5	2.7
2008	851	948.4	16.6	10.3
2009	948	1026.4	15.6	0.9
2010	948	816.8	16.0	5.9
2011	919	1130.2	16.0	16.4
2012	1577	1766.8	15.1	0.3
2013	3155	933.7	15.1	3.5
2014	3341	1625.2	14.7	−3.0
2015	2687	850.2	15.3	−10.1

12

Performance of oysters selected for dermo resistance compared to wild oysters in northern Gulf of Mexico estuaries

Sandra Casas[1], William Walton[2], Glen Chaplin[2], Scott Rikard[2], John Supan[1,3], Jerome La Peyre[1,*]

[1]School of Renewable Natural Resources, Louisiana State University Agricultural Center, Baton Rouge, LA 70803, USA
[2]Auburn University Shellfish Laboratory, Dauphin Island, AL 36528, USA
[3]Louisiana Sea Grant College Program, Louisiana State University, Baton Rouge, LA 70803, USA

ABSTRACT: The performance of the progeny of eastern oysters *Crassostrea virginica* from Louisiana selected for resistance to dermo, caused by *Perkinsus marinus* (referred to as 'OBOY') and of wild oysters collected from Louisiana (Calcasieu Lake) and Alabama (Cedar Point, Perdido Pass), USA, estuaries was compared for their potential use in aquaculture. Seed oysters from each stock were deployed in September 2011 at 2 dermo-endemic sites, Dauphin Island and Sandy Bay, Alabama, using an adjustable longline system, and their survival and shell heights were monitored bimonthly. *P. marinus* infection intensity and condition index were measured at deployment and in March, July and September 2012. The OBOY stock showed lower mortality than the unselected stocks (Cedar Point, Perdido Pass, Calcasieu Lake) at Dauphin Island, and both Louisiana stocks had lower mortality than the Alabama stocks at Sandy Bay, a slightly more saline site. Mortality increased in summer, especially between July and September, concomitant with increasing *P. marinus* infection intensities at the higher temperatures and favorable salinities. At the higher salinity site, both Louisiana stocks had lower *P. marinus* infection intensities than the Perdido Pass stock, the stock with the highest percentage of oysters with moderate and heavy infection and cumulative mortality. The OBOY stock reached greater mean shell height than Calcasieu Lake and Perdido Pass stocks. Condition index of the oyster stocks decreased by more than half between March and July following expected spawning. Differences in stock performance highlight the importance of stock selection for aquaculture in dermo-endemic estuaries of the northern Gulf of Mexico.

KEY WORDS: *Crassostrea virginica* · Off-bottom aquaculture · Mortality · Growth · Dermo disease · *Perkinsus marinus*

INTRODUCTION

Off-bottom oyster aquaculture is a nascent and fast-expanding industry in the northern Gulf of Mexico (Maxwell et al. 2008, Walton et al. 2013a). Since 2010, at least 20 commercial oyster farms have begun operation in Louisiana, Alabama and Florida, USA, collectively, with interest in Mississippi where regulations are being developed to allow this type of oyster farming (Petrolia et al. 2017). Off-bottom oyster aquaculture could play an important role in stabilizing production and increasing the sales of premium oysters from Gulf States.

In northern Gulf of Mexico estuaries, most traditional eastern oyster *Crassostrea virginica* production occurs on-bottom between salinities of 5 and 15 because of excessive mortality due to dermo—a disease caused by the protistan parasite *Perkinsus marinus* and directly transmitted from one oyster to another through the water—and predation from southern oyster drill

*Corresponding author: jlapeyre@agcenter.lsu.edu

snails *Stramonita haemastoma* at salinities >15 (Breithaupt & Dugas 1979, Craig et al. 1989, La Peyre et al. 2016). Off-bottom oyster culture virtually eliminates predation, can improve growth and shell shape and provides opportunities to control fouling and to farm vast areas of previously unsuitable bottom (e.g. mud) and waters at higher salinity where oyster growth is greatest. Moreover, using hatchery-produced seed enables selection of brood stocks for increased resistance to dermo and which are adapted to local environmental conditions (Frank-Lawale et al. 2014, Proestou et al. 2016, Leonhardt et al. 2017).

The need to develop stocks of locally adapted eastern oysters that are resistant to disease has long been recognized (Haskin & Ford 1979, Matthiessen et al. 1990). Previous research studies in the northeastern and mid-Atlantic regions have successfully bred stocks of eastern oysters with increased resistance to the effects of *Haplosporidium nelsoni* (MSX) (Haskin & Ford 1979, Matthiessen et al. 1990) as well as to the effects of both MSX and dermo (Ragone Calvo et al. 2003a) and *Roseovarius* oyster disease (Davis & Barber 1999). Following these findings, selective breeding programs were organized and have significantly contributed to the expansion and success of oyster aquaculture in those regions (Guo et al. 2008, Frank-Lawale et al. 2014, Proestou et al. 2016).

In the northern Gulf of Mexico, a Louisiana eastern oyster stock named 'OBOY' has been selectively bred for dermo resistance since 1999 (Stickler et al. 2001), as *P. marinus* is the only parasite causing significant mortalities in this region. The objective of the current study was to compare the performance of this locally selected stock to wild oysters from Louisiana and Alabama and to determine the true potential of the OBOY stock for use in the rapidly expanding numbers of oyster farms in Alabama estuaries. Specifically, the mortality, growth, *P. marinus* infection intensity, and condition index of the progeny of wild oysters collected in Louisiana and Alabama, and of the OBOY stock, were compared in a common garden experimental design at 2 dermo-endemic Alabama sites. The wild oysters were collected from waters of intermediate to high salinities where dermo is typically endemic.

MATERIALS AND METHODS

Oysters

The eastern oyster *Crassostrea virginica* stock named 'OBOY' consists of the descendants of large oysters,

collected in 1999 from a dermo-endemic area (i.e. Oyster Bayou, Cameron Parish, Louisiana, USA, 29° 47′ 39″ N, 93° 23′ 14″ W) and whose progeny have been challenged in the field (i.e. generations F0 and F3) and in the laboratory (i.e. F1 and F2) by *Perkinsus marinus*. Wild market-size eastern oysters from Calcasieu Lake (Cameron Parish, 29° 51′ 00″ N, 93° 18′ 40″ W, salinity yearly mean ± SD of 20.4 ± 2.2, USGS recorder 08017118) were collected in the fall of 2010. OBOY (F3) and Calcasieu Lake oysters were maintained in labeled aquaculture bags (50 oysters per bag) hanging on an adjustable longline system (ALS; BST Oyster Co.) at the Oyster Research and Demonstration Farm in Grand Isle, Louisiana (29° 14′ 20″ N, 90° 00′ 11″ W, 2003–2011 salinity yearly mean ± SD of 20.9 ± 1.9, USGS recorder 073802516) prior to spawning.

The Louisiana stocks were spawned at the Louisiana Sea Grant Oyster Hatchery in Grand Isle in May 2011 to produce a F0 generation of Calcasieu Lake and a F4 generation of the OBOY stock. Each stock (2 separate spawning events) was naturally spawned (150 oysters per stock) and the resulting larvae reared using methods similar to Dupuy et al. (1977). Pediveliger (~280 μm) larvae were set on micro-cultch material (~500 μm ground oyster shell) to produce single oyster spats that were shipped to the Auburn University Shellfish Laboratory (AUSL), Dauphin Island, Alabama, for rearing.

Alabama Cedar Point stock was produced from a cross of a F1 Cedar Point brood stock (CP09; 21 females, 3 males) line with a F0 Heron Bay brood stock (HB09; 17 females, 6 males); Heron Bay, 30° 19′ 55″ N, 88° 8′ 25″ W, is in close proximity to Cedar point, 30° 18′ 33″ N, 88° 8′ 14″ W. No effort was made to select for dermo resistance in these brood stocks, which were 24 mo (CP09) and 22 mo (HB09) old at the time of spawning. Alabama Perdido Pass stock was produced from Perdido Pass wild-stock oysters (32 females, 14 males) collected from Perdido Pass (30° 16′ 25″ N, 87° 33′ 20″ W) in Orange Beach, Alabama. Yearly mean salinity ± SD for 2011–2015 was 14.5 ± 6.9 in Cedar Point and 28.1 ± 5.3 in Perdido Pass (Mobile Bay National Estuary Program, www.mymobilebay.com, data collection for both sites started in 2011). The brood stock oysters were spawned at AUSL in May 2011 and the larvae were reared and set to produce single spat as described above.

Spat from all 4 oyster stocks were reared in individual flow-through upwellers at a density of ~5000 oysters per upweller at AUSL. Once the majority of oysters were retained on a 6 mm grading screen, all stocks were placed in 4.5 mm mesh oyster bags and reared in off-bottom floating cages (OysterGro®) in Sandy Bay (Fig. 1), Alabama, until experimental deployment.

Study sites

The study was conducted in Dauphin Island waters (Billy Goat Hole, 30° 15′ 4″ N, 88° 4′ 46″ W) and in Sandy Bay (30° 22′ 59″ N, 88° 18′ 45″ W), both located in Mobile Bay National Estuary, Alabama (Fig. 1). Daily mean ± SD water salinity and temperature at Dauphin Island Sea Lab station, located ~150 m from our Dauphin Island study site, were 20.3 ± 7.7 and 21.9 ± 7.0°C for the 2007–2011 period (Dauphin Island Sea Lab station 30° 15′ 4″ N, 88° 4′ 40″ W, Mobile Bay National Estuary Program, www.mymobilebay.com). Point Aux Chenes is the closest station to the Sandy Bay study site with daily monitoring (30° 20′ 54″ N, 88° 25′ 6″ W, NOAA NERR National Estuarine Research Reserve System Centralized Data Management Office), and its salinity data showed a strong correlation (r = 0.904, p < 0.001) with ADPH 176 Alabama Department of Public Health station, which is even closer to our Sandy Bay site but does not continuously record environmental conditions (Fig. 1). Point Aux Chenes daily mean water salinity and temperature for the 2007–2011 period were 21.8 ± 6.1 and 22.1 ± 6.9°C, respectively.

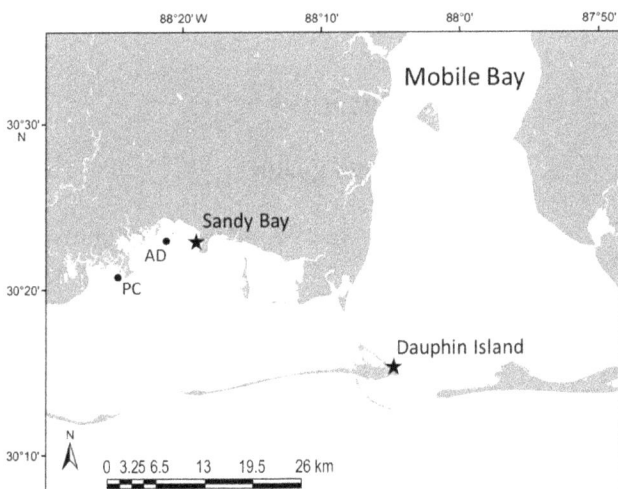

Fig. 1. Mobile Bay estuary, Alabama, USA, with Dauphin Island and Sandy Bay study sites (stars) where oysters were deployed. Water salinity and temperature recorders were located at the Dauphin Island Sea Lab station located ~150 m from our Dauphin Island study site, and at (black dots) Point Aux Chenes (PC) and Alabama Department of Public Health station 176 (AD)

Experimental design

In September 2011, 4 ALS aquaculture bags of each stock containing 75 oysters per bag were deployed at each study site, for a total of 300 oysters per stock and site. ALS bags were fully enclosed to prevent the risk of predation mortality and suspended beneath the water surface. Since predation was largely removed, mortality could be more readily attributed to stressful abiotic conditions and *P. marinus* infection. At both sites, oyster growth (shell height) and mortality (counts of live/dead) data were collected every other month, starting in November 2011 and ending in November 2012, for a total of 7 sampling periods. Condition index and *P. marinus* infection intensity data were collected at the time of deployment to establish pre-deployment baselines for each stock (N = 15 oysters per stock × 4 stocks, total of 60) and in March, July, and September 2012 (N = 15 oysters × 4 stocks × 2 sites, total of 120 per sampling time).

Water quality

Daily mean water salinity and temperature data from September 2011 to November 2012 were obtained from the Dauphin Island Sea Lab station and Point Aux Chenes monitoring stations. Mean water temperature and salinity for each sampling interval were calculated for use as predictor variables in multiple linear regression analyses to examine the relationships between water quality and interval mortality or growth of oysters.

Mortality

At each sampling time the numbers of live and dead oysters in each bag were recorded and the dead oysters were discarded. Interval mortality and cumulative mortality were calculated as described by Ragone Calvo et al. (2003b).

Growth

Shell height, the greatest distance between hinge and growth edge, was measured with digital calipers (Absolute Coolant Proof Calipers, series 500, resolution 0.01 mm, Mituyoto America Corporation) in 100 oysters selected haphazardly per stock, site, and sampling time (25 oysters per bag, 4 bags). Monthly interval growth rate was calculated as the increment in mean shell height between 2 consecutive sampling times divided by the number of days between samplings and normalized to a 30 d period. Mean growth rate was calculated using the mean shell height of each bag (N = 4).

P. marinus infection intensity

The number of *P. marinus* parasites per gram of wet oyster tissue was determined using the whole-oyster procedure as described by Fisher & Oliver (1996) and modified by La Peyre al. (2003). Oysters were further classified as either uninfected, lightly infected (i.e. $<10^4$ parasites g^{-1} wet tissue), moderately infected (i.e. 1×10^4 to 5×10^5 parasites g^{-1} wet tissue), or heavily infected (i.e. $>5 \times 10^5$ parasites g^{-1} wet tissue) (Bushek et al. 1994).

Condition index

Condition index was calculated as the ratio of dry tissue weight to the whole oyster weight minus its shell weight (i.e. filled cavity weight) multiplied by 100 using a variation of Hopkins' (1949) formula as recommended by Lawrence & Scott (1982). For each oyster, a 10 ml aliquot of oyster tissue homogenate that had been prepared to determine *P. marinus* infection intensity, was dried at 65°C for 48 h, and the dry weight for the whole oyster was calculated based on the total volume of homogenized tissue as described by La Peyre et al. (2003).

Statistical analyses

All statistical analyses were done using SigmaStat version 3.5 (Systat Software). Results for daily salinity and temperature from continuous data recorders were analyzed with a 1-factor (site) ANOVA followed by Tukey's multiple comparison procedure. Interval salinity and temperature were compared with Kruskal-Wallis 1-factor (sampling interval) ANOVA on ranks followed by Dunn's multiple comparison procedure. Cumulative mortalities (%) at the time oysters reached commercial size and at the end of the study were compared using a series of chi-square analyses to determine differences between stocks at each site. Interval mortality (%) at each sampling interval was analyzed with a 2-factor (stock, site) ANOVA followed by Tukey's multiple

comparison procedure. Interval mortality (%) of each stock was analyzed with a 2-factor (site, sampling interval) ANOVA followed by Tukey's multiple comparison procedure. Interval mortality data of all stocks and sites were pooled together and analyzed using a multiple linear regression with interval temperature and interval salinity as predictor variables. Shell height was compared with a 2-factor (stock, site) ANOVA at time of deployment (September 2011) and at the end of the study (November 2012) followed by Tukey's multiple comparison procedure. Shell height was also compared by a 1-factor (stock) ANOVA at the time oysters reached commercial size at each site. Interval growth rate (mm mo^{-1}) of each stock was analyzed with a 2-factor (site, sampling interval) ANOVA followed by Tukey's multiple comparison procedure. Interval growth rate data of all stocks and sites were pooled together and analyzed using a multiple linear regression with interval temperature, interval salinity and interval initial shell height as predictor variables. *P. marinus* body burden and CI were compared at the beginning of the study with 1-factor (stock) ANOVA and data collected in March, July and September were compared with a 2-factor (stock, sampling interval) ANOVA followed by Tukey's multiple comparison procedure. The number of oysters with heavy and moderate infections in the 4 stocks at each site was also compared with a Chi-square test. To achieve normality and homogeneity of variance, shell height and *P. marinus* body burden were log transformed. All data are reported as mean ± SD.

RESULTS

Water quality

Mean daily salinity was significantly higher at Sandy Bay (i.e. Point Aux Chenes, 23.4 ± 4.5, range 5.3–31.4) than at Dauphin Island (21.3 ± 4.9, range 7.6–31.7). Daily salinity at both sites throughout the year tended to be >18 except for the January–March interval. (Fig. 2). Interval salinity at Dauphin Island was significantly lower than at Sandy Bay during the November–January, January–March, and July–September intervals (Fig. 3).

Temperature followed expected seasonal trends and ranges for this region (Fig. 2), and differences between both sites were not found (Dauphin Island: 22.4 ± 5.7°C, range 9.5–31.3°C; Sandy Bay: 23.1 ± 5.7°C, range 10.3–31.8°C). The lowest interval temperatures were during the November–January and

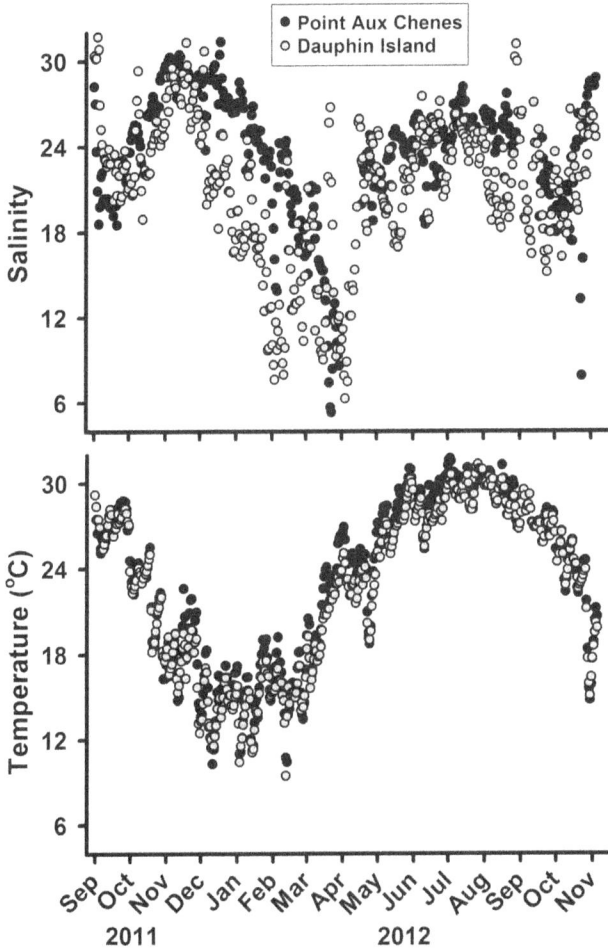

Fig. 2. Daily water temperature and salinity from September 2011 to November 2012 from continuous recorders at Dauphin Island Sea Lab and Point Aux Chenes stations (see Fig. 1)

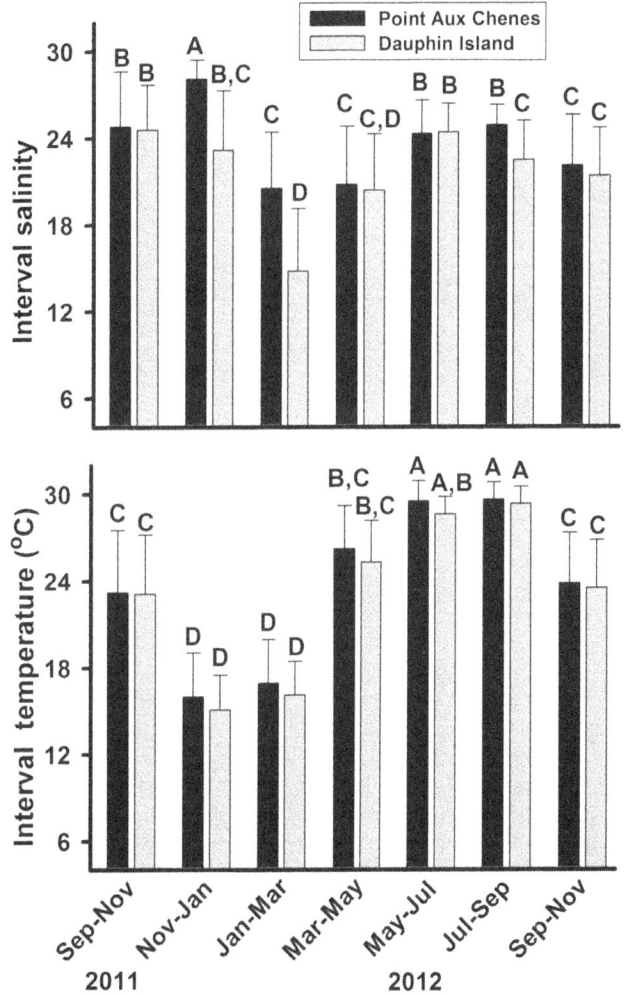

Fig. 3. Mean (±SD) interval water salinity and temperature from September 11, 2011 to November 12, 2012 calculated from daily salinity and temperature recorded at Dauphin Island Sea Lab and Point Aux Chenes stations. Groups that do not share letters have means that are significantly different (p < 0.05)

January–March intervals while the highest interval temperatures were during the May–July and July–September intervals (Fig. 3).

Mortality

At the end of the study, significant differences in cumulative mortality were found between stocks at Sandy Bay (p < 0.001) and Dauphin Island (p < 0.001). At Sandy Bay, the cumulative mortalities of both Louisiana stocks (Calcasieu Lake: 13%, OBOY: 11%) were significantly lower than the cumulative mortalities of either Alabama stocks (Perdido Pass: 29%, Cedar Point: 20%), and Cedar Point stock had significantly lower cumulative mortality than Perdido Pass stock (Fig. 4). At Dauphin Island, the cumulative mortality of the OBOY stock (9%) was significantly

lower and the cumulative mortality of Perdido Pass stock (26%) was significantly higher than the cumulative mortalities of either Cedar Point (18%) or Calcasieu Lake (17%) stocks.

In March 2012, when oysters reached commercial size at Sandy Bay, no difference in cumulative mortality between stocks could be shown (0.7–1.0% cumulative mortality, Fig. 4). At the time point when oysters reached commercial size at Dauphin Island in May 2012, cumulative mortality was significantly higher (p = 0.001) in Perdido Pass stock (6.2%) than in Cedar Point (1.1%) and OBOY (1.1%) stocks (Fig. 4).

Interval mortality (%) at each sampling interval was only significantly different between stocks during the March–May (p < 0.05) and July–September

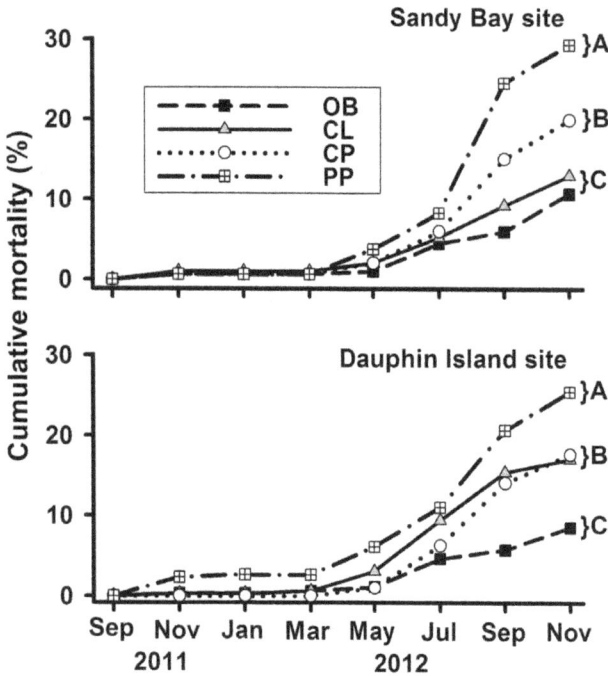

Fig. 4. Cumulative mortality of Louisiana eastern oysters from the OBOY (OB) and Calcasieu Lake (CL) stocks, and Alabama oysters from the Cedar Point (CP) and Perdido Pass (PP) stocks at the Sandy Bay and Dauphin Island study sites. Stocks that do not share a letter have cumulative mortalities that are significantly different at the end of the study (p < 0.05)

(p < 0.0001) intervals and differences between sites were not found. Interval mortality of the Perdido Pass stock was higher than the OBOY stock during the March–May interval, and than the OBOY, Calcasieu Lake and Cedar Point stocks in the July–September interval. The interval mortality of Cedar Point stock was also higher than of OBOY stock in the July–September interval.

When analyzed by stock, interval mortality was only significantly different between intervals (p < 0.001) and differences between sites were not found (Fig. 5). Overall mortalities of 4 mo old oysters deployed in September 2011 was very low (< 5%) during fall and winter intervals and increased significantly in summer intervals at both sites (Fig. 5).

Multiple linear regression analysis was used to determine the relationship between interval mortality and the potential predictors of interval temperature and interval salinity. Interval temperature (p < 0.001) but not interval salinity (p = 0.944) contributed significantly and produced a multiple regression model with interval mortality = −7.054 + (0.413 × interval temperature) + (0.009 × interval salinity) and an R^2 of 0.397.

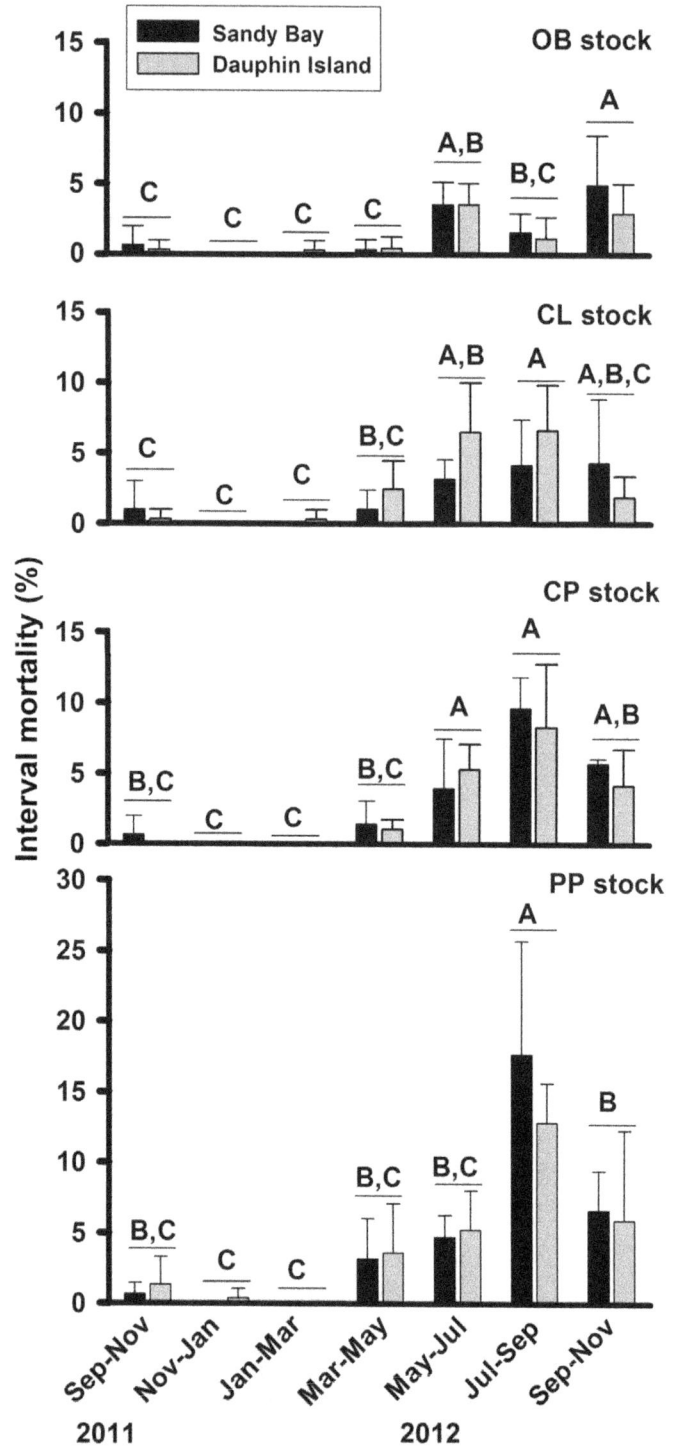

Fig. 5. Mean (±SD) interval mortality of Louisiana eastern oysters from the OBOY (OB) and Calcasieu Lake (CL) stocks, and Alabama oysters from the Cedar Point (CP) and Perdido Pass (PP) stocks at the Sandy Bay and Dauphin Island study sites. Interval mortality was significantly affected by sampling interval but not site. Groups that do not share letters have means that are significantly different (p < 0.05). Lines over bars indicate that no significant differences in means could be shown between sites

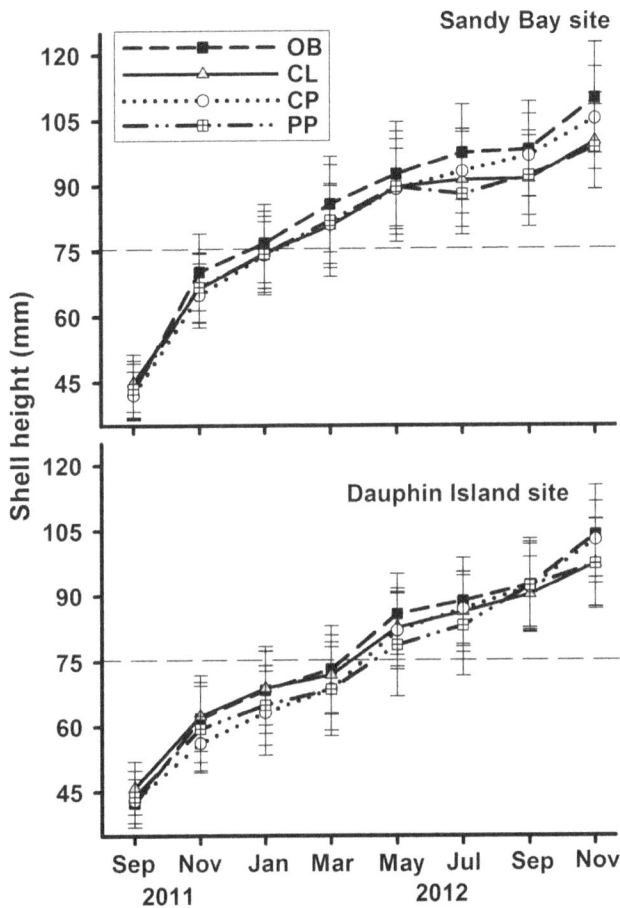

Fig. 6. Mean (±SD) shell heights of Louisiana eastern oysters from the OBOY (OB) and Calcasieu Lake (CL) stocks, and Alabama oysters from the Cedar Point (CP) and Perdido Pass (PP) stocks at the Sandy Bay and Dauphin Island study sites. Dashed lines at 75 mm represent harvest threshold shell height

Growth

At the time of deployment, in September 2011, no differences in shell height were found between stocks (in mm; OBOY: 42.7 ± 6.0, Calcasieu Lake: 45.4 ± 6.3, Cedar Point: 42.5 ± 5.2, Perdido Pass: 43.7 ± 6.3) or sites (in mm; Sandy Bay: 43.3 ± 6.3, Dauphin Island: 43.8 ± 5.8). Stocks reached commercial size (75 mm) in March 2012 at Sandy Bay and May 2012 at Dauphin Island, when the oysters were

Fig. 7. Mean (±SD) interval growth rates of Louisiana eastern oysters from the OBOY (OB) and Calcasieu Lake (CL) stocks, and Alabama oysters from the Cedar Point (CP) and Perdido Pass (PP) stocks at the Sandy Bay and Dauphin Island study sites. Interval growth rate of each stock was significantly affected by the interaction of site and sampling interval. Groups that do not share letters have means that are significantly different (p < 0.05)

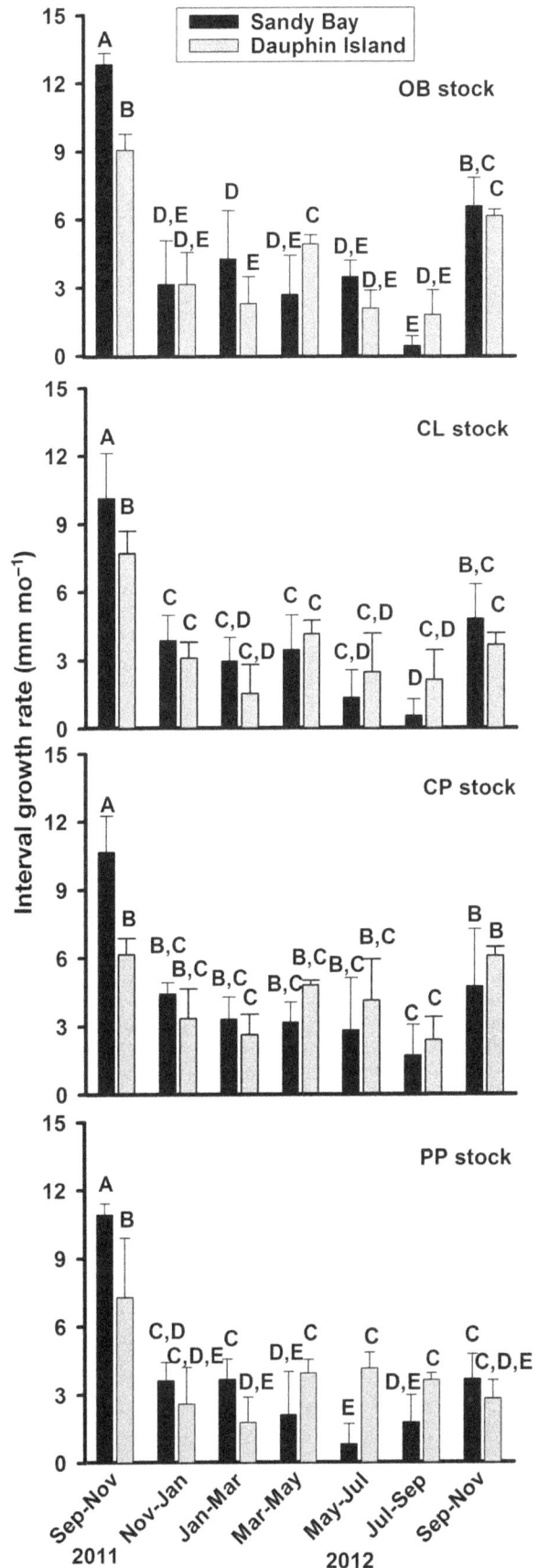

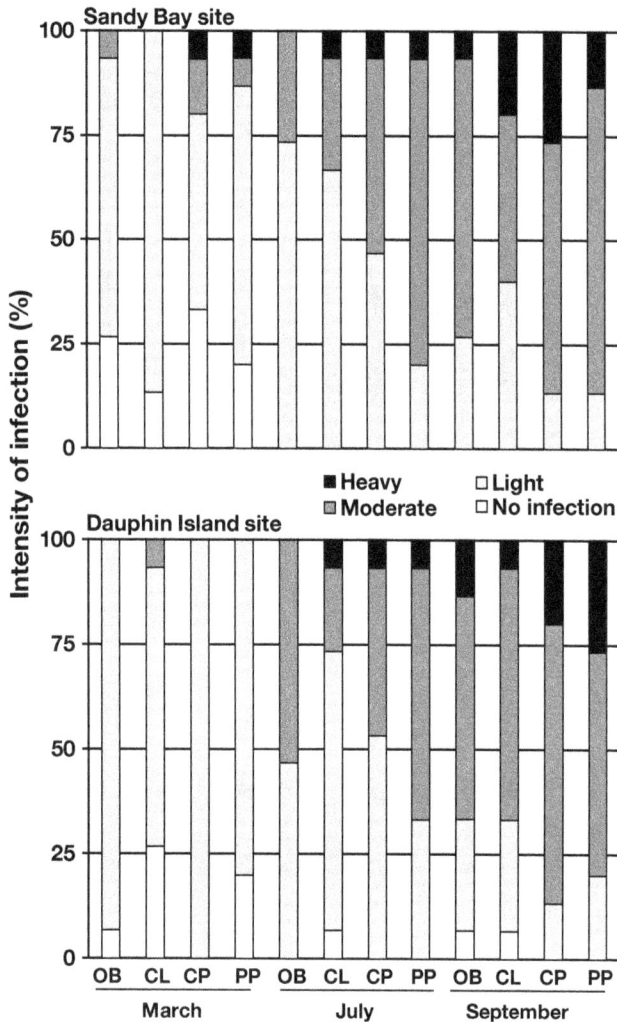

Fig. 8. Percentage of OBOY (OB), Calcasieu Lake (CL), Cedar Point (CP) and Perdido Pass (PP) oysters with no *Perkinsus marinus* infection and with light ($<10^4$ parasites g^{-1} wet tissue), moderate (1×10^4 to 5×10^5 parasites g^{-1} wet tissue) and heavy ($> 5 \times 10^5$ parasites g^{-1} wet tissue) infections sampled in March, July, and September 2012 at the Sandy Bay and Dauphin Island study sites

10 and 12 mo old, respectively (Fig. 6). In March 2012, OBOY stock (85.8 ± 10.9 mm) was significantly larger than Calcasieu Lake (80.8 ± 9.6 mm) and Cedar Point (81.1 ± 8.9 mm) stocks at Sandy Bay (p = 0.004). In May 2012, OBOY stock (85.8 ± 9.3 mm) was significantly larger than Cedar Point (82.0 ± 8.8 mm) and Perdido Pass (78.8 ± 11.8 mm) stocks at Dauphin Island, and Calcasieu Lake stock (82.7 ± 8.9 mm) was also larger than Perdido Pass stock (p < 0.001). At the end of the study in November 2012, there was a significant effect of stock (p < 0.001) and site (p < 0.001) with no stock × site interaction (p = 0.149); OBOY (106.8 ± 12.4 mm) and Cedar Point (104.1 ± 10.5 mm) oysters were significantly larger than Calcasieu Lake

(98.8 ± 11.1 mm) and Perdido Pass (98.1 ± 10.0 mm) oysters, and oysters at Sandy Bay (103.7 ± 12.4 mm) were significantly larger than at Dauphin Island (100.3 ± 10.5 mm).

Interval growth rate of each stock was significantly affected by the interaction of site and sampling interval (OBOY, Cedar Point, Perdido Pass: site × sampling interval interaction; p < 0.001, Calcasieu Lake: site × sampling interval interaction, p = 0.017; Fig. 7). Interval growth rates in OBOY stock were the highest (>6 mm mo^{-1}) in the September–November 2011 and September–November 2012 intervals and the lowest in the July–September interval. In the other stocks, there was only 1 high growth rate interval (>6 mm mo^{-1}) at the beginning of the study in the September–November 2011 interval, and the lowest growth rates were also in the July–September interval. Interval oyster growth rate was significantly higher at Sandy Bay than at Dauphin Island during the September–November 2011 interval and tended to be also higher in the January-March interval, while it tended to be lower than at Dauphin Island in the period between March and September 2012 (Fig. 7).

Multiple linear regression analysis was used to determine the relationship between interval growth rate and the potential predictors of interval temperature, interval salinity and initial shell height at the beginning of each interval. Interval temperature (p < 0.056) and initial shell height (p < 0.001), but not interval salinity (p = 0.337), contributed significantly and produced a multiple regression model with interval growth rate = 7.969 + (0.120 × interval temperature) + (0.0895 × interval salinity) − (0.117 × initial shell height) and an R^2 of 0.474.

P. marinus infection intensity

No significant differences in *P. marinus* infection intensity between stocks were detected at the time of deployment (in $\log_{10}$ parasites g^{-1} wet tissue; OBOY: 1.7 ± 1.4, Calcasieu Lake: 1.0 ± 1.2, Cedar Point: 1.5 ± 1.2, Perdido Pass: 1.3 ± 1.3). At Sandy Bay, there were significant effects of sampling times (p < 0.001) and stocks (p = 0.001) on *P. marinus* infection intensities after deployment. Infection intensities increased significantly from March (1.5 ± 1.5) to July (3.7 ± 1.6) and again from July to September (4.5 ± 1.5). Infection intensities of Calcasieu Lake (2.7 ± 1.9) and OBOY stocks (2.9 ± 1.8) were significantly lower than Perdido Pass stock (3.8 ± 2.0). Calcasieu Lake stock also had significantly lower infection intensity than Cedar Point stock (3.6 ± 2.1). Moreover, the combined percentage of oysters with moderate and heavy infections was significantly lower in the Louisiana stocks, OBOY (p = 0.035) and Calcasieu Lake (p = 0.011) stocks, than in the Perdido Pass stock. At Dauphin Island, *P. marinus* infection intensity was only significantly affected by sampling time (p = 0.047), with higher infection intensities in September (4.5 ± 1.6) compared with March (1.5 ± 1.0). No differences in the combined percentage of oysters with moderate and heavy infections could be shown between stocks at Dauphin Island (Fig. 8).

Condition index

No differences in mean condition indices between stocks were found at the time of deployment (OBOY: 16.9 ± 1.8, Calcasieu Lake: 15.1 ± 2.8, Cedar Point: 17.2 ± 3.1, Perdido Pass: 15.7 ± 1.8). After deployment, there was a significant effect of stock (p = 0.036) and sampling time (p < 0.001) in Dauphin Island. Specifically, Perdido Pass (10.5 ± 5.0) stock had a significantly greater condition index than Calcasieu Lake (9.5 ± 5.0) stock, and condition index in March (16.5 ± 1.5) was significantly greater than in July (7.0 ± 1.5) and September (7.0 ± 1.9). Oyster condition index in Sandy Bay was only affected by sampling time, with a significant decrease, by more than half, in condition index of all stocks from March (15.7 ± 2.0) to July (5.8 ± 1.3) and September (7.1 ± 2.3).

DISCUSSION

The mortality, growth, *Perkinsus marinus* infection intensity and condition index of 1 oyster stock selected for dermo resistance in Louisiana and 3 unselected stocks, 1 from Louisiana and 2 from Alabama estuaries, were compared at 2 Alabama sites. The Louisiana stock selected for dermo resistance showed significantly lower mortality than the unselected stocks over the course of the study at Dauphin Island, and both Louisiana stocks had lower mortality than the Alabama stocks at Sandy Bay, a slightly more saline site. Mortality peaked in summer concomitant with increasing *P. marinus* infection intensities and higher temperatures. The selected stock reached greater mean shell height than the unselected Louisiana stock and the Alabama oyster stock with the highest mortality and *P. marinus* infection intensity (i.e. Perdido Pass stock). The shell height of oysters at the higher salinity site was slightly but significantly greater than at the lower salinity site.

Growth rates decreased during intervals with higher temperatures and as shell heights at the beginning of each interval increased. Condition indices of the oyster stocks decreased by more than half between March and July, following expected oyster spawning, and stayed depressed throughout the summer. Differences in stock performance especially in dermo-related mortality highlight the importance of stock selection for intensive aquaculture in dermo-endemic estuaries of the northern Gulf of Mexico.

Less than 10% of the OBOY stock oysters selectively bred for dermo-resistance died by the end of the study compared to >25% of the worst-performing Alabama stock oysters (i.e. Perdido Pass). The greater oyster mortality in the worst-performing stock was associated with higher *P. marinus* infection intensities, similarly described in previous studies (Ray 1954, Mackin 1962). Specifically, in Sandy Bay, Perdido Pass stock oysters had significantly higher infection intensity and a greater percentage of moderately and heavily infected oysters with *P. marinus* than Louisiana stock oysters. In Dauphin Island the performance of Perdido Pass stock oysters followed similar trends but significant differences with the other stocks were not found. Mean daily salinity at Sandy Bay (23.4) was significantly greater than at Dauphin Island (21.3); still, differences were small and average salinities at both sites were well within the range favorable for parasite propagation (Chu et al. 1993, Dungan & Hamilton 1995, La Peyre et al. 2006). The salinities prevalent at Dauphin Island between the end of January and the end of March, however, were much lower than at Sandy Bay and lower than the optimum conditions for *P. marinus* proliferation (>15). Those salinities impeded the proliferation of *P. marinus*, as indicated by finding all oysters in March lightly or not-infected at Dauphin Island while some Alabama stock oysters (7–20%) had moderate and heavy infections at Sandy Bay.

The highest susceptibility to dermo of Perdido Pass stock oysters was unexpected as this population inhabits a high-salinity area and was anticipated to have developed some dermo resistance. For all the stocks, but especially for Perdido Pass stock, mortality peaked in summer concomitant with increasing *P. marinus* infection intensities (7–27% heavily infected oysters) at the higher temperatures. High temperatures are well known to promote the propagation of this parasite *in vitro* and *in vivo* in laboratories studies (Chu & La Peyre 1993, Dungan & Hamilton 1995, La Peyre et al. 2008). In field studies, increased infection intensities of *P. marinus* in eastern oysters have consistently been recorded when temperatures exceed 20°C and at salinity >12–15, leading to peak mortalities following summer high temperatures (Ray 1954, Mackin 1962, Ragone Calvo et al. 2003b).

The unselected Louisiana oyster stock, consisting of the progeny of wild oysters collected from Calcasieu Lake public oyster grounds, showed lower mortalities than the Perdido Pass stock at Dauphin Island and both Alabama stocks at the Sandy Bay site. The progeny of Calcasieu Lake oysters also showed the lowest mortality when compared to the progeny of oysters collected on oyster public grounds from other Louisiana estuaries (Stickler et al. 2001, Leonhardt et al. 2017). This is the reason why oysters from Oyster Bayou on the southern edge of Calcasieu Lake were used as the founding brood stock for the OBOY line. Oysters from this location are exposed to higher salinities than oysters of other Louisiana public grounds, which may have favored selection for increased survival upon *P. marinus* infection. Differences in disease susceptibility between oyster populations along both the Atlantic and Gulf of Mexico coasts have been reported and should continue to serve as the basis for brood stock development for local use (Bushek & Allen 1996, Stickler et al. 2001, Brown et al. 2005, Frank-Lawale et al. 2014).

At the end of the study, the selected OBOY stock which showed the lowest mortality also reached the highest shell height while the Perdido Bay stock with the highest mortality had the lowest shell height. As with mortality, the mean *P. marinus* infection intensities and percentage of oysters with moderate and heavy infection intensities which were greatest in Perdido Pass stock oysters than in OBOY stock oysters might explain the lower shell height; increase in *P. marinus* infection intensities can lead to a decrease in oyster growth rates (Menzel & Hopkins 1955, Paynter & Burreson 1991). Oyster growth rates were the lowest during the July–September interval and might be due to increased *P. marinus* infection intensities and the elevated temperatures, which often exceeded 30°C. High temperatures may have contributed directly to the decrease in growth rate because of reduction or cessation of pumping and shell closure as reported in past studies (Collier 1954, Loosanoff 1958). Interestingly, temperatures tended to be higher and oyster growth rate lower at Sandy Bay compared to Dauphin Island during the July–September interval. Overall, however, temperature and initial shell height were the major predictors of growth rates in this and previous studies (reviewed in Kraeuter et al. 2007).

The differences in shell heights and in salinities between sites were small but significant. Although salinity is a major factor controlling growth rates (Kraeuter

et al. 2007), the favorable salinity (generally ≥20) at both sites, along with relatively small differences in interval salinities between sites explain why salinity was not identified as a predictor of growth rates under our study's environmental conditions. While salinities at both sites were >20 during most intervals, it is interesting to note that when salinity was lower (~15) at Dauphin Island compared to Sandy Bay (~20) during the January–March interval, the oyster growth rates at Dauphin Island were also significantly lower than at Sandy Bay. Other factors such as phytoplankton quantity and quality or suspended sediment, which were not measured, may have also had some differential impact on oyster growth at both sites.

The condition index is a prime indicator of how well an oyster uses the shell cavity available for somatic and gonadal tissue growth and reflects physiological or nutritive status (Haven 1961, Rainer & Mann 1992). Oyster condition index is often used to estimate meat quality and yield (Lawrence & Scott 1982). The only significant difference found in condition index was between Perdido Pass and Calcasieu Lake stock oysters, but it is unlikely that the small difference impacted oyster meat quality. Quantitative comparison of oyster meat quality between stocks will be needed to confirm this assertion (Zhang et al. 2016)

The most significant differences between oyster stocks in our study were their mortalities, specifically between July and September during peak *P. marinus* infection intensities. It is therefore recommended that local growers preferably use stocks with greater dermo resistance at sites with comparable or higher salinity regime or harvest oysters prior to this time. In our current study, oysters could have been harvested before summer because they were spawned early (i.e. May) and reached market size within 12 mo, but this strategy may not always work in years where local environmental conditions may not be as favorable for growth as exemplified in an earlier study (Walton et al. 2013b), or where harvest is not allowed due to environmental conditions (e.g. prolonged closures due to rainfall or harmful algal blooms). Better characterization or understanding of the processes involved in disease resistance to dermo may lead in the future to the development of specific markers for use in candidate gene or marker-assisted selection (Cancela et al. 2010, La Peyre et al. 2010). This is important for the development of off-bottom aquaculture at high salinity, an environmental condition that is well known to increase eastern oyster growth rate but also dermo-related mortalities (Ragone Calvo et al. 2003b, Kraeuter et al. 2007, Bushek et al. 2012).

Acknowledgements. We thank Julie Davis at Auburn for field help and Jacqueline Tai, Jaren Lee, Alexis Allen, and April Chow at the Louisiana State University for laboratory help. This research was funded by the National Oceanic and Atmospheric Administration Aquaculture Program.

LITERATURE CITED

Breithaupt RL, Dugas RJ (1979) A study of the southern oyster drill (*Thais haemastoma*) distribution and density on the oyster seed grounds. Louisiana Wildlife and Fisheries Commission, Tech Bull 30, New Orleans, LA

Brown BL, Butt AJ, Shelton SW, Meritt D, Paynter KT (2005) Resistance of dermo in eastern oysters, *Crassostrea virginica* (Gmelin), of North Carolina but not Chesapeake Bay heritage. Aquacult Res 36:1391–1399

Bushek D, Allen SK Jr (1996) Host–parasite interactions among broadly distributed populations of the eastern oyster *Crassostrea virginica* and the protozoan *Perkinsus marinus*. Mar Ecol Prog Ser 139:127–141

Bushek D, Ford SE, Allen SK Jr (1994) Evaluation of methods using Ray's fluid thioglycollate medium for diagnosis of *Perkinsus marinus* infection in the eastern oyster, *Crassostrea virginica*. Annu Rev Fish Dis 4:201–217

Bushek D, Ford SE, Burt I (2012) Long-term patterns of an estuarine pathogen along a salinity gradient. J Mar Res 70:225–251

Cancela ML, Bargelloni L, Boudry P, Boulo V and others (2010) Genomic approaches in aquaculture and fisheries. In: Cock JM, Viard F, Tessamar-Raible K, Boyen C (eds) Introduction to Marine Genomics. Springer-Verlag, Heidelberg, p 213–286

Chu FLE, La Peyre JF (1993) *Perkinsus marinus* susceptibility and defense-related activities in eastern oysters *Crassostrea virginica*: temperature effects. Dis Aquat Org 16:223–234

Chu FLE, La Peyre JF, Burreson CS (1993) *Perkinsus marinus* infection and potential defense related activities in eastern oysters, *Crassostrea virginica*: salinity effects. J Invertebr Pathol 62:226–232

Collier A (1954) A study of the response of oysters to temperature, and some long range ecological interpretations. In: Papers delivered at the Convention of the National Shellfisheries Association, New Orleans, LA, June 22–25, 1953. Fish and Wildlife Service, US Department of Interior, Washington, DC, p 13–38

Craig A, Powell EN, Fay RR, Brooks JM (1989) Distribution of *Perkinsus marinus* in gulf-coast oyster populations. Estuaries 12:82–91

Davis CV, Barber BJ (1999) Growth and survival of selected lines of eastern oysters, *Crassostrea virginica* (Gmelin 1791) affected by juvenile oyster disease. Aquaculture 178:253–271

Dungan CF, Hamilton RM (1995) Use of a tetrazolium based cell proliferation assay to measure the effects of *in vitro* conditions on *Perkinsus marinus* (Apicomplexa) proliferation. J Eukaryot Microbiol 42:379–388

Dupuy JL, Windsor NT, Sutton CE (1977) Manual for design and operation of an oyster seed hatchery for the American oyster *Crassostrea virginica*. Spec Rep No 142, Virginia Institute of Marine Science, Gloucester Point, VA

Fisher WS, Oliver LM (1996) A whole-oyster procedure for diagnosis of *Perkinsus marinus* disease using Ray's fluid thioglycollate culture medium. J Shellfish Res 15:109–118

Frank-Lawale A, Allen SK Jr, Degremont L (2014) Breeding and domestication of eastern oyster (*Crassostrea virginica*) lines for culture in the mid-Atlantic, USA: line development and mass selection for disease resistance. J Shellfish Res 33:153–165

Guo X, Wang Y, DeBrosse G, Bushek D, Ford SE (2008) Building a superior oyster for aquaculture. Jersey Shoreline 25:7–9

Haskin HH, Ford SE (1979) Development of resistance to *Minchinia nelsoni* (MSX) mortality in laboratory-reared and native oyster stocks in Delaware Bay. Mar Fish Rev 41:54–63

Haven D (1961) Seasonal cycle of condition index of oysters in the York and Rappahannock rivers. Proc Natl Shellfish Ass 52:42–66

Hopkins AE (1949) Determination of condition of oysters. Science 110:567–568

Kraeuter JN, Ford S, Cummings M (2007) Oyster growth analysis: a comparison of methods. J Shellfish Res 26:479–491

La Peyre JF, Xue QG, Itoh N, Li Y, Cooper RK (2010) Serine protease inhibitor cvSI-1 potential role in the eastern oyster host defense against the protozoan parasite *Perkinsus marinus*. Dev Comp Immunol 34:84–92

La Peyre MK, Nickens AD, Volety AK, Tolley GS, La Peyre JF (2003) Environmental significance of freshets in reducing *Perkinus marinus* infection in eastern oysters *Crassostrea virginica*: potential management applications. Mar Ecol Prog Ser 248:165–176

La Peyre M, Casas S, La Peyre J (2006) Salinity effects on viability, metabolic activity and proliferation of three *Perkinsus* species. Dis Aquat Org 71:59–74

La Peyre MK, Casas SM, Villalba A, La Peyre JF (2008) Determining the effects of temperature on two *Perkinsus* species viability, metabolic activity and proliferation and its significance to understanding seasonal cycles of perkinsosis. Parasitology 135:505–519

La Peyre MK, Geaghan J, Decossas G, La Peyre JF (2016) Analysis of environmental factors influencing salinity patterns, oyster growth and mortality in lower Breton Sound Estuary, Louisiana using 20 years of data. J Coast Res 32:519–530

Lawrence DR, Scott GI (1982) The determination and use of condition index of oysters. Estuaries 5:23–27

Leonhardt JM, Casas S, Supan JE, La Peyre JF (2017) Stock assessment for eastern oyster seed production and field grow-out in Louisiana. Aquaculture 466:9–19

Loosanoff VL (1958) Some aspects of behavior of oysters at different temperatures. Biol Bull 114:57–70

Mackin JG (1962) Oyster disease caused by *Dermocystidium marinum* and other microorganisms in Louisiana. Publ Inst Mar Sci Univ Tex 7:132–229

Matthiessen GC, Feng SY, Leibovitz L (1990) Patterns of MSX (*Haplosporidium nelsoni*) infection and subsequent mortality in resistant and susceptible strains of the eastern oyster *Crassostrea virginica* (Gmelin, 1971), in New England. J Shellfish Res 9:359–366

Maxwell VJ, Supan J, Schiavinato LC, Showalter S, Treece GD (2008) Aquaculture parks in the coastal zone: a review of legal and policy issues in the Gulf of Mexico state waters. Coast Manage 36:241–253

Menzel RW, Hopkins SH (1955) The growth of oysters parasitized by the fungus *Dermocystidium marinum* and by the trematode *Bucephalus cuculus*. J Parasitol 41:333–342

Paynter KT, Burreson EM (1991) Effects of *Perkinsus marinus* infection in the eastern oyster, *Crassostrea virginica* II. Disease development and impact on growth rate at different salinities. J Shellfish Res 10:425–431

Petrolia DR, Walton WC, Yehouenou L (2017) Is there a market for branded Gulf of Mexico oysters? J Agric Appl Econ 49:45–65

Proestou DA, Vinyard BT, Corbett RJ, Piesz J and others (2016) Performance of selectively-bred lines of eastern oyster, *Crassostrea virginica*, across eastern US estuaries. Aquaculture 464:17–27

Ragone Calvo LM, Calvo GW, Burreson EM (2003a) Dual disease resistance in a selectively bred eastern, *Crassostrea virginica*, strain tested in Chesapeake Bay. Aquaculture 220:69–87

Ragone Calvo LM, Dungan CF, Roberson BS, Burreson EM (2003b) Systematic evaluation of factors controlling *Perkinsus marinus* transmission dynamics in lower Chesapeake Bay. Dis Aquat Org 56:75–86

Rainer JS, Mann R (1992) Comparison of methods for calculating condition index in eastern oysters, *Crassostrea*

virginica (Gmelin, 1791). J Shellfish Res 11:55–58

Ray SM (1954) Biological studies of *Dermocystidium marinum*. Rice Institute Pamphlet 41, Rice Institute, Houston, TX

Stickler S, Wagner E, Supan J, Allen S, La Peyre J (2001) Natural Dermo resistance and its role in the development of hatcheries for the Gulf of Mexico. J Shellfish Res 20: 557 (Abstract)

Walton WC, Davis JE, Supan JE (2013a) Off-bottom culture of oysters in the Gulf of Mexico. SRAC Publication-Southern Regional Aquaculture Center No. 4308, Stoneville, MS

Walton WC, Rikard FS, Chaplin GI, Davis JE, Arias CR, Supan JE (2013b) Effects of ploidy and gear on the performance of cultured oysters, *Crassostrea virginica*: survival, growth, shape, condition index and *Vibrio* abundances. Aquaculture 414–415:260–266

Zhang J, Walton WC, Wang Y (2016) Quantitative quality evaluation of eastern oyster (*Crassostrea virginica*) cultured by two different methods. Aquac Res, doi:10.1111/are.13126

13

Simulation of mussel *Mytilus galloprovincialis* growth with a dynamic energy budget model in Maliakos and Thermaikos Gulfs (Eastern Mediterranean)

Yannis Hatzonikolakis[1,2], Kostas Tsiaras[2], John A. Theodorou[3], George Petihakis[4], Sarantis Sofianos[1], George Triantafyllou[2,*]

[1]Department of Environmental Physics, University of Athens, 15784 Athens, Greece

[2]Hellenic Centre for Marine Research (HCMR), Athens-Sounio Avenue, Mavro Lithari, 19013 Anavyssos, Greece

[3]Department of Fisheries and Aquaculture Technology, Technological Educational Institute of Western Greece, Nea Ktiria, Mesolonghi 30200, Greece

[4]Hellenic Centre for Marine Research (HCMR), 71003 Heraklion, Greece

ABSTRACT: A dynamic energy budget (DEB) model was developed to investigate the growth and reproduction of cultured bivalve species raised under different environmental conditions (varying phytoplankton carbon biomass [Phyto-C], particulate organic carbon [POC] and temperature) and tuned against field data for *Mytilus galloprovincialis* from the Maliakos and Thermaikos Gulfs (Aegean Sea, Greece). Values of most DEB model parameters were adopted from the literature, while half saturation constant (X_k) and initial values of energy reserves (E) and reproductive buffer (R) were calibrated. Different values have been found for X_k in the 2 areas (Maliakos: $X_k = 36$ mg C m^{-3}; Thermaikos: $X_k = 28$ mg C m^{-3}), suggesting that X_k should be treated as a site-specific parameter. Food density (X) was adapted to include not only Phyto-C but also POC in the diet of *M. galloprovincialis* and only when Phyto-C density was low compared to POC density. Results showed a small contribution of POC during spring in the Maliakos Gulf and almost none at Thermaikos Gulf. The simulated mussel growth showed good agreement with field data. Sensitivity tests on the calibrated parameters (E, R and X_k) were performed to investigate model uncertainty. The standard deviation of simulations with perturbed parameter/initial values remained relatively small and appeared to increase as the modeled mussel grew, in agreement with observations.

KEY WORDS: Dynamic energy budget · DEB model · Mussel culture · *Mytilus galloprovincialis* · Growth · Eastern Mediterranean · Uncertainty · Ensemble forecasting

INTRODUCTION

According to the Food and Agriculture Organization of the United Nations (FAO), the world's population will reach 8 billion people in 2030. This increase is not reflected in wild fisheries or oyster production, which have been practically steady since 1989. On the other hand, world aquaculture production has shown an enormous growth in the last 3 decades, increasing from 14% of total seafood production in 1988 to 44.1% in 2014 (FAO 2016). A significant amount of this aquaculture production comes from shellfish farms. For example, in Europe in 2009, 57% of total aquaculture production came from mussel, clam and oyster farming (Eurostat 2016). Considering that shellfish farming takes place in coastal zones which are affected both by anthropogenic pressures and climate change, it is important to analyze and understand the processes that affect production. Among the most important cultured bivalve species is the Mediterranean mussel *Mytilus galloprovincialis*, which has significant global production (116 262 metric tonnes [t] in 2014

*Corresponding author: gt@hcmr.gr

according to FAO[1]) and is mainly cultured on the northern shores of the Mediterranean Sea (Rodrigues et al. 2015). Greece contributes significantly to Mediterranean *M. galloprovincialis* farming, with an estimated maximum farming carrying capacity up to 35 000 or 40 000 t gross weight (although production levels are currently lower: 18 000 t in 2014; FGM 2015). Because of the overall oligotrophic characteristics of the Mediterranean, major farming areas are only found close to estuarine systems. In Greece, these are mainly located in the northern part of the country where major rivers discharge (Theodorou et al. 2011, 2015a), and are more scattered in other areas of continental Greece (Fig. 1).

In the present study, the growth of *M. galloprovincialis* in the Maliakos and Thermaikos Gulfs was investigated. A model describing the growth of *M. galloprovincialis* was developed based on dynamic energy budget (DEB) theory (Kooijman 1986, 2000). The model simulates the growth of an individual cultured mussel, assuming that the simulated individual represents the average state of the farm's population. DEB models have been applied successfully for several bivalve species (Casas & Bacher 2006, Pouvreau et al. 2006, Zaldívar 2008, Bourlès et al. 2009, Troost et

---

[1]China and Spain mussel production is not reported as *Mytilus galloprovincialis* production by FAO

al. 2010, Handa et al. 2011, Thomas et al. 2011, Wijsman & Smaal 2011, Sarà et al. 2012, among others).

DEB models show important benefits in describing the growth of an individual organism and have the powerful aspect of being generic: DEB theory assumptions represent physiological processes that are common among different species, with their differences reflected only in the values of the parameters. Moreover, DEB models can be used as a basis for modeling other processes, e.g. concentrations of contaminants in an individual (Zaldívar 2008) or bioaccumulation of trace metals (Casas & Bacher 2006). Larsen et al. (2014) compared the results of bio-energetic growth (BEG), scope for growth (SFG) and DEB models on growth data of blue mussels from Danish waters, and concluded that the DEB model provided the best results and predictions regarding mussel somatic growth. Brigolin et al. (2009) chose a model based on the dynamic estimation of SFG as being more appropriate to study nutrient, carbon and phosphorus fluxes related to ingestion and the production of feces and pseudofeces in *M. galloprovincialis*. In their study, they also compared the results between their model and the DEB model produced by Casas & Bacher (2006), concluding that both models can give good simulations regarding the somatic growth of *M. galloprovincialis*.

The primary objective of this work was to develop a model describing the growth and reproduction of the Mediterranean mussel *M. galloprovincialis* that can be used to investigate processes affecting production on shellfish farms, offering a useful tool for the study of mussel farming in the Maliakos and Thermaikos Gulfs and estimating the carrying capacity of the study areas.

MATERIALS AND METHODS

Study area and mussel growth data

Maliakos Gulf is a semi-enclosed, shallow (14 m mean depth) estuarine embayment in the central western part of the Aegean Sea which covers a total surface of 110 km². It receives fresh water discharge from the Spercheios River at an average rate of 68 m³ s⁻¹; there is also some inflow of more saline water from the N. Evoikos Gulf through an anti-clockwise circulation in the

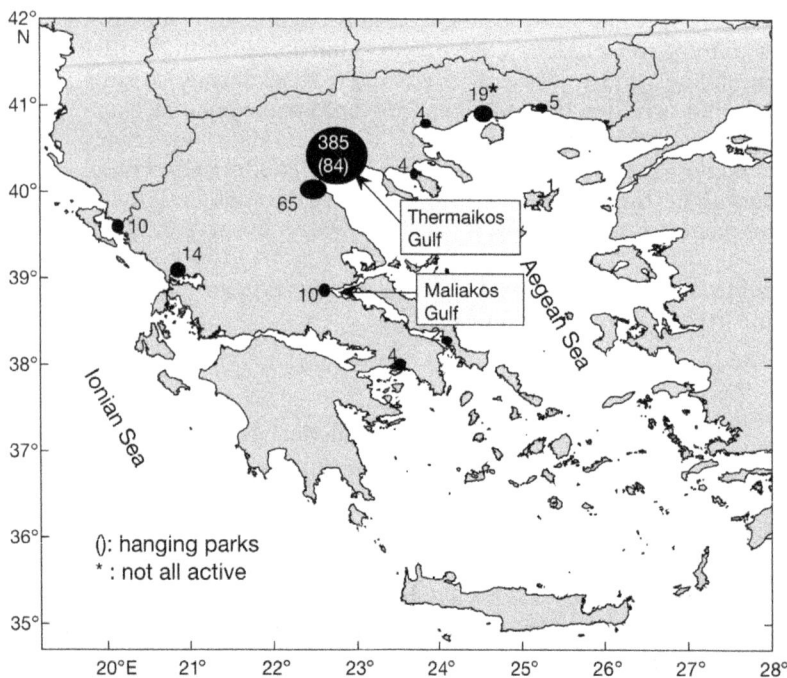

Fig. 1. Mussel farms (black circles) in Greece, Eastern Mediterranean. Numbers of floating longlines (no. of hanging parks in brackets) in the area are indicated. (★) Licensed but not yet active longlines. Study areas (Thermaikos and Maliakos Gulfs) are indicated (adapted from Theodorou et al. 2011, 2015a)

northern part of the gulf (Christou et al. 1995). Mussel farming was established in late 1980s, and today there are 10 farms with an estimated total annual production of around 1500 to 1700 t yr^{-1} (Dimitriou et al. 2015, Theodorou et al. 2015b).

Monthly mussel growth data (i.e. fresh tissue mass [g] and shell length [cm]) were derived from a farm (CalypsoSeafood/Aqua-Consulting) in the area of Molos (Southern Maliakos). Specifically, 3 pergolaries (mussel 'socks' made of plastic cylindrical nets) 3 m in length were filled with mussel seed (n = 30) of average (±SD) length (3.68 ± 0.53 cm) and weight (1.43 ± 0.21 g). The pergolaries were attached to a 'mother' longline rope at the edge of the farm in September 2004. To estimate animal growth, on a monthly basis 30 to 40 mussels were randomly collected from this batch for morphometric analysis during the period from October 2004 to June 2005.

Thermaikos Gulf is a semi-enclosed basin with a total surface area of 5100 km^2, located in the northwest Aegean. Depth varies from 10 to 75 m and the tidal range is 0.25 m. The inner Thermaikos Gulf is one of the few areas in Greece that can be characterized as eutrophic (Pagou 2005, Papakonstantinou et al. 2007), and contains one of the most extensive and productive mussel aquacultures, reaching 90% of total production in Greece (Konstantinou et al. 2015).

Mussel growth data, provided by Kravva (2000) from the coastal area of Chalastra, were used for tuning the Thermaikos Gulf model. This area is influenced by 2 of the most important rivers (Axios, Aliakmon) in the northern Aegean. It receives significant inputs of particulate matter and dissolved constituents (Price et al. 2005), creating favorable conditions for phytoplankton growth and thus is an ideal area for the cultivation of mussels.

Environmental data

Near-surface temperature and phytoplankton carbon biomass (Phyto-C) data (Fig. 2) were used as forcing functions for the DEB mussel model implementations in the Maliakos and Thermaikos Gulfs.

Data provided for Maliakos (August 2004 to August 2005) are outcomes of the Project ARCHIMIDES I –EPEAEK II-EU (contract no. 10012-00004) 'Environmental Interactions of the Mussel farming' (2004-2007). Relevant data used in the present study are from the deliverables of this effort presented in Theodorou et al. (2006a,b 2007) and Kakali et al. (2006). Monthly data samples at 4 depths (0.5, 2, 3 and 6 m) were used to calculate mean near-surface

chlorophyll *a* (chl *a*) and temperature. Phyto-C was obtained from chl *a* data, assuming a constant carbon:chl *a* ratio (50:1) that can be considered as a mean value of ratio's seasonal variability (Malone & Chervin 1979, Geider & Piatt 1986, Kormas et al. 2002). These datasets were used to force the model, and their values at each model time step were obtained by linear interpolation from the monthly data. Particulate organic carbon (POC) used in model simulations was obtained from Kormas et al. (2002).

In Thermaikos Gulf, input data (temperature, Phyto-C, POC) for the period during which mussel growth data were available (May 1995 to July 1996) were obtained from a 3-dimensional (3-D) hydrodynamic/biochemical long-term model simulation over the 1980 to 2000 period (Tsiaras et al. 2014). The hydrodynamic model was based on the Princeton Ocean Model (POM; Blumberg & Mellor 1983) while the biochemical model was based on the European Regional Seas Ecosystem Model (ERSEM; Baretta et al. 1995). The 3-D model results were validated against available SeaWiFS chl *a* and *in situ* data (Tsiaras et al. 2014).

The 2 coastal environments show some differences, mostly related to the maximum values and variability of Phyto-C. In Maliakos Gulf, Phyto-C reaches its highest value in winter (277 mg C m^{-3}), when the phytoplankton bloom takes place. This bloom is attributed to the supply of nutrients from Sperchios River in the early winter and is characterized by rapid sedimentation of phytoplankton cells to the shallow seafloor (Kormas et al. 1998). In Thermaikos Gulf (Chalastra), Phyto-C peaks in late April with a value of 138 mg C m^{-3}. Although Phyto-C in Maliakos Gulf exhibits a significantly higher peak compared to Thermaikos, the annual mean values are similar in the 2 areas (Maliakos: 94.5 mg C m^{-3}; Thermaikos: 90 mg C m^{-3}). In Maliakos Gulf, Phyto-C was high throughout August 2004 to February 2005, but afterward decreased to very low levels between March 2005 and August 2005. Phyto-C in Thermaikos Gulf showed more moderate fluctuations. This can be attributed to the fact that the 3-D hydrodynamic/biochemical model of Thermaikos Gulf adopts a climatologic seasonal variability for river discharge and therefore cannot capture high-frequency variability.

DEB model

Description of the DEB mussel model

The basic assumption of a DEB model is that the assimilated food first enters a reserve pool and then is

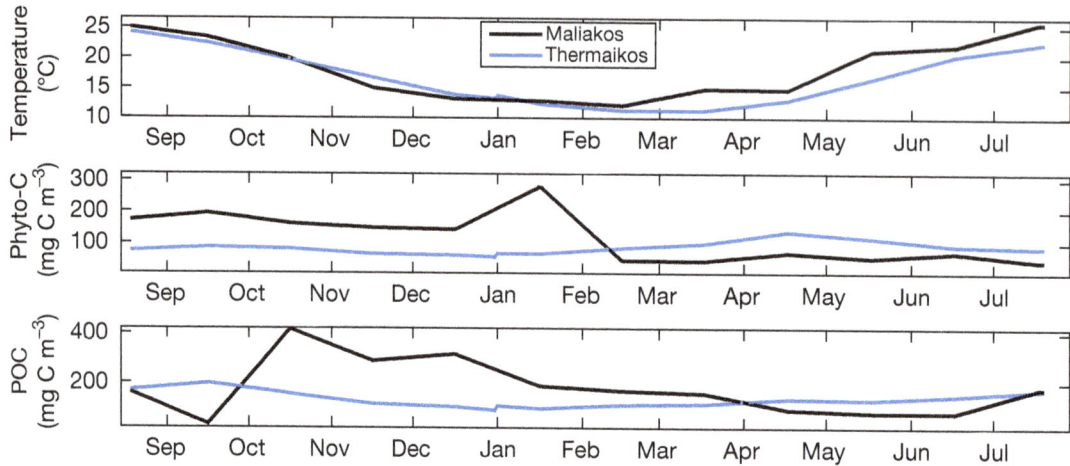

Fig. 2. Environmental data used for the forcing of the dynamic energy budget model in the Maliakos (black line) and Thermaikos (blue line) Gulf simulations, showing temperature (top), phytoplankton carbon biomass (middle), and particulate organic carbon (bottom)

allocated between the other compartments: a fixed part, κ, is spent on somatic maintenance and growth, while the remaining, 1 − κ, on maturity maintenance and reproduction. This rule is known as the κ-rule. The individual is characterized by 3 state variables: structural volume V (cm^3), energy reserves E (joules) and energy allocated to development and reproduction R (joules), while the environment of the individual is described by food density and temperature. All model equations describing the time evolution of feeding, maintenance, growth, development and reproduction are shown in Table 1; descriptions of model variables are provided in Table 2. The interested reader may refer to Zaldívar (2008) and Casas & Bacher (2006) for a full description of the model equations.

Parameter values

Most of the model parameters used in the present study for *Mytilus galloprovincialis* are adapted from those estimated by Van der Veer et al. (2006) for the blue mussel *M. edulis* L. in the northeast Atlantic (see Table 3 for exceptions). The 2 mussel species are closely related and are very similar 'with no single morphological or genetic character being clearly diagnostic' (Gosling 1984, p. 554). Preliminary experimentation showed that this specific parameterization resulted in good agreement of *M. galloprovincialis* model-simulated growth with observations. A similar approach was used by Casas & Bacher (2006) for *M. galloprovincialis* along the French Mediterranean shoreline. The half-saturation coefficient (X_k)

was tuned in the current study in order to obtain a better fit of model-simulated growth with observations. Widdows et al. (1984) and later Camacho et al. (1995) showed that the differences in physiological responses among populations, which are mainly

Table 1. Dynamic energy budget model: equations. See Table 2 for model variables, Table 3 for parameters and Table 4 for initial values

$$\frac{dE}{dt} = \dot{p}_a - \dot{p}_c \tag{1}$$

$$\frac{dV}{dt} = \frac{\kappa \cdot \dot{p}_c - [\dot{p}_M] \cdot V}{[E_g]} \tag{2}$$

$$\frac{dR}{dt} = (1-k) \cdot \dot{p}_c - [\frac{1-\kappa}{\kappa}] \cdot \min(V, V_p) \cdot [\dot{p}_M] \tag{3}$$

$$\dot{p}_a = \{\dot{p}_{Am}\} \cdot f \cdot k(T) \cdot V^{2/3} \tag{4}$$

$$f = \frac{X}{X + X_k} \tag{5}$$

$$\dot{p}_c = \frac{[E]}{[E_g] + \kappa \cdot [E]} \cdot \left(\frac{[E_g] \cdot \{\dot{p}_{AM}\} \cdot k(T) \cdot V^{2/3}}{[E_m]} + [\dot{p}_M] \cdot V \right) \tag{6}$$

$$[E] = \frac{E}{V} \tag{7}$$

$$[\dot{p}_M] = k(T) \cdot [\dot{p}_M]_m \tag{8}$$

$$k(T) = \frac{\exp(\frac{T_A}{T_1} - \frac{T_A}{T})}{1 + \exp(\frac{T_{AL}}{T} - \frac{T_{AL}}{T_L}) + \exp(\frac{T_{AH}}{T_H} - \frac{T_{AH}}{T})} \tag{9}$$

$$L = \frac{V^{1/3}}{\delta_m} \tag{10}$$

$$W = d \cdot \left(V + \frac{E}{[E_g]} \right) + \frac{R}{\mu_E} \tag{11}$$

Table 2. Dynamic energy budget model: variables

Variable	Description	Units
V	Structural volume	cm^3
E	Energy reserves	J
R	Energy allocated to development and reproduction	J
$\dot{p}_a$	Assimilation energy rate	$J\ d^{-1}$
$\dot{p}_c$	Energy utilization rate	$J\ d^{-1}$
f	Functional response function	–
X	Food density	$mg\ C\ m^{-3}$
$[\dot{p}_M]$	Maintenance costs	$J\ cm^{-3}\ d^{-1}$
T	Temperature	K
$k(T)$	Temperature dependence	–
L	Shell length	cm
W	Fresh tissue mass	g

responsible for differences in growth rate, are mainly the result of environmental conditions and to a lesser extent, genetic differences. Therefore, a good modeling approach would be to capture the differences in physiological responses among the 2 similar species and populations (Hilbish et al. 1994, Fly & Hilbish 2013) by the site-specific parameter, X_k (Troost et. al. 2010). X_k is the amount of food (Phyto-C, POC) where food uptake is at half its maximum value (see 'Food density' below) and may be considered representative of the environment to which the organism has adapted. A higher value of X_k is thus expected in more productive environments, such as the French Mediterranean shoreline as in Casas & Bacher (2006) ($X_k = 3.88\ \mu g\ l^{-1}$; ~194 mg C m^{-3}) and a

lower value in less productive areas, such as the Aegean coastal areas. Specific density (d) was set to 1 g cm^{-3} as suggested by Kooijman (2000). The values of all DEB model parameters used are summarized in Table 3.

Initial values

The initial values used for each simulation run are shown in Table 4. For both study areas, the initial value of shell length (L) was set from the field data, while initial V was calculated from Eq. (10) (see Table 1). For the Thermaikos simulation, initial R was chosen to be 0; the initial mussel body volume, V, indicates that the individual is in the juvenile phase ($V < V_p$; see Table 3) and thus it was assumed that the animal has no energy allocated for reproduction yet, following the same approach as Thomas et al. (2011). In Maliakos Gulf, the initial V, obtained from the field data, suggests that the individual is mature and thus some energy has to be allocated for reproduction (R). In the absence of available data, a model simulation initialized as in the Thermaikos Gulf ($V < V_p$ and $R = 0$) was used to estimate the initial R at the observed mussel length ($L = 3.68$ cm) and weight ($W = 1.43$ g) at Maliakos. For both study areas, the initial value of E was calibrated (Table 4), so that the computed initial W (Eq. 11) shows the best fit with field data. Regarding initial allocation between E and R, Rosland et al. (2009) performed sensitivity experiments and demonstrated that it has little impact on the results.

Table 3. Dynamic energy budget model: parameters

Parameter	Units	Description	Value	Reference
$\{\dot{p}_{Am}\}$	$J\ cm^{-2}\ d^{-1}$	Maximum surface area-specific assimilation rate	147.6	Van der Veer et al. (2006)
X_k	$mg\ C\ m^{-3}$	Half saturation coefficient	Calibrated	–
T_A	K	Arrhenius temperature	5800	Van der Veer et al. (2006)
T_I	K	Reference temperature	293	Van der Veer et al. (2006)
T_L	K	Lower boundary of tolerance rate	275	Van der Veer et al. (2006)
T_H	K	Upper boundary of tolerance rate	296	Van der Veer et al. (2006)
T_{AL}	K	Rate of decrease of lower boundary	45430	Van der Veer et al. (2006)
T_{AH}	K	Rate of decrease of upper boundary	31376	Van der Veer et al. (2006)
$[\dot{p}_M]_m$	$J\ cm^{-3}\ d^{-1}$	Volume specific maintenance costs	24	Van der Veer et al. (2006)
$[E_G]$	$J\ cm^{-3}$	Volume specific costs of growth	1900	Van der Veer et al. (2006)
$[E_m]$	$J\ cm^{-3}$	Maximum energy density	2190	Van der Veer et al. (2006)
κ	–	Fraction of utilized energy spent on maintenance/growth	0.7	Van der Veer et al. (2006)
V_p	cm^3	Volume at start of reproductive stage	0.06	Van der Veer et al. (2006)
δ_m	–	Shape coefficient	0.25	Casas & Bacher (2006)
d	$g\ cm^{-3}$	Specific density	1.0	Kooijman (2000)
μ_E	$J\ g^{-1}$	Energy content of reserves	6750	Casas & Bacher (2006)

Table 4. Dynamic energy budget model: initial values. L: shell length; W: fresh tissue mass; V: structural volume; E: energy reserves; R: energy allocated to development and reproduction

Maliakos Gulf		Thermaikos Gulf	
Variable	Value	Variable	Value
Start date	28 Sep 2004	Start date	15 May 1995
L	3.68 cm	L	0.84 cm
W	1.43 g	W	0.054 g
V	0.7787 cm^3	V	0.0093 cm^3
E	700 J	E	350 J
R	300 J	R	0 J

This conclusion has been verified with the method of Rosland et al. (2009) on preliminary tests. Allocating all initial energy to E increased final L by 0.87 % and final W by 2.84 %. Allocating all initial energy to R decreased final L by 2.18 % and W by 6.32 %.

Simulation of reproduction

To simulate the loss of mussel weight at spawning (Van Haren et al. 1994), the buffer R was completely emptied ($R = 0$) on the spawning day, which was set at the time of the year when the field data and literature indicate that spawning events occur. The same method was applied by Handa et al. (2011) and Zaldívar (2008). Spawning events for *M. galloprovincialis* in Maliakos and Thermaikos Gulfs occur between December and March (Fasoulas & Fantidou 2008, Theodorou et al. 2011). Thus, for the simulated mussel individual the spawning day was set at about the middle of the spawning season.

Food density

The relation between food uptake and food density is described by a Holling Type II (Holling 1959) functional response, f (Eq. 5), which can vary between 0 and 1. As a first approach, only Phyto-C was considered in the food density (X): X = [Phyto-C]. As a second method, the functional response was adjusted to include not only Phyto-C but also POC in the simulated mussel diet. In this case, the food density X is given by:

$$X = \frac{a \cdot [\text{Phyto-C}] + b \cdot [\text{POC}]}{a_f + b_f} \tag{12}$$

where a and b are given by:

$$a = a_f \cdot \frac{[\text{Phyto-C}]}{[\text{Phyto-C}] + X_k} \tag{13}$$

$$b = b_f \cdot \frac{[\text{POC}]}{[\text{POC}] + X_k} \tag{14}$$

where [Phyto-C] is the density of available Phyto-C and [POC] is the density of available POC. Parameters a_f and b_f describe the mussels' relative preference for Phyto-C and POC, which is related to food quality. Troost et al. (2010) investigated the importance of detritus as a food source between different shellfish species and concluded that the contribution of detritus to shellfish's diet might differ among different environments. For example, cockles prefer phytoplankton as a food source but in those areas where phytoplankton concentrations are low, cockles can also assimilate detritus (Troost et al. 2010). To simulate this behavior, variable preference weights (a, b) were adopted, depending on the availability of food resources in terms of Phyto-C and POC, as given by Eqs. (13) & (14). In this way, the contribution of POC will be significant only when Phyto-C is low compared to POC density. The seasonal contribution of POC for *M. galloprovincialis* is discussed below. Parameters a_f and b_f were determined by calibration: a_f = 0.55, and b_f = 0.45 (see 'Discussion' for more details regarding these values).

Simulation of starvation

Following Rosland et al. (2009) and Handa et al. (2011), when the growth rate according to Eq. (2) is negative, it is assumed that the energy utilization rate is not enough to cover somatic maintenance. In this case, the mussel is assumed to be in a starvation state and stops growing (dV/dt is set to 0). Energy is also withdrawn from the reproductive buffer to cover the maintenance deficit, following Handa et al. (2011); thus the reproduction equation changes to:

$$\frac{dR}{dt} = \kappa \cdot \dot{p}_c - \dot{p}_M \tag{15}$$

RESULTS

Model simulation

Simulations of the growth of *Mytilus galloprovincialis* for the same period as the experimental data were performed first with food density X = [Phyto-C]. Results are shown in Fig. 3 for Maliakos Gulf and Fig. 4 for Thermaikos Gulf. X_k was tuned to different

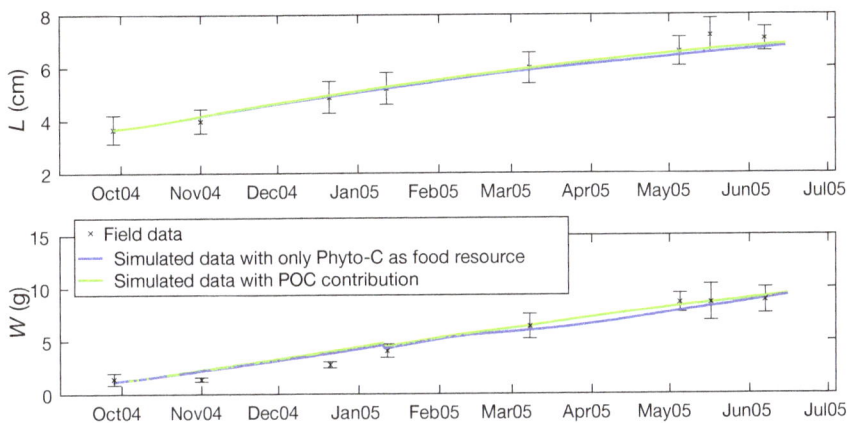

Fig. 3. Simulated mussel shell length (L) (top) and fresh tissue mass (W) against Maliakos data (mean ± SD), using phytoplankton carbon biomass (Phyto-C) (X = [Phyto-C]; blue line) and both Phyto-C and particulate organic carbon (POC) (Eq. 12; green line) in the mussel diet

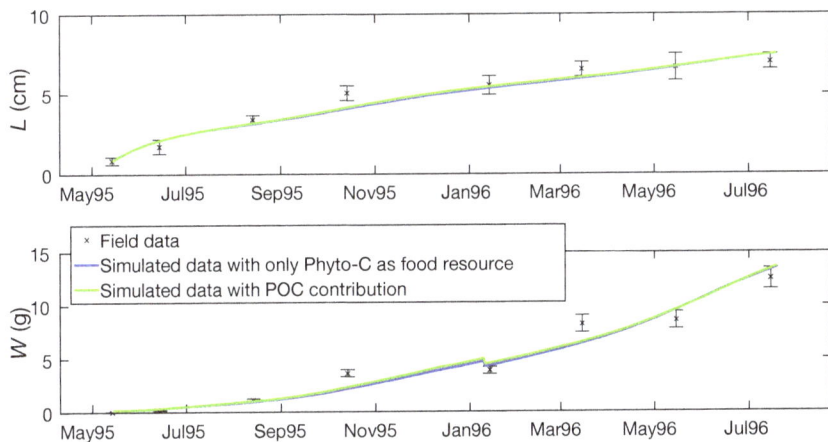

Fig. 4. Simulated mussel shell length (L) (top) and fresh tissue mass (W) against Thermaikos data (mean ± SD), using phytoplankton carbon biomass (Phyto-C) (X = [Phyto-C]; blue line) and both Phyto-C and particulate organic carbon (POC) (Eq. 12; green line) in the mussel diet

To achieve a more realistic simulation of mussel growth, POC was added to the diet of *M. galloprovincialis*, with food function, X, given by Eq. (12). In Figs. 3 & 4, the results can be compared with those obtained with food density X = [Phyto-C]. At Maliakos Gulf, including POC in the diet resulted in a better fit between simulated and field data. This area is characterized by low Phyto-C concentrations from the middle of February until late April. During this period, POC values are higher relative to Phyto-C and this appears to significantly contribute to mussel growth. On the other hand, POC appears to have no significant role in the *M. galloprovincialis* diet at Thermaikos Gulf, as in this area Phyto-C is characterized by much weaker variability, with POC concentrations always lower than Phyto-C.

Comparison between field and simulated data: model performance

As a useful index of the model skill, simulated mussel growth was plotted against field data in Figs. 5 & 6. Points along the $x = y$ line indicate a perfect fit. In most occasions, points are very close to that line. Moreover, the model bias and unbiased root-mean-square-deviation (RMSD) against field data were calculated and plotted on target diagrams (Jolliff et al. 2009) in Figs. 7 & 8. The marks are very close to the center of the diagram, indicating a successful simulation. The mean model bias, indicated on the y-axis, shows that referring to the overall mean value, the simulated shell length is slightly overestimated ($y > 0$), while the fresh mass tissue is slightly underestimated ($y < 0$) at Maliakos Gulf and in a very good agreement with field data at Thermaikos Gulf ($y \sim 0$). Additionally, the unbiased RMSD suggests that the standard deviation of the model is larger ($x > 0$) except for the fresh mass tissue at Thermaikos Gulf ($x < 0$).

values for the 2 areas: X_k = 36 mg C m^{-3} for Maliakos Gulf and X_k = 28 mg C m^{-3} for Thermaikos Gulf; X_k not only depends on species but is also site-specific (Troost et al. 2010). The different values of X_k (36 mg C m^{-3} or 0.72 µg chl a l^{-1} at Maliakos and 28 mg C m^{-3} or 0.56 µg chl a l^{-1} at Thermaikos) that were fitted for the 2 study areas can be attributed to differences in Phyto-C variability between the 2 environments (see 'Environmental data' above). These fitted values of X_k were, as expected, slightly lower compared to those found in more productive areas. Casas & Bacher (2006) found X_k = 3.88 µg l^{-1} (~194 mg C m^{-3}) on the French Mediterranean shoreline, while Troost et al. (2010) found X_k = 2.23 µg l^{-1} (~116.5 mg C m^{-3}) in southwest Netherlands.

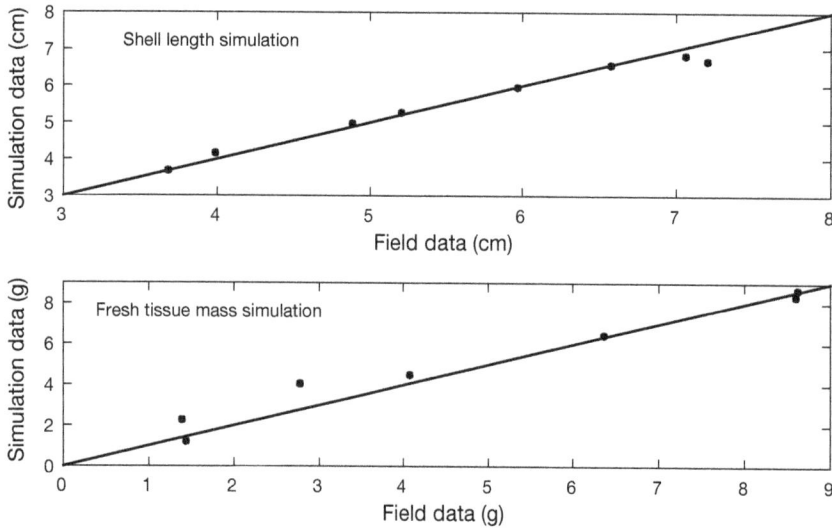

Fig. 5. Field data against simulation data (stars) at Maliakos Gulf plotted against a
$y = x$ line

Model uncertainty and ensemble forecasting

The initial values of E and R and the values of the calibrated parameters (X_k, a_f, b_f) may be considered relatively unknown. To examine the model's sensitivity relative to the uncertainty of initialization/calibration, a series of sensitivity simulations were performed, adopting a representative envelope of 5 different values (Table 5) for each initial value and calibrated parameter, with the exception of the preference coefficients a_f and b_f, which did not show a significant sensitivity in preliminary tests. X_k was perturbed by 15 and 30% of its standard value, while initial E and R, to which the model shows less sen-

sitivity (Bacher & Gangnery 2006), were perturbed by 50 and 100%. In the Thermaikos Gulf simulation, where initial R was set at zero, 5 values from 0 to 100 J were adopted. Model runs were executed with each possible combination of the 3 parameter values, giving a total of 125 runs. The mean and standard deviation from all model results were then calculated, as shown in Figs. 9 & 10 for the Maliakos and Thermaikos study cases, respectively. Araújo & New (2007) demonstrated the advantages of an ensemble forecasting in biological models. Instead of selecting the best tuning values, a better procedure is to present a range of possible model states within an envelope concerning the representative parameterization/initialization values.

The uncertainty of the model, as represented by the standard deviation, appears to increase as the mussel grows, particularly regarding its wet weight. This does not indicate a decrease in the model skill, as a similar increase of standard deviation with growing mussels was also found in the wet weight field data, considering that each individual of the farm may grow in a different way, ultimately reaching a different final weight. This does not apply to the shell length evolution, which seems to be a more standard process. It is also noticeable that a weakness of the model is in efficiently simulating the individual's production of fresh tissue mass during the period before spawning (from early November 2004 to early January 2005 for Maliakos, and during October 1995 for Thermaikos Gulf). This could imply errors in the simulated reproduction of the individual that are either related to the adopted spawning day/period or to the assumed weight loss due to spawning. According to Van Haren et al. (1994), mussels lose 40 to 70% of their wet weight during spawning. This does not occur in the presented simulations, where weight loss is around 10%. However, if the

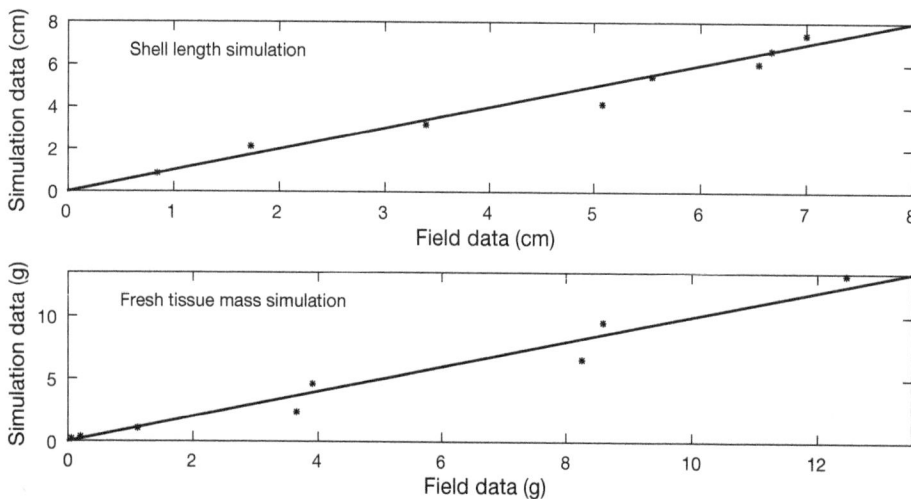

Fig. 6. Field data against simulation data (stars) at Thermaikos Gulf plotted against a $y = x$ line

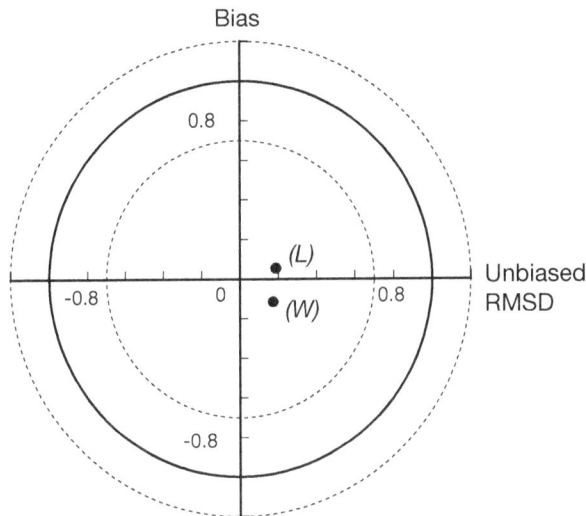

Fig. 7. Target diagram of simulated shell length (L) and fresh mass tissue weight (W) against field data from the Maliakos Gulf. The model bias is indicated on the y-axis while the unbiased root-mean-square-deviation (RMSD) is indicated on the x-axis

simulation continues for a second or a third year, the percentage of wet weight loss increases to 20% on subsequent spawning events.

DISCUSSION

A DEB model was developed and tuned against data from Maliakos and Thermaikos Gulfs to study the growth of cultured mussel *Mytilus galloprovincialis*.

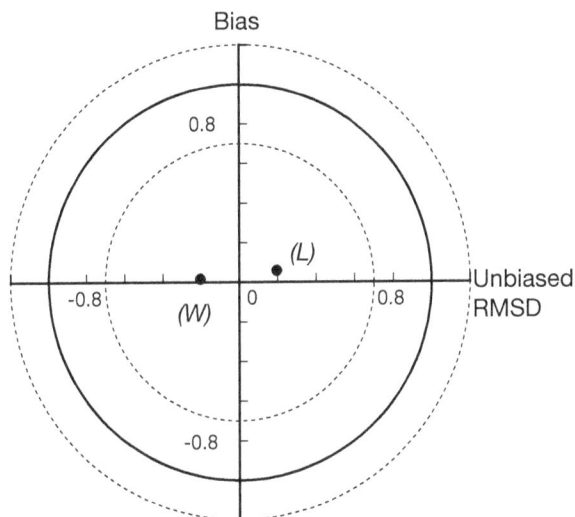

Fig. 8. Target diagram of simulated shell length (L) and fresh mass tissue weight (W) against field data from the Thermaikos Gulf. The model bias is indicated on the y-axis while the unbiased root-mean-square deviation (RMSD) is indicated on the x-axis

The 2 environments showed differences mostly in terms of Phyto-C and POC fluctuations and maximum values, which has an impact on the diet of *M. galloprovincialis*. At first, model simulations were performed accounting for only Phyto-C as a food resource for the mussel. In a second approach, available food density was modified to also include POC, aiming to investigate the contribution of POC to mussel growth. In this case, following the available literature, it was assumed that *M. galloprovincialis* assimilates POC only when the density of Phyto-C is not enough for its needs. This was formulated, adopting different preference coefficients for Phyto-C ($a_f = 0.55$) and POC ($b_f = 0.45$) in the mussel food density (Eq. 12). The results showed a small contribution of POC at Maliakos Gulf and almost none at Thermaikos Gulf. This is in agreement with Troost et al. (2010), who found a site-dependent contribution of detritus. The values of a_f and b_f could be parameterized in a way that would allow POC to have a more significant role in the mussel growth. Simulations with higher b_f (0.55 to 0.8) and lower a_f (0.45 to 0.2), combined with higher values of X_k, showed that the model could produce similar results to those presented; however, this would be in conflict with most of the literature, where phytoplankton is considered the principal food source for mussels (Williams 1981, Langdon & Newell 1990, Garen et al. 2004). In general, the simulations are satisfactory when considering only Phyto-C as available food for the mussel. Under this assumption, the model is simplified, with X_k being the only parameter that has to be tuned. However, the model appears more stable when POC is included in the mussel's diet.

The 2 study areas are very similar with regard to sea surface temperature, and thus differences in the

Table 5. Parameter values used for the estimation of model uncertainty. **Bold** numbers are the standard values of initial energy reserves (E), initial energy allocated to development and reproduction (R) and half saturation coefficient (X_k)

Initial E (J)	Initial R (J)	X_k (mg C m^{-3})
Maliakos Gulf		
0	0	25.2
350	150	30.6
700	**300**	**36**
1050	450	41.4
1400	600	46.8
Thermaikos Gulf		
0	**0**	19.6
175	20	23.8
350	50	**28**
500	80	32.2
700	100	36.4

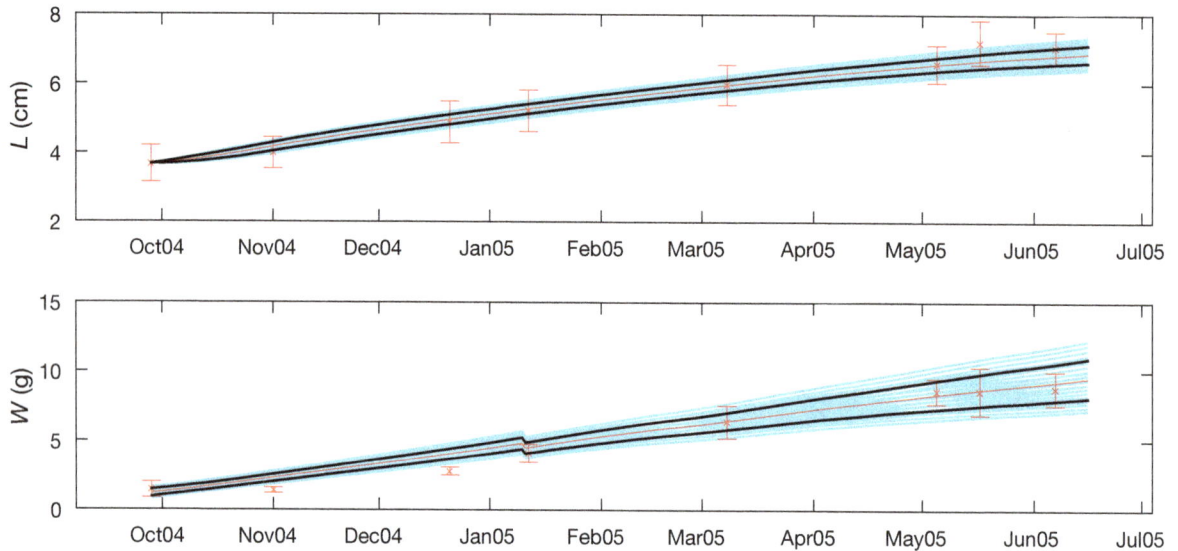

Fig. 9. Model uncertainty for the Maliakos simulation. Blue lines: the 125 model runs with each possible combination among the perturbed values of the initial energy reserves (E), reproductive buffer (R) and half-saturation coefficient (X_k) given in Table 5. Red line: computed mean value; black lines: ± SD. Red crosses and bars are for field data (mean ± SD)

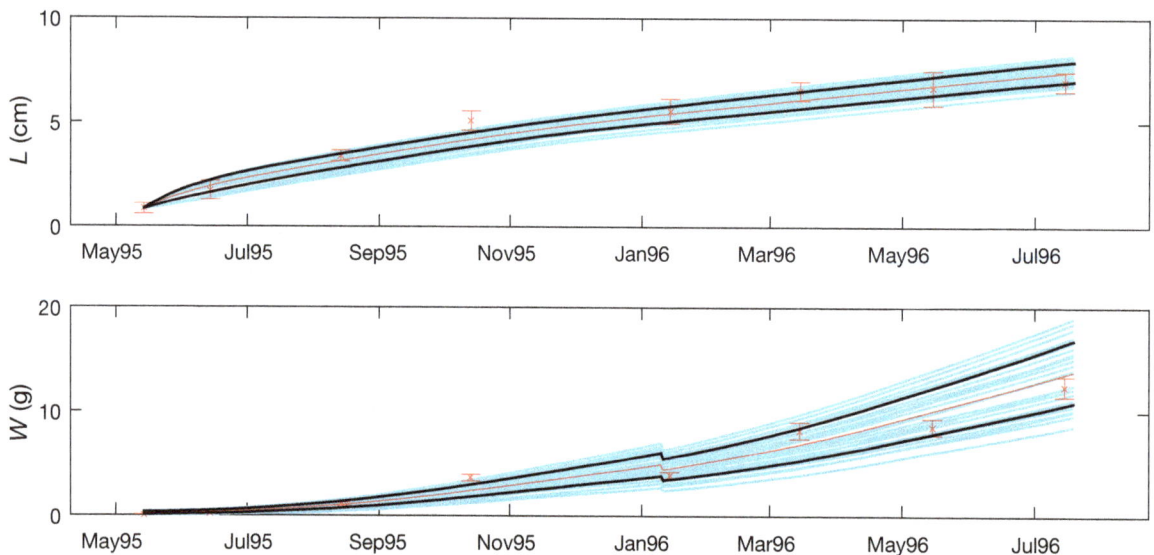

Fig. 10. Model uncertainty for the Thermaikos simulation. Blue lines: the 125 model runs with each possible combination among the perturbed values of initial energy reserves (E), reproductive buffer (R) and half saturation coefficient (X_k) given in Table 5. Red line: computed mean value; black lines: ± SD. Red crosses and bars are for field data (mean ± SD)

growth of cultured mussels may be attributed only to differences in food resources. Temperature affects the growth of the simulated mussel through the temperature dependence $k(T)$, which is multiplied by each physiological rate (i.e. $\dot{p}_a$, $\dot{p}_c$, $[\dot{p}_M]$). Increasing the temperature time series by 1°C in the Maliakos Gulf simulation resulted in an increase of 0.86 % for final shell length and 2.83 % for final fresh mass tissue. Although this response to temperature does not seem significant for the purpose of the present study, it could be interesting in the con-

text of a climate change scenario investigating the effect of global warming on Aegean Sea mussel farms.

To quantify the uncertainty related to the fitted model parameters (initial values of E and R and X_k), different values of these parameters were tested and a series of simulations was performed with all possible combinations, leading to a representation of the DEB model uncertainty. The results showed that in most cases, the uncertainty related to the different simulations is within the limits of the field data standard deviation, suggesting that small perturbations to the calibrated values of E, R and X_k do not have a strong influence on model outputs and thus the model remains valid.

In general, the model performed well in simulating the growth of the cultured mussel *M. galloprovincialis*. While the simulation of shell length growth was satisfactory, the agreement between observed and simulated weights was not as good, suggesting that there is a need for optimization of parameters related to the weight–length relations, such as the energy content of reserves (μ_E) and the shape coefficient (δ_m), for which many different estimates can be found in the literature (Casas & Bacher 2006, Van der Veer et al. 2006, Thomas et al. 2011, Sarà et al. 2012). In the present study, the best fit with the field data was obtained with parameter values adopted from Casas & Bacher (2006). Representation of model uncertainty due to initial E, R and X_k led to a more reliable simulation, as in most occasions there is no easy way to estimate those parameters with satisfactory accuracy. More work in this direction needs to be done in the future, as there is still room for optimization of the DEB parameters. In his work on the estimation of DEB parameters for bivalve species, Van der Veer et al. (2006) concluded that there was a standard error of about 30 % on his estimates. Different values can be found among different studies on the same species; i.e. Troost et al. (2010) and Thomas et al. (2011) used a fraction of utilized energy spent on maintenance/growth $\kappa = 0.45$ for *M. edulis*, while Picoche et al. (2014) used $\kappa = 0.67$. Other relevant examples involve δ_m and $\{\dot{p}_{Am}\}$; Sarà et al. (2012) and Rinaldi et al. (2014) used $\delta_m = 0.2254$ and $\{\dot{p}_{Am}\} = 173.184$ J cm^{-2} d^{-1} (= 7.216 J cm^{-2} h^{-1}) for *M. galloprovincialis*, while Casas & Bacher (2006) and Zaldívar (2008) used $\delta_m = 0.25$ and $\{\dot{p}_{Am}\} = 147.6$ J cm^{-2} d^{-1}, which worked better with the presented data. Building an envelope of a targeted set of the most sensitive model parameters with different values found in the literature (such as κ, δ_m and $\{\dot{p}_{Am}\}$) could lead to more reliable and general ensemble simulations, and also provide an estimate of the model uncertainty related to these parameters. This could be strengthened by including in the ensemble other bio-energetic individual models (such as a SFG model; Brigolin et al. 2009 among others). Furthermore, a full description of the model uncertainties should include uncertainty due to environmental forcing, along with uncertainties due to parameterization and initialization. This complete representation of a DEB model uncertainties could lead to a better understanding of the model dynamics. Additionally, ensemble forecasting (Araújo

& New 2007) in the DEB model could be used to apply climate change scenarios and investigate climate change effects on species.

Although the model worked well at the 2 specific study areas and ensemble forecasting provided a quantification of uncertainties due to parameters and initial value calibrations, the model still suffers from some limitations. In the present study, the simulated individual is considered to be representative of the mussel farm's mean state. However, one should also take into account the farm's population effect. Generally, in a predator–prey system the rate of an individual's consumption is affected by the density of predators (Kratina et al. 2009). Kratina et al. (2009) tested different functional responses, concluding that those taking into account predator density provide better results. In the context of the present study, one way to take into account the effect of the farm's population on individual mussel growth with the DEB model would be to modify the Hollings Type II functional response function (f), expressing X and X_k in terms of food resources per mussel density. This approach could be tested in future work.

Another simplification that can be regarded as a limitation of the model is the assumption that *M. galloprovincialis* filtrates phytoplankton and POC of every size. Many studies emphasize the selectivity of *Mytilus* spp. with respect to size and species of phytoplankton (Bayne et al. 1987, Raby et al. 1997 among others). Other studies (Lehane & Davenport 2006, Prato et al. 2010, Ezgeta-Bali et al. 2012) suggest that *Mytilus* spp. can consume even zooplankton in certain periods and areas. Therefore, in order to have a detailed description of mussel diet, food density X should consist of each proxy food (different size groups of phytoplankton and POC) separately adjusted by a suitable preference weight; an effort that should be supported by field data diet analyses. On the other hand, such a model would be more difficult to tune, and uncertainties due to parameter calibration would be higher.

Due to its generic character, the model developed in the present study can easily be adapted to simulate the growth of other bivalve species, such as native oysters *Ostrea edulis* and European clams *Tapes decussatus*, with potential farming interest in Greek coastal waters.

Acknowledgements. The authors thank George Verriopoulos, professor of marine biology at the University of Athens for his helpful advice on biological matters. Data provided for Maliakos are outcomes of the Project ARCHIMIDES I –EPEAEK II-EU funded (contract no 10012-00004) 'Environmental Interactions of the Mussel farming' (2004–2007).

LITERATURE CITED

Araújo MB, New M (2007) Ensemble forecasting of species distributions. Trends Ecol Evol 22:42–47

Bacher C, Gangnery A (2006) Use of dynamic energy budget and individual based models to simulate the dynamics of cultivated oyster populations. J Sea Res 56: 140–155

Baretta JW, Ebenhoh W, Ruardij P (1995) The European regional seas ecosystem model, a complex marine ecosystem model. Neth J Sea Res 33:233–246

Bayne B, Hawkins A, Navarro E (1987) Feeding and digestion by the mussel *Mytilus edulis* L. (Bivalvia: Mollusca) in mixtures of silt and algal cells at low concentrations. J Exp Mar Biol Ecol 111:1–22

Blumberg AF, Mellor GL (1983) Diagnostic and prognostic numerical circulation studies of the South Atlantic Bight. J Geophys Res 88:4579–4592

Bourlès Y, Alunno-Bruscia M, Pouvreau S, Tollu G and others (2009) Modelling growth and reproduction of the Pacific oyster *Crassostrea gigas*: advances in the oyster-DEB model through application to a coastal pond. J Sea Res 62:62–71

Brigolin D, Maschio G, Rampazzo F, Giani M, Pastres R (2009) An individual-based population dynamic model for estimating biomass yield and nutrient fluxes through an off-shore mussel (*Mytilus galloprovincialis*) farm. Estuar Coast Shelf Sci 82:365–376

Camacho A, Labarta U, Beiras R (1995) Growth of mussels (*Mytilus edulis galloprovincialis*) on cultivation rafts: influence of seed source, cultivation site and phytoplankton availability. Aquaculture 138:349–362

Casas S, Bacher C (2006) Modelling trace metal (Hg and Pb) bioaccumulation in the Mediterranean mussel, *Mytilus galloprovincialis*, applied to environmental monitoring. J Sea Res 56:168–181

Christou ED, Pagou K, Christianidis S, Papathanassiou E (1995) Temporal and spatial variability of plankton communities in a shallow embayment of the Eastern Mediterranean. In: Eleftheriou A, Ansell AD, Smith CJ (eds) Biology and ecology of shallow coastal waters. Olsen and Olsen, Fredensborg, p 3–10

Dimitriou PD, Karakassis I, Pitta P, Tsagaraki TM and others (2015) Effects of mussel farming on quality indicators of the marine environment: good benthic below poor pelagic ecological status. Mar Pollut Bull 101:784–793

Eurostat (2016) Eurostat. http://ec.europa.eu/eurostat (accessed 12 Sep 2016)

Ezgeta-Bali D, Najdek M, Peharda M, Blažina M (2012) Seasonal fatty acid profile analysis to trace origin of food sources of four commercially important bivalves. Aquaculture 334-337:89–100

FAO (Food and Agriculture Organization of the United Nations) (2016) The state of world fisheries and aquaculture 2016. Contributing to food security and nutrition for all. FAO, Rome

Fasoulas TA, Fantidou ES (2008) Spat dynamic pattern of cultured mussel *Mytilus galloprovincialis* (Lamarck, 1819) in NW Gulf of Thessaloniki. Abstract presented at the 4th International Congress on Aquaculture, Fisheries Technology and Environmental Management, November 21–22, 2008, Athens, Greece

FGM (Federation of Greek Maricultures) (2015) The Greek aquaculture. Federation of Greek Maricultures, Athens

Fly EK, Hilbish TJ (2013) Physiological energetics and biogeographic range limits of three congeneric mussel species. Oecologia 172:35–46

Garen P, Robert S, Bougrier S (2004) Comparison of growth of mussel, *Mytilus edulis*, on longline, pole and bottom culture sites in the Pertuis Breton, France. Aquaculture 232:511–524

Geider J, Piatt T (1986) A mechanistic model of photoadaptation in microalgae. Mar Ecol Prog Ser 30:85–92

Gosling EM (1984) The systematic status of *Mytilus galloprovincialis* in western Europe: a review. Malacologia 25:551–568

Handa A, Alver M, Edvardsen C, Halstensen S and others (2011) Growth of farmed blue mussels (*Mytilus edulis* L.) in a Norwegian coastal area; comparison of food proxies by DEB modeling. J Sea Res 66:297–307

Hilbish TJ, Bayne BL, Day A (1994) Genetics of physiological differentiation within the marine mussel genus *Mytilus*. Evolution 48:267–286

Holling C (1959) Some characteristics of simple types of predation and parasitism. Can Entomol 91:385–398

Jolliff J, Kindle J, Shulman I, Penta B, Friedrichs M, Helber R, Arnone R (2009) Summary diagrams for coupled hydrodynamic-ecosystem model skill assessment. J Mar Syst 76:64–82

Kakali F, Vildou I, Theodorou JA, Tzovenis I, Rizos D, Kagalou I (2006) Trophic state evaluation of Maliakos bay (Greece), for potential development of the mussel *Mytilus galloprovincialis* farming. Abstract presented at Aqua2006: annual meeting of the World Aquaculture Society, 9–13 May 2006, Florence, Italy

Konstantinou Z, Kombiadou K, Krestenitis Y (2015) Effective mussel-farming governance in Greece: testing the guidelines through models, to evaluate sustainable management alternatives. Ocean Coast Manage 118: 247–258

Kooijman SALM (1986) Energy budgets can explain body size relations. J Theor Biol 121:269–282

Kooijman S (2000) Dynamic energy and mass budgets in biological systems. Cambridge University Press, Cambridge

Kormas KA, Kapiris K, Thessalou-Legaki M, Nicolaidou A (1998) Quantitative relationships between phytoplankton, bacteria and protists in an Aegean semi-enclosed embayment (Maliakos Gulf, Greece). Aquat Microb Ecol 15:255–264

Kormas K, Garametsi V, Nicolaidou A (2002) Size-fractionated phytoplankton chlorophyll in an Eastern Mediterranean coastal system (Maliakos Gulf, Greece). Helgol Mar Res 56:125–133

Kratina P, Vos M, Bateman A, Anholt B (2009) Functional responses modified by predator density. Oecologia 159: 425–433

Kravva N (2000) Genetic composition and growth in *Mytilus galloprovincialis* populations in the Thermaikos Gulf. PhD dissertation, Aristotle University of Thessaloniki

Langdon CJ, Newell RIE (1990) Utilization of detritus and bacteria as food sources by two bivalve suspension-feeders, the oyster *Crassostrea virginica* and the mussel *Geukensia demissa*. Mar Ecol Prog Ser 58:299–310

Larsen P, Filgueira R, Riisgård H (2014) Somatic growth of mussels *Mytilus edulis* in field studies compared to predictions using BEG, DEB, and SFG models. J Sea Res 88: 100–108

Lehane C, Davenport J (2006) A 15-month study of zooplankton ingestion by farmed mussels (*Mytilus edulis*)

in Bantry Bay, Southwest Ireland. Estuar Coast Shelf Sci 67:645–652

Malone T, Chervin M (1979) The production and fate of phytoplankton size fractions in the plume of the Hudson River, New York Bight. Limnol Oceanogr 24:683–696

Pagou K (2005) Eutrophication in Hellenic coastal areas. In: Papathnasiou E, Zenetos A (eds) State of the Hellenic marine environment. Hellenic Centre for Marine Research (HCMR), Anavissos, p 311–317

Papakonstantinou C, Zenetos A, Vassilopoulou V, Tserpes G (2007) State of Hellenic fisheries. Hellenic Centre for Marine Research, Institute of Marine Biological Resources, Athens

Picoche C, Le Gendre R, Flye-Sainte-Marie J, Françoise S, Maheux F, Simon B, Gangnery A (2014) Towards the determination of Mytilus edulis food preferences using the dynamic energy budget (DEB) theory. PLOS ONE 9: e109796

Pouvreau S, Bourlès Y, Lefebvre S, Gangnery A, Alunno-Bruscia M (2006) Application of a dynamic energy budget model to the Pacific oyster, Crassostrea gigas, reared under various environmental conditions. J Sea Res 56:156–167

Prato E, Portacci G, Biandolino F (2010) Effect of diet on growth performance, feed efficiency and nutritional composition of Octopus vulgaris. Aquaculture 309:203–211

Price N, Karageorgis A, Kaberi H, Zeri C and others (2005) Temporal and spatial variations in the geochemistry of major and minor particulate and selected dissolved elements of Thermaikos Gulf, Northwestern Aegean Sea. Cont Shelf Res 25:2428–2455

Raby D, Mingelbier M, Dodson J, Klein B, Lagadeuc Y, Legendre L (1997) Food-particle size and selection by bivalve larvae in a temperate embayment. Mar Biol 127: 665–672

Rinaldi A, Montalto V, Manganaro A, Mazzola A, Mirto S, Sanfilippo M, Sarà G (2014) Predictive mechanistic bioenergetics to model habitat suitability of shellfish culture in coastal lakes. Estuar Coast Shelf Sci 144:89–98

Rodrigues LC, van den Bergh JCJM, Massa F, Theodorou JA, Ziveri P, Gazeau F (2015) Sensitivity of Mediterranean bivalve mollusc aquaculture to climate change and ocean acidification: results from a producers' survey. J Shellfish Res 34:1161–1176

Rosland R, Strand Ø, Alunno-Bruscia M, Bacher C, Strohmeier T (2009) Applying dynamic energy budget (DEB) theory to simulate growth and bio-energetics of blue mussels under low seston conditions. J Sea Res 62: 49–61

Sarà G, Reid G, Rinaldi A, Palmeri V, Troell M, Kooijman S (2012) Growth and reproductive simulation of candidate shellfish species at fish cages in the Southern Mediterranean: dynamic energy budget (DEB) modelling for integrated multi-trophic aquaculture. Aquaculture 324–325:259–266

Theodorou JA, Nathanailidis K, Kagalou I, Rizos D, Georgiou P, Tzovenis I (2006a) Spat settlement pattern of the cultured Mediterranean mussel Mytilus galloprovincialis in the Maliakos Bay (Greece). Abstract presented at Aqua2006: annual meeting of the World Aquaculture Society, 9–13 May 2006, Florence, Italy

Theodorou JA, Nathanailidis K, Makaritis P, Kagalou I, Negas I, Anastasopoulou G, Alexis M (2006b) Gonadal maturation of the cultured mussel Mytilus galloprovincialis in the Maliakos Bay (Greece): preliminary results. Abstract presented at Aqua2006: annual meeting of the World Aquaculture Society, 9–13 May 2006, Florence, Italy

Theodorou J, Makartis P, Tzovenis I, Fountoulaki E, Nengas I, Kagalou I (2007) Seasonal variation of the chemical composition of the farmed mussel Mytilus galloprovincialis in Maliakos Gulf. Abstract presented at 13th Hellenic Congress of Ichthyologists, Mytilini, Greece

Theodorou J, Viaene J, Sorgeloos P, Tzovenis I (2011) Production and marketing trends of the cultured Mediterranean mussel Mytilus galloprovincialis Lamarck 1819, in Greece. J Shellfish Res 30:859–874

Theodorou JA, Perdikaris C, Filippopoulos NG (2015a) Evolution through innovation in aquaculture: the case of the Hellenic mariculture industry (Greece). J Appl Aquacult 27:160–181

Theodorou JA, James R, Tzovenis I, Hellio C (2015b) The recruitment of the endangered fan mussel (Pinna nobilis, Linnaeus 1758) on the ropes of a Mediterranean mussel long line farm. J Shellfish Res 34:409–414

Thomas Y, Mazurié J, Alunno-Bruscia M, Bacher C and others (2011) Modelling spatio-temporal variability of Mytilus edulis (L.) growth by forcing a dynamic energy budget model with satellite-derived environmental data. J Sea Res 66:308–317

Troost TA, Wijsman JWM, Saraiva S, Freitas V (2010) Modelling shellfish growth with dynamic energy budget models: an application for cockles and mussels in the Oosterschelde (southwest Netherlands). Philos Trans R Soc Lond B Biol Sci 365:3567–3577

Tsiaras K, Petihakis G, Kourafalou V, Triantafyllou G (2014) Impact of the river nutrient load variability on the North Aegean ecosystem functioning over the last decades. J Sea Res 86:97–109

Van der Veer HW, Cardoso JFMF, van der Meer J (2006) The estimation of DEB parameters for various Northeast Atlantic bivalve species. J Sea Res 56:107–124

Van Haren RJF, Schepers HE, Kooijman SALM (1994) Dynamic energy budgets affect kinetics of xenobiotics in the marine mussel Mytilus edulis. Chemosphere 29: 163–189

Widdows J, Donkin P, Salkeld PN, Cleary JJ, Lowe DM, Evans SV, Thomson PE (1984) Relative importance of environmental factors in determining physiological differences between two populations of mussels (Mytilus edulis). Mar Ecol Prog Ser 17:33–47

Wijsman J, Smaal A (2011) Growth of cockles (Cerastoderma edule) in the Oosterschelde described by a dynamic energy budget model. J Sea Res 66:372–380

Williams P (1981) Detritus utilization by Mytilus edulis. Estuar Coast Shelf Sci 12:739–746

Zaldívar J (2008) A general bioaccumulation DEB model for mussels. JRC Scientific and Technical Reports. European Commission, Institute for Health and Consumer Protection, Joint Research Centre (JSR), Luxembourg

14

Aquaculture's struggle for space: the need for coastal spatial planning and the potential benefits of Allocated Zones for Aquaculture (AZAs) to avoid conflict and promote sustainability

P. Sanchez-Jerez[1,*], I. Karakassis[2], F. Massa[3], D. Fezzardi[3], J. Aguilar-Manjarrez[4], D. Soto[4], R. Chapela[5], P. Avila[6], J. C. Macias[7], P. Tomassetti[8], G. Marino[8], J. A. Borg[9], V. Franičević[10], G. Yucel-Gier[11], I. A. Fleming[12], X. Biao[13], H. Nhhala[14], H. Hamza[15], A. Forcada[1], T. Dempster[16]

[1]Department of Marine Science and Applied Biology, University of Alicante, PO Box 99, 03080 Alicante, Spain

[2]Institute of Marine Biology of Crete, PO Box 2214, 71003 Heraklion, Greece

[3]General Fisheries Commission for the Mediterranean, Food and Agriculture Organisation of the United Nations, Via Vittoria Colonna 1, 00193 Rome, Italy

[4]Fisheries and Aquaculture Department, Food and Agriculture Organization of the United Nations, Viale delle Terme di Caracalla, 00153 Rome, Italy

[5]CETMAR, C/ Eduardo Cabello s/n, 36208 Bouzas-Vigo Spain

[6]Junta de Andalucía, Agencia de Gestión Agraria y Pesquera de Andalucía, C/ Severo Ochoa 38, Parque Tecnológico de Andalucía, 29590 Campanillas, Málaga, Spain

[7]Aquaculture Consultant, C/ Crucero 2ªF n1, Sanlúcar de Barrameda, 11540 Cadiz, Spain

[8]Institute for Environmental Protection and Research (ISPRA), Via Vitaliano Brancati 48, 00144 Rome, Italy

[9]Department of Biology, Faculty of Science, University of Malta, 20810 Msida, Malta

[10]Ministry of Agriculture, Department of Fisheries, I. Mažurani a 30, 23000 Zadar, Croatia

[11]Institute of Marine Sciences and Technology, Dokuz Eylül University, 35340 Izmir, Turkey

[12]Department of Ocean Science, Memorial University of Newfoundland, St. John's, NL A1C 5S7, Canada

[13]School of Geography Sciences, Nanjing Normal University, 1 Wenyuan Road, Nanjing 210023, PR China

[14]Centre Aquacole Institut National de Recherche Halieutique, BP n°31, M'diq, Morocco

[15]Direction Générale des Pêches et de l'Aquaculture Ministère de l'Agriculture, 30 Rue Alain Savary, 1002 Tunis, Tunisia

[16]Sustainable Aquaculture Laboratory – Temperate and Tropical (SALTT), Department of Zoology, University of Melbourne, Victoria 3010, Australia

ABSTRACT: Aquaculture is an increasingly important food-producing sector, providing protein for human consumption. However, marine aquaculture often struggles for space due to the crowded nature of human activities in many marine coastal areas, and because of limited attention from spatial planning managers. Here, we assess the need for coastal spatial planning, emphasising the establishment of suitable areas for the development of marine aquaculture, termed Allocated Zones for Aquaculture (AZAs), in which aquaculture has secured use and priority over other activities, and where potential adverse environmental impacts and negative interactions with other users are minimised or avoided. We review existing examples of marine aquaculture spatial development worldwide and discuss the proper use of site selection in relation to different legal and regulatory requirements. National or regional authorities in charge of coastal zone management should carry out spatial planning defining optimal sites for aquaculture to promote development of sustainable marine aquaculture and avoid conflict with other users, following a participatory approach and adhering to the principles of ecosystem-based management.

KEY WORDS: Site selection · Spatial planning · Aquaculture

*Corresponding author: psanchez@ua.es

INTRODUCTION

Aquaculture, including marine aquaculture (mariculture), plays a major role in meeting the rising demand for fish products and supplying healthy and nutritious protein to a growing global population (Larsen & Roney 2013). Proponents of aquaculture have called for increased fish production from mariculture to increase fish delivery for human consumption, to reduce the use of scarce natural resources such as freshwater (Duarte et al. 2009) and wild fish (FAO 2012a), and because of the economic benefits. If mariculture is to expand, access to adequate areas for production will be a key determinant of future production levels because, despite calls to 'conquer the ocean' with offshore mariculture (Marra 2005), for the moment coastal waters remain the most desirable location for marine aquaculture.

However, available space for new mariculture development in coastal zones is now becoming increasingly limited. This trend can be expected to continue because, along with aquaculture, other human activities in coastal and oceanic waters will increase significantly over the next 20 yr, including tourism (Hall 2001) and offshore renewable energy generation (Douvere & Ehler 2009). Traditional activities, such as capture fisheries, are deeply embedded in many coastal communities and they will continue to play an important role in the use of nearshore coastal seascapes. This competition for space has resulted in negative interactions with traditional and new coastal users (Dempster & Sanchez-Jerez 2008), and negative effects of other activities on aquaculture (e.g. agriculture and sewage discharges; Díaz et al. 2012). The complex interactions among users of coastal areas often leave little space for aquaculture, particularly since marine aquaculture requires coastal waters with specific environmental and water quality characteristics. The result is that suitable coastal areas for aquaculture are severely limited in many areas and this has become a major barrier to expansion.

The seventh session of the Sub-Committee on Aquaculture (SCA) of the FAO Committee on Fisheries expressed a strong interest in spatial planning to ensure the allocation of space for aquaculture growth. The SCA acknowledged that spatial planning needs to follow an ecosystem approach and to define socioeconomic, environmental and governance objectives that result in the integrated management of land, water and living resources for the development and expansion of the sector in a sustainable and equitable way (FAO 2013a). Therefore, to ensure a balance between conserving ecosystem services and other legitimate uses of marine resources, spatial planning of aquaculture activities through optimal site selection is necessary.

The purpose of this paper is to review some examples of spatial planning in marine waters across the world to assist aquaculture development. We discuss the potential benefits of establishing suitable areas for the development of marine aquaculture, Allocated Zones for Aquaculture (AZAs), in which aquaculture has priority over other activities, and where adverse environmental and social impacts, as well as negative interactions with other users, are minimised or avoided, and propose key considerations for the successful implementation and management of AZAs.

SPATIAL PLANNING OF MARINE AQUACULTURE

Identifying optimal sites

The success of an aquaculture project depends to a large extent on the selection of an appropriate site to establish the farm; this entails an intricate multicriteria decision-making process. For site selection, in addition to consideration of physical and environmental factors, other crucial aspects concerning the efficiency and economy of the aquaculture operations are central (see Ross et al. 2013 for an extensive description of the area and site selection process). The producer must identify the marketability of the product, after making a proper evaluation of existing demand and its stability, and of potential future markets (Knapp 2008). Once a market has been identified, a suitable site for production must be found which is in accordance with the biology of the cultivated organism and the projected volume of production. Haphazard development of aquaculture without adequate planning and regulation can lead to adverse environmental impacts, lack of economic feasibility, and/or social conflicts. Negative environmental and social conflicts have emerged at broad scales previously where planning has been inadequate, such as seen in the early development of prawn farming in tropical regions (Primavera 1997, Pattanaik & Prasad 2011).

Spatial planning of aquaculture activities should be a public process aimed at achieving ecological, economic, and social objectives that usually have been specified through a governance process, with a broad participatory approach (IUCN 2009, FAO 2013b, Ross et al. 2013, Hishamunda et al. 2014). Essentially, spatial planning contributes to the process of selecting suitable areas and sites for aquaculture when it is integrated and forward-looking with respect to regulation, management and protection of the marine environment; the allocation of space should address the multiple, cumulative, and potentially conflicting uses of the sea (FAO 2013b). The shared use of public domain areas and the conservation policies adopted for coastal areas reduce the availability of sites. At the same time, however, the demand for aquaculture products is increasing, especially because this activity can supply a constant stream of quality products at stable prices (FAO 2010a).

Preliminary site selection processes should define the geographical location and extent of aquaculture in a determined region (Ross et al. 2013). As part of this process, physical, ecological and socio-economic criteria should be taken into account. Next, a location that minimises conflict with the other users of coastal waters, such as shipping, fishing, recreational activities and the energy industry (Dempster & Sanchez-Jerez 2008), should be identified to site the farm. As aquaculture is often the 'new kid on the block' in terms of its use of space, in many coastal areas it will be difficult to find suitable sites which do not conflict with pre-existing uses that may well be considered more important socio-economically for the region (Toledo-Guedes et al. 2014). As this scenario of multiple pre-existing uses of coastal spatial resources is widespread, we suggest that the AZA concept should be considered to provide space for marine aquaculture, and avoid environmental degradation and negative interaction with other traditional users. Here, we define an AZA as: 'a marine area where the development of aquaculture has priority over other uses, and therefore will be primarily dedicated to aquaculture. Identification of an AZA will result from zoning processes through participatory spatial planning, whereby administrative bodies legally establish that specific spatial areas within a region have priority for aquaculture development'.

FAO (2010b) defines an Ecosystem Approach to Aquaculture (EAA) as a strategy for the integration of the activity within the wider ecosystem context that promotes sustainable development, equity and resilience of interlinked socio-ecological systems. A key principle of this approach is that aquaculture development and management should take account of the full range of ecosystem functions and services, and should not threaten the sustained delivery of these to society. Aquaculture should be developed in the context of ecosystem functions, which entails estimating assimilative and production carrying capacities, and adapting farming practices considering both ecological and social components. The EAA provides the

strategy including steps, process and tools to facilitate the definition of AZAs. This was recently proposed by Ross et al. (2013).

Another consideration is that aquaculture should help improve human well-being and equity for local communities. Following the IUCN (2009) and FAO recommendations (Ross et al. 2013), selection of AZAs should be based on a participatory approach. It should promote social acceptability by applying the precautionary principle, and consider the potential of adaptive management. Based on these principles, political decisions on the use of coastal space, following spatial planning, should favour aquaculture in coastal regions with reduced social confrontation. Therefore, AZAs should be identified in all coastal areas where a potential space for aquaculture exists, and be included in national and regional administrative structures alongside other sectors, serving as a tool for the integration of aquaculture into coastal zone management (Chapela Pérez 2009). It is clear, however, that the designation of an AZA is not enough to guarantee sustainable aquaculture. Within an AZA, the specific site selection and the production per site must match ecosystem carrying capacities and a permanent monitoring programme within the relevant water body (which could be an AZA) is needed to assess the impacts of each farm. Other potential issues, such as diseases, can be spread in AZAs and therefore biosecurity aspects must also be considered (Ross et al. 2013).

Allocation of mariculture in different regions

Spatial planning for selection of AZAs is already widespread globally, although they exist under very diverse legislative and administrative arrangements and are referred to by an array of names. Here, we provide key examples from each global region to demonstrate how spatial planning is implemented and how the concept of AZA has been flexibly implemented in regions with different legal and regulatory frameworks.

Europe

In the Mediterranean region, the development of European seabass *Dicentrarchus labrax* and gilthead seabream *Sparus aurata* culture is of high economic importance, but this been hindered by existing heavy use of the coastal zone and competition with many other users. Consequently, spatial planners have established AZAs across the region. Chapela & Ballesteros (2011) review the procedures for site selection, regulatory schemes and environmental impact assessment (EIA) processes undertaken by different Mediterranean countries.

In Italy, the legislative framework for marine aquaculture is heterogeneous across regions. This is because of the transfer of most administrative, regulative and control functions related to agriculture, fisheries and aquaculture from the Italian Ministry of Agricultural, Food and Forestry Policies and the Italian Ministry for Environment and Territory and Sea (MATTM) to the Regional Authorities (Decree 143, 4 Jun 2007). Only a few Italian regions have developed rules and guidelines for aquaculture development and monitoring. In 2008, the Sicilian region issued 'Guidelines for the setting-up of marine fish farms' (ARPA Sicilia 2008), which includes site selection criteria and a monitoring programme based on the relationship between the hydrodynamic regime and other environmental conditions, and the farmed fish biomass. The Liguria region issued 'Criteria and directives for marine aquaculture' (Committee Decree 1415, 30 Nov 2007), while the Toscana region provided guidelines and provisions for the application of EIA to fish farms (Regional Law 68, 18 Apr 1995). All these new regional regulations adopt an AZA approach and include guidelines for an environmental monitoring programme. The Emilia Romagna region set up the Information System for the Sea and the Coasts (SIC) as a decision-making tool to support integrated coastal zone management (ICZM) activities. In this framework, SICs support multiple uses of coastal areas (e.g. fisheries, mussel culture, oil platforms), with an interactive cartography based on geographic information system (GIS) tools. The Marches Regional Council issued a decision (18 Oct 2005) to identify public areas for aquaculture mainly based on socio-economic and environmental aspects. The Puglia Regional Council approved a specific law (no. 17, 23 Jun 2006) for the 'Plan of Regional Coasts', where, for further development of aquaculture activities, fish farms need to be certified by the European Eco-Management and Audit Scheme or as organic farms.

At the national level, the Protocol on Mediterranean ICZM (EU Official Journal L34/19, 4 Feb 2009) has been implemented by MATTM, through the institutional cooperation between the ministry and local public authorities. The protocol includes some recommendations on aquaculture, related to identifying allocated areas for aquaculture where these activities are under development, and the management of aquaculture wastes and the production process. The national Legislative Decree 152/2006 on the Protection of Waters against Pollution includes Article 111, which is specifically related to aquaculture. Pursuant to this article, a specific set of rules will be issued, that include the application of AZA principles. In June 2012 a new Legislative Decree (83/2012) transferred responsibility for issuing fish farm licenses to the Ministry of Agricultural, Food and Forestry Policies. It is expected that future integration of these 2 new pieces of legislation will provide a clearer legislative framework for aquaculture development in coastal areas and for implementation of AZAs.

In Spain, as in Italy, regional communities are responsible for coastal spatial planning. Thus, aquaculture site selection programmes are at various stages of development depending on region. As an example, in the region of Andalucía, planning and management of aquaculture activities fall under the responsibility of the Regional Ministry of Agriculture and Fisheries. In this region, studies for site selection since 2001 have been based on data collection and analysis using GIS for aquaculture planning and development. The first step in the process is to identify all users of the coastal zone and the roles of the different administrative bodies that have jurisdiction over the activities within the zone (Macías et al. 2006). This is followed by an environmental assessment of the identified areas (Macías et al. 2006). Studies undertaken in the final stages of the process address socio-economic aspects and consider the need for specific regulations for given species (Order 2006-BOJA/76, 10 Apr 2006). These studies and regulations are then applied to identify areas suitable for aquaculture within the whole Maritime Public Domain in the Region.

In Malta, the concept of 'Offshore Aquaculture Zones' (a synonym of AZA) originated in the mid-2000s from the need to translocate existing tuna farms offshore and to provide additional sites for new tuna ranches and other fish farms. The first such site, located 6 km off the southeastern coast of Malta, was established in 2006, and is managed by the Fisheries and Aquaculture Department of the Ministry for Sustainable Development, the Environment and Climate Change. Establishment of the Maltese AZA required an environmental impact statement (EIS) as per the requirements of the Malta Environment and Planning Authority (MEPA). The EIS included studies on aspects of feasibility and socio-economics, and of the marine environment, notably the benthic habitats present in the area. The Maltese AZA currently has 2 tuna farms operating within it, but the site has been designated to support a maximum of 6 farms. Allocations are on a lease basis and the fish farms operating within the site are required to undertake regular environmental monitoring as per the MEPA's guidelines.

Although Croatia has optimal natural conditions and a long tradition in aquaculture, the development of aquaculture has not proceeded as rapidly as in neighbouring Mediterranean countries. One of the main obstacles for its development has been delays in the selection of locations for aquaculture activities. Over the past decade, Croatia has developed a legal framework in order to define basic spatial requirements for different aquaculture activities, as well as requirements for EIA procedures. These measures have resulted in significant development of the sector. There are 7 administrative regions on Croatia's coastline. In these regions, AZAs are identified within physical plans. Basic biophysical criteria for AZAs are defined by specific national legislation. In addition to biophysical criteria, this legislation prescribes criteria regarding the availability of the necessary infrastructure. Based on the analysis of natural conditions and anthropogenic activities, areas that are excluded for aquaculture are defined (e.g. urban centres, military zones and areas of intensive maritime traffic). Aquaculture in protected areas, including NATURA 2000 sites (European Commission 2014), requires special authorisation. The permits for locations of aquaculture farms within an AZA must align with the physical plan. A permit is issued following an EIA that takes account of the species, production technique and production quantities. The EIA also defines future monitoring procedures. Based on the location permits, regional administrations then allocate concessions for aquaculture production for a maximum period of 20 yr. Final aquaculture licenses are issued by the Ministry of Agriculture based on the concession agreement and results of the EIA.

In Greece, the long size of the coastline, low level of development and lack of adequate funding and human resources resulted in a rather inadequate national plan for the development of the coastal zone. Since the beginning of the 1980s, fish farming has grown rapidly in sheltered, shallow coastal sites. Since the vast majority of fish farmers have applied for sites in remote, undeveloped areas, most licenses awarded before the 2000s were assigned without broader examination of the suitability of the area or assessment of the cumulative effects of multiple farms. After the Council of State cancelled all the acts of the administration regarding new licenses, a new planning system for fish farming was implemented. This includes 3 types of planning schemes for designated Areas for Development of Aquaculture: (1) the

POAY (acronym of the respective Greek terminology: areas of organised development of aquaculture) which follows the typical AZA concept, with a leased area of at least 10 ha and with coordinated management and monitoring of the farms; (2) the PASMI (acronym of the respective Greek terminology: areas of scattered concentrations of fish farming) with less than 5 farms, a farmed area less than 10 ha and a distance between farms of between 500 and 2000 m; and (3) individual farms mainly in remote areas which have specific limits placed on area and maximum production. The new regulation in Greece requires that farms in AZAs should be located at least 1000 m away from urban sites, tourist facilities, harbours, industrial areas, mining and other incompatible activities, and at least 500 m from diving centres and major swimming beaches. Establishment of a POAY requires extensive comprehensive EIA, including measurement of sea currents, mapping of habitats and sampling of biological and physico-chemical variables. POAYs are established by a Presidential Decree and require a management operator and a plan for the management, monitoring and restoration of the area.

Cage fish farming of European seabass and gilthead seabream began in the Turkish Aegean in 1986. Initially, farming took place in sheltered, shallow coastal areas between Izmir and Mugla along the Aegean coastline. In parallel with an increase in aquaculture production, there was a marked development of urbanisation and tourism. This generated conflicts among different uses of the area. New legislation was developed which led to relocation of fish farming offshore. This new legislation prohibited the siting of farms in locations shallower than 30 m depth and closer than 0.6 nautical miles from the shore. Furthermore, it considered eutrophication values using TRIX Index as a guideline for selection of AZAs (MEF 2007). According to the legal framework, 2 large AZAs at Milas and Bodrum were defined by the Inter-Ministerial consortium in 2007. Before the establishment of the 2 AZAs, an EIA was undertaken. Annually, each farm must be reassessed by the Ministry of Environment and Forestry for water column and sediment attributes within the Environmental Monitoring Programme (MEF 2009).

Oceania and Asia

A similar approach to spatial management to that of AZA is exemplified by the adoption of Aquaculture Management Areas (AMA) in New Zealand.

The aquaculture industry has experienced rapid recent growth, with more than 500 greenshell mussel *Perna canaliculus* farms in operation that cover a combined total area of 30 km^2. Rapid industry growth, coupled with the near saturation of traditional sites and advances in culture technologies, have led the industry offshore. The industry's desire to explore offshore areas, together with recent central government requirements, has created the need for environmental managers to designate zones for aquaculture through the creation of AMAs (Longdill et al. 2008). In November 2010, the Minister of Conservation released a revised New Zealand Coastal Policy Statement, providing stronger direction and recommendations on how coastal environment must be managed in different regions, including requiring the spatial identification of important elements. An example of the application of this policy is the Hauraki Gulf Marine Park Act 2000. This management plan includes integrated management of the gulf, including aquaculture, protection and enhancement of the resources and life-supporting capacity of the gulf's environment (Hauraki Gulf Forum 2011). This example underscores that AMAs go beyond AZAs, since after the zoning there is the development of a management plan for the AMA, and this is followed up by permanent monitoring, periodic evaluation and adaptive management measures. This is in full agreement with the recommended steps for EAA implementation (FAO 2010b).

In Australia, a recent example of AZA implementation can be seen in the development of a regional marine aquaculture management plan for coastal waters of the Sandy Straits region in Queensland (State of Queensland 2011). The Great Sandy Regional Marine Aquaculture Plan established guidelines and identified suitable zones for marine aquaculture development, which lie within a larger area of a marine park, where there was an array of zoned areas with different levels of protection. The AZAs were chosen after a consultation process between industry and government to minimise conflict with other user groups and ensure that social and environmental assets of the marine park were not compromised. Management controls were developed during the process, to reduce the risk of adverse impacts on these assets. These controls provide rules to be applied to aquaculture activities and specify the conditions under which aquaculture farms can operate, including infrastructure and design requirements, environmental requirements, specifications for ongoing environmental monitoring programmes and measures to ensure biosecurity.

In China, with the establishment of a research group by the State Aquaculture Administration in 1978, technological development allowed the intensification of aquaculture and production grew dramatically, from less than 1.3 million t in 1970 to approximately 47.8 million t in 2010, by which time it accounted for 61.2 % of the total global aquaculture production by weight, and representing 49.3 % of the total global aquaculture production by value (FAO 2012b). Aquaculture in China is a diverse industry, which includes production of a variety of fish, crustaceans, shellfish and algae. Marine aquaculture covers 2.08 million ha (MABF 2011), it is mostly carried out in shallow seas, mud flats and sheltered bays and also includes land-based installations (Cao et al. 2007). Geographically, China's marine aquaculture activities are divided in 3 sectors: the Bohai Sea and Yellow Sea culture zone, the southeast coastal culture zone and the Yangtze Valley culture zone. In general, 2 main factors affect aquaculture site selection in China: the local functional zoning plan and the Regulation on Water Quality and Effluent from Aquaculture. Thus, the coastal areas must fit the local functional zoning plan, considering that ownership of land and water areas in China is public. For example, the Functional Zoning Scheme of the Coastal Areas of Guangdong region was issued in 1999, which specified division of the coastal area into different 'function zones', including aquaculture. In 2004, the 'Aquaculture Plan for Inland Water Areas and the Coastal Zone of Guangdong' was approved by the provincial government, establishing guidelines for aquaculture development and management by local authorities.

North and South America

While Canada has a number of policy documents designed to provide a framework for sustainable aquaculture (reviewed in VanderZwaag et al. 2012), it lacks specific federal legislation to address it. Among the various policy documents, the National Aquaculture Strategic Action Plan Initiative 2011–2015 is the most overarching and is endorsed by federal and regional governments. It is designed to provide a strategic vision for aquaculture development based on environmental protection, social well-being and economic prosperity. Despite such a policy framework, Canada continues to rely on a complex patchwork of more than 70 pieces of federal and regional legislation to regulate the industry. Thus, Canada needs to develop and enact modern integrated legislation along the lines of a sustainable aquaculture act that specifies requirements and guidance on national objectives and procedures for all aquaculture operations (Hutchings et al. 2012). This is recognised and called for by industry, stakeholders and non-governmental organisations alike (VanderZwaag et al. 2012).

On the Pacific coast, Fisheries and Oceans Canada is in the process of developing Integrated Management of Aquaculture Plans for aquaculture licensing that include spatial planning of aquaculture sites. The region of British Columbia remains involved in assessing land tenure applications for aquaculture facilities during the transition to federal regulation. While more of the regulation of finfish aquaculture falls within the purview of regional governments on Canada's Atlantic coast, the federal government retains the authority to be consulted on and to provide recommendations regarding the issue of leases or extension of those already issued by regions. As a result, the regulatory frameworks have been largely site-specific, developed locally or regionally, and have not been reconciled with each other around the country (DFO 2005). To date, however, integrated management efforts and accomplishments have been limited (Jessen 2011, VanderZwaag et al. 2012). It remains for Canada to develop marine spatial planning with clear geographical priorities, explicit timelines and transparent measures for public reporting (Hutchings et al. 2012).

In United States of America, the 2011 Department of Commerce and National Oceanic and Atmospheric Administration aquaculture policies as well as the National Ocean Policy, provide federal guidance for marine aquaculture activities (NOAA 2011). Projects that are sited in USA waters must meet a suite of federal, state, and local regulations that ensure environmental protection, water quality, food safety, and protection of public health. Currently, the first comprehensive regional approach to authorising aquaculture in federal waters is being developed through a Fishery Management Plan for Regulating Offshore Marine Aquaculture in the Gulf of Mexico (GMFMC & NOAA 2009). The purpose of this 'Gulf Aquaculture Plan' plan is to maximise benefits to the nation through the development of an environmentally sound and economically sustainable aquaculture industry in federal waters of the Gulf of Mexico. The plan establishes a regional permit process and includes a first approach to marine spatial planning. The plan would allow up to 20 offshore aquaculture operations to be permitted in federal waters of the gulf over a 10 yr period. At a regional scale, there are a few examples of marine spatial planning. In Florida, for example, the process of site selection for farms growing *Mercenaria* spp. has been described by Arnold et al. (2000).

In Ecuador, which supports an important crustacean culture industry, the national government has adopted a law on fisheries and aquaculture planning called the Management Plan of Fisheries and Aquaculture (Official Register 14, 4 Feb 2003; Arriaga & Martinez 2003). This management plan enforces the zoning of marine and coastal areas, including the selection of AZAs.

The General Law of Fisheries and Aquaculture of the Chilean Government (Article 67) defines Suitable Areas for Aquaculture (AAA, from the Spanish term 'Areas Aptas para la Acuicultura'). AAAs are areas under public ownership that, following consultation with competent authorities, are entitled to contain aquaculture production (SUBPESCA 2014). The AAAs are selected by regions along the Chilean coast, and their designation should be published in the Official Journal of the Republic of Chile. Other activities can be carried out in these AAAs, and all types of aquaculture can be developed (from shore based to open water). However, the Chilean salmon farming industry was strongly affected by infectious salmon anaemia (ISA) disease, which is spread within aquaculture areas (Niklitschek et al. 2013). This might indicate that the selection of AZAs was not correctly carried out.

Africa

In Morocco, shellfish culture (the oyster *Crassostrea* spp.) first started on the Atlantic coast (Lagoon of Oualidia) in the 1950s, while fish culture (gilthead seabream and European seabass) began first on the Mediterranean coast (Lagoon of Nador) in 1986. Aquaculture took place in sheltered sites until the 1990s, after which coastal aquaculture installations began to proliferate. Floating cages were used for fish culture in the Mediterranean, and rafts and floating and sub-floating long-lines for mussel (*Mytilus galloprovincialis*) culture, both in the Mediterranean and Atlantic. However, marine aquaculture in Morocco developed less rapidly than in neighbouring European countries. The National Institute of Marine Fisheries undertakes studies and experiments to support aquaculture development and diversification in terms of space, species and techniques. Spatial planning for marine site selection was initiated in 2000 to allow for aquaculture to integrate and enhance its development at national level.

There is a 3-step procedure for aquaculture planning in Morocco: (1) characterise the current environmental, social and economic states of the target area; (2) define suitable zones for aquaculture inside the area using GIS tools and a participative approach, including the selection of suitable species and techniques; and (3) assessment the environmental impact of the defined aquaculture project. Criteria and requirements taken into consideration are those adopted by the European Union. This procedure allows definition of a specific zone to be dedicated exclusively for aquaculture purposes, without any conflicts of space or interest. It also allows local integration of aquaculture activities into maritime public domain, with a concession for a determined period (generally 20 yr and renewable). Aquaculture farmers are required to undertake regular environmental monitoring of their farm site annually. The first aquaculture plan was established for Dakhla Bay in early 2000 and since then, 4 other aquaculture plans were established (3 for lagoons and 1 for the coastal zone).

In Tunisia, the long coastline hosts numerous competitive and sometimes conflicting activities, including tourism, fishing, and aquaculture. This pressure on the coastal marine space has directly impacted procedures to assign sites to aquaculture activity. Indeed, the selection of aquaculture sites in sea areas remains a private initiative in Tunisia. The proposer must submit a preliminary study to obtain approval by the Interdepartmental Commission of Set Net. This committee (Decision of the Minister of Agriculture, 28 Sep 1995, Article 43) brings together the main ministries responsible for the use, management and control of maritime space. Representatives of local communities, chosen by local stakeholders, also take part in the work of the committee. The first approval granted by the committee reserves the requested area for aquaculture activity for 6 mo, to allow the development of technical proposals and appraisal of economic viability. At a later stage, it is necessary to obtain agreement by the National Agency for the Protection of the Environment based on the conclusions of the EIA. Once a final acceptance is granted, the area is allocated exclusively for aquaculture and becomes an aquaculture concession. Any activity other than aquaculture is prohibited in the concession area and in a 500 m buffer zone around the fish farm (Decision of the Minister of Agriculture, 28 Sep 1995, Article 49). Even the concession holder is not allowed to fishing in this area. Any modification or technical change must obtain prior an approval from a supervisory administrative body (Article 6). The concession holder is also held responsible for all irreversible changes of the environment (Article 13).

POTENTIAL BENEFITS FROM ALLOCATED ZONES FOR AQUACULTURE (AZA)

Appropriate AZA selection

As described above, mariculture is part of marine spatial planning in many regions of the world, based on different administrative approaches. The best scenario is that the spatial expansion of marine aquaculture is nested within the broader context of marine spatial planning to minimise adverse impacts on the environment and biodiversity, and to preserve ecosystem services at regional scale. Alternatively, selection of AZAs could be carried out by specific aquaculture spatial planning bodies, or site selection could even be made on a case by case basis, considering environmental and social aspects at local scale. In general, the selection of AZAs within the framework of coastal planning programmes, following different administrative and technical approaches, seems to be the optimal tool for a rational use of the marine space. However, sometimes the results are the opposite from what was intended, with unexpected negative effects on natural resources or on the aquaculture industry itself. It is necessary to balance properly demands for aquaculture development with the need to protect marine ecosystems (Douvere & Ehler 2009). In order to avoid future problems, selection of AZAs must take account of current and future uses of ecosystem services and must be done in a participative way, under the ecosystem-approach to aquaculture, and within the framework of marine spatial planning to secure sustainable benefits from aquaculture without exceeding the resilience of natural systems (FAO 2010b).

In any case, conservation of habitats and of species of high ecological value, and maintenance of ecological biodiversity and water quality, will improve social acceptance of aquaculture (GFCM 2011). In addition to natural limiting factors, it is important to note that the social interactions with other human activities, together with conflicts on use and partitioning of resources in the much-exploited coastal zone, are constraints to be considered when aquaculture facilities are set up (IUCN 2009). Moreover, aquaculture is an economic activity more dependent on the maintenance of environmental quality than many others. Therefore spatial planning of the aquaculture sector should avoid economical losses due to environmental degradation (e.g high mortality because of diseases in Chile; Bustos-Gallardo 2013) or selection of sites with unsuitable oceanic conditions (e.g. bankruptcy due to escapes in La Palma Island; Toledo-Guedes et al. 2014).

Role of management in AZA implementation

It is important to note that, following development of the spatial planning procedure for selection of the AZA, adoption of sustainable aquaculture is not necessarily guaranteed, as this depends on effective enforcement of regulations by the relevant authorities. The salmon industry crisis that resulted in widespread infection of farmed fish by ISA in 2007, serves as a recent example of bad management. This disease, first detected in Norway in 1984 (Asche et al. 2010), led to the collapse of production and of employment, and devastation of the environment. Official information for Chile, available from the national fisheries service (SERNAPESCA), records verified outbreaks due to the ISA virus at a total of 134 fish farms between August 2007 and July 2009. The spread of ISA across farms was very rapid because of the inaction of the authorities and farm companies. Bustos-Gallardo (2013) argues that, at the time of prevalence of the disease, the Chilean environmental authorities did not consider the available scientific information, demonstrating an inability of the political system to address the crisis. This failure had 2 facets: firstly, an investigative phase during which scientific explanations were sought but not found in the form needed for policy-making; and secondly an intervention phase when the tentative solutions focused on economic restructuring and financial considerations while failing to address long-term ecological implications. Spatial planning and AZA selection (AMA in Chile) were probably the most relevant and important factors for curbing the ISA crises, since some areas that had been identified as optimal for aquaculture had a very high density of farms, and the risk of outbreaks of disease had not been taken into consideration. This was evidenced by the failure to allocate safety distances between farms to mitigate the spread of disease. Criticism from the scientific community focused on the degree to which national environmental policy was subordinated to the economic strategy pursued by Chile over the previous 30 yr, which had allowed excessive farm densities in some coastal regions. One of the most important and debated aspects of the recommendations from a joint working group was the establishment of 'barrios' (i.e. groups of farming activities in the same neighbourhood). Wherever there were at least 3 contiguous concessions, a group of owners could request to be declared a 'barrio', so as to coordinate their seeding and harvesting operations, application of antibiotics or other chemicals, and most importantly, to ensure better control and containment of future disease outbreaks. Governmental incentives were offered to the farm owners to relocate at a greater distance from other farms and operations; but the need for adequate spacing between farms, following the long experience of Norwegian companies growing Atlantic salmon, should have been included as a criterion for AZA selection in advance. Currently there are concerns because of changes made to the previous Fisheries and Aquaculture Regulations of 2010, which allow establishment of new AZAs in ISA-free sites for farm relocation. This would result in salmon farming activities moving south (into the Magallanes region); however, there is no robust scientific evidence to support such a measure (Vega-Salinas 2013). A panel of fish health experts weighted the risk factors predictive of the spread of the ISA virus among barrios in Chile, and defined fish density and site selection as 2 of the most important aspects to consider (Gustafson et al. 2014). The salmon crisis in Chile is an example of bad environmental governance, which raises important questions about the science–policy interface, and the extent to which current arrangements have the capacity to lead to appropriate spatial planning for the sustainable management of natural resources.

Another example of unintended adverse outcomes resulting from inappropriate spatial planning and implementation of AZAs is the new Turkish legislation for site selection, which requires fish farms to relocate offshore to 2 designated AZAs. Prior to this legislation, there were 127, mostly small farms, located in small semi-enclosed inlets in Gulluk Bay. As a result of a further amendment to the Turkish law, the fish farms were unable to fulfil new criteria and were compelled to relocate to the offshore AZA during the period 2008–2009. Following relocation, an additional 81 new larger fish farms began to operate in Gulluk Bay and the annual production capacity was increased to 88000 t. Fish farms that were found to be in contravention of this new law, mostly small fish farms, were closed down by the authorities. Therefore, the new Turkish legislation negatively affected the small fish farms, which could not afford the financial investment required for reallocation, and led to concentration of farming activities in a few large-scale fish farms.

The review of AZAs makes it clear that in countries where aquaculture is a strategic sector, spatial planning, and/or environmental management for aquaculture have been implemented in diverse ways (Fig. 1). In simple terms, spatial aquaculture management might be classified based on: (1) the length of the coastline of the country; (2) the governance level

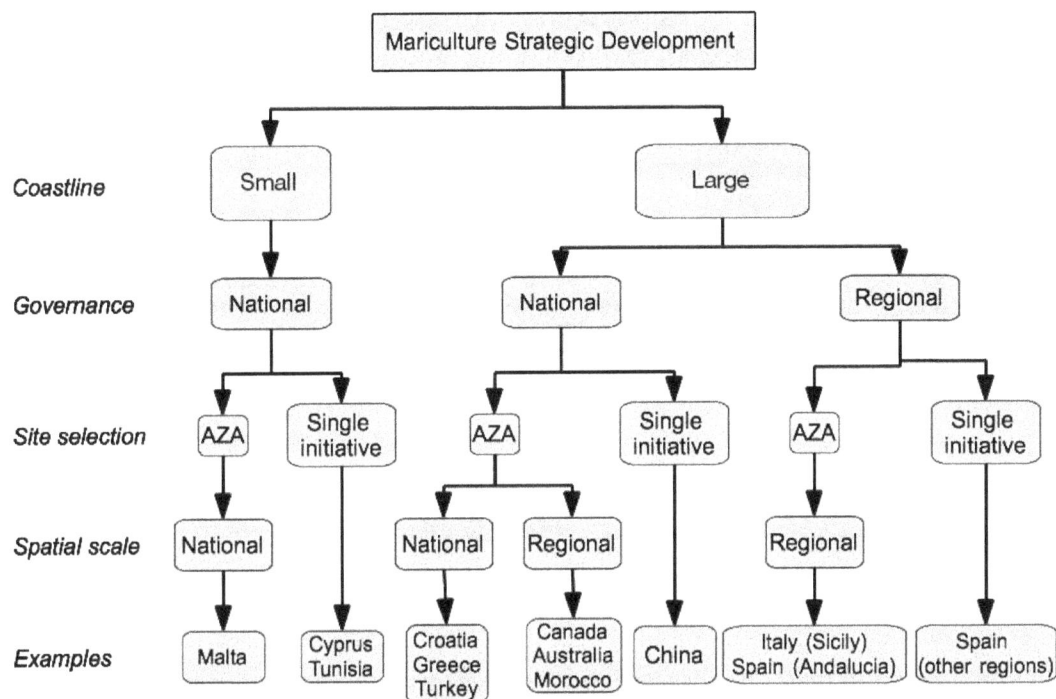

Fig. 1. Classification of case studies of Allocated Zones for Aquaculture (AZAs) included in the present study, relative to (1) the length of the coastline of the country, (2) the governance level, (3) the site selection procedure (based on previous spatial planning or relying on private single initiative) and, in the event of spatial planning, (4) the spatial scale of AZA selection

(national or regional); (3) whether site selection is framed by previous spatial planning or relies on private initiative; and, if a spatial planning process exists, (4) the spatial scale of AZA selection (the entire national coastline or by regions). Logically, the management strategy depends on the size of the coastline. Small countries (with less than 2000 km of coastline) can more easily develop AZA-based site selection for the total available marine space (74 countries from a total of 149, following the World Resources Institute, e.g. Malta), because governance is at the national level. However, this does not always occur, as noted for Cyprus and Tunisia, small countries where site selection depends very much on individual initiatives. For larger countries, site selection is strongly dependent on the administrative structure (i.e. the distribution of central and regional legal authorities, among states, autonomous regions, provinces and counties). By contrast, some countries with a very long coastline, such as Turkey, Croatia and Greece, have developed spatial planning at the national level. At this level, success depends principally on how different administrative bodies coordinate AZA implementation within the total national territory (for an example, see Karka et al. 2011). It is common for a regional approach to be adopted for selection of an AZA, especially in regions in which

aquaculture is a priority sector (e.g. British Columbia in Canada, Queensland in Australia or the Mediterranean coast of Morocco). Despite the considerable efforts at both national and regional levels to developing guidelines for the management of aquaculture activities, there are many examples where spatial planning is not correctly implemented at the regional scale; these include cases in which site selection depends on single business initiatives (e.g. the Valencian Community in Spain). Some large Asian countries, for example China, which has the highest aquaculture production growth worldwide, have general marine spatial planning guidelines that are embedded in aquaculture planning regulations, and which are based on a 'site by site' management plan. It appears that China considers the physical and production aspects of aquaculture as driving forces for spatial planning, without taking account of environmental aspects, social acceptance or potential user conflicts. Such guidelines and regulations could lead to enhanced user conflict and environmental degradation at the regional scale. In such cases, creation of 'social spaces' to allow space for users and stakeholders to interact is essential for optimal AZA selection and designation. After top-down site selection, local issues including farming activities that exceed the carrying capacity, environmental pressure due to

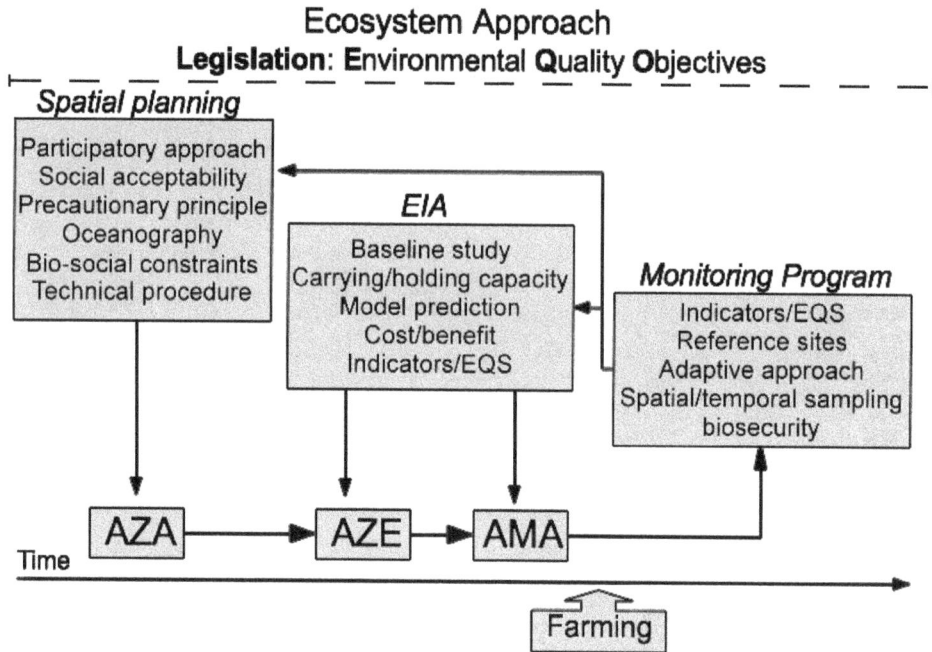

Fig. 2. Technical procedure for correct full environmental management of aquaculture, which must include the definition of an Allocated Zone for Aquaculture (AZA), an Environmental Impact Assessment (EIA), definition of an Allowable Zone of Effect (AZE) and an Aquaculture Management Area (AMA) and, following the initiation of farming activity, development of monitoring programmes (modified from Sanchez-Jerez & Karakassis 2011)

problems with area zoning scheme enforcement, and lack of effective monitoring and of legislation on effluent discharge, are the main bottlenecks that are currently limiting appropriate aquaculture site selection and carrying capacity management in China (Chen et al. 2011).

In all these scenarios, successful development of aquaculture in the coastal zone requires the establishment of a regulatory framework for coastal and offshore aquaculture, and consideration of important aspects such as licensing, site selection and monitoring, to keep possible harmful effects on the marine environment under control. The challenge for governments is to ensure that appropriate governance measures are implemented. Without effective governance, there will be misallocation of resources or stagnation of production, and this affects all businesses, whether aquaculture or any other. Communication across administrative levels must be effective and transparent and with mechanisms for societal participation. Regulatory procedures can be conducive to investment or may instead hinder all entrepreneurial initiatives in aquaculture. Without the rule of law, there will be little predictability and security. In such situations, farmers have no incentive to take risks or to invest. Rent-seeking rather than efficiency becomes rational behaviour for resource users, with a resulting loss of productivity (Hishamunda et al. 2014).

Development of full environmental management of aquaculture, beyond marine spatial planning, should include EIAs and monitoring programmes and, for each AZA, definition of the corresponding Allowable Zone of Effect (AZE) and Aquaculture Management Area (AMA) (Fig. 2). Under the umbrella of an ecosystem approach, national and/or regional legislation should define the Environmental Quality Objectives (EQOs), which ensure the safeguarding of ecosystem services. For the selection of AZAs, it is important to establish a regulatory process that clearly identifies where aquaculture facilities can be located and for how long. A technical procedure of site selection, using biological and oceanographic information and taking into consideration ecological and social constraints (i.e. carrying capacities), must also be applied in defining the AZA. This process must follow a participatory approach, which considers the social acceptability of aquaculture in the area. For each AZA, an EIA will: (1) forecast the potential environmental and socio-economic impact of aquaculture; (2) define the spatial extent of the AZE (Sanchez-Jerez & Karakassis 2011) for a determinate level of production and type of mariculture; (3) estimate the area of influence around the AZE; and (4) define the AMA. The EIA will also facilitate collection of baseline environmental information, enable estimation of the holding/carrying capacity (see a review in Ross et al. 2013) and define the indicators to be used by the monitoring programme, as well as values for Environmental Quality Standards (EQS). Therefore, establishment of the AMA should be the last step, incorporating a management plan which includes a monitoring programme, and considers biosecurity aspects and social and environmental constraints. Estimation of carrying/holding capacity and setting a limit to maximum production would also be fundamental. Monitoring results should be made public and communicate the status of the AZA to the public in an accessible way (e.g. using a traffic light system). When farming is initiated, monitoring programmes should ensure that the EQS are within optimal limits using pre-defined indicators of environmental change, with an appropriate spatial and temporal sampling design. Fortunately, all these aspects are considered to a greater or lesser extent by most countries where aquaculture is an important sector.

CONCLUSIONS

Because management of marine aquaculture should be ecosystem-based, balancing ecological, economical, and social objectives for sustainable development, we suggest that national or regional authorities in charge of coastal zone management must use spatial planning for identifying suitable AZAs, and implement them more widely. Development of aquaculture spatial plans as a framework for the establishment of AZAs, will help avoid negative externalities, provide business opportunities, and decouple environmental degradation and aquaculture development. It is essential that, for AZAs to be effective, they should be accompanied by a clear national/subnational (regional or other national administrative subdivision) regulatory framework and regulated under optimal governance, in which the different procedures for licensing and leasing aquaculture activities are clearly outlined. Finally, site selection and definition of AZAs are only the first step towards sustainable aquaculture. To follow up, one or several AMAs should be defined within an AZA, each with a management plan in accordance with the EIA, ongoing monitoring, and provisions for adaptive management. There should be appropriate stakeholder participation, and information about the AZA should be made available to the public.

The General Fisheries Commission for the Mediterranean (GFCM), a regional fisheries management organisation with a specific mandate for aquaculture, adopted a specific resolution on AZAs in 2012, in which marine spatial planning, EIA and a monitoring programme are considered essential for the implementation of a regional strategy for the creation of AZAs, and as a priority for the responsible development and management of aquaculture activities in the Mediterranean and Black Sea (GFCM 2012). Further to this resolution, the Committee on Fisheries of FAO (2013b) emphasised the need for spatial planning to ensure the allocation of adequate space for aquaculture. It highlighted the benefits of spatial planning for multiple outcomes, including: (1) a more coordinated and integrated approach to the use and management of the environment; (2) accountability and transparency by involving relevant stakeholders at all levels; (3) a better understanding of the cumulative and combined effects of different uses of coastal resources and the interactions among resource users and between users and the environment; (4) a more effective mechanism for governments to deliver their commitments to sustainable development; (5) greater clarity on policy and decision-making; and (6) a better understanding of the changes required to improve different enabling policy and regulatory frameworks.

In a future, with improved technology, aquaculture systems will move away from inshore to open waters. Offshore mariculture may be an option to avoid issues of competition for space, reducing environmental impact and negative public perceptions (Lovatelli et al. 2013). In this emerging scenario, AZAs also need to be selected by spatial planning processes within the exclusive economic zones of each country; or even further, beyond the 200 nautical miles belt of national jurisdiction (Hishamunda et al. 2014). Therefore new analytical frameworks will be necessary to define appropriate aquaculture site selection, closely linked to aquaculture technological developments. Selection of an AZA should be an adaptive process, in order to respond to the potential beneficial and negative effects of climate change and weather uncertainty, some that must be taken into consideration for future marine spatial planning.

LITERATURE CITED

ä Arnold WS, White MW, Norris HA, Berigan ME (2000) Hard clam (*Mercenaria* spp.) aquaculture in Florida, USA: geographic information system applications to lease site selection. Aquacult Eng 23:203–223

ARPA (Agenzia Regionale per la Protezione Dell'Ambiente) Sicilia (2008) Linee guida per la realizzazione di impianti di maricoltura in Sicilia. Regione Siciliana, Palermo

Arriaga L, Martinez J (2003) Subsecretaria de Recursos Pesqueros. Plan de ordenamiento de la pesca y acuicultura del Ecuador. (accessed 20 Sep 2014)

Asche F, Hansen H, Tveteras R, Tveterås S (2010) The salmon disease crisis in Chile. Mar Resour Econ 24: 405–411

ä Bustos-Gallardo B (2013) The ISA crisis in Los Lagos Chile: a failure of neoliberal environmental governance? Geoforum 48:196–206

ä Cao L, Wang WM, Yang Y, Yang CT, Yuan ZH, Xiong SB, James D (2007) Environmental impact of aquaculture and countermeasures to aquaculture pollution in China. Environ Sci Pollut R 14:452–462

Chapela Pérez R (2009) Review on cage aquaculture licensing procedures: a focus on Chile, Greece, Norway, Spain and the United States of America. In: FAO Regional Commission for Fisheries. Report of the Regional Technical Workshop on Sustainable Marine Cage Aquaculture Development, Muscat, 25–26 January 2009. FAO Fish Aquacult Rep 892, FAO, Rome, p 95–130

Chapela R, Ballesteros M (2011) Procedures for site selection, regulatory schemes and EIA procedures in the Mediterranean. In: Report of the 35th session of the General Fisheries Commission for the Mediterranean (GFCM). Site selection and carrying capacity in Mediterranean marine aquaculture: key issues (WGSC-SHoCMed). Unpublished document (GFCM:XXXV/2011/Dma.9), p 91–136. http://bit.ly/GFCM-XXXV-2011-Dma9 (accessed 15 Feb 2014)

Chen LX, Zhu CB, Dong SL (2011) Aquaculture site selection and carrying capacity management in China. Guangdong Agricult Sci 21:1–7

Dempster T, Sanchez-Jerez P (2008) Aquaculture and coastal space management in Europe: an ecological perspective. In: Holmert M, Black K, Duarte CM, Marbà N, Karakassis I (eds) Aquaculture in the ecosystem. Springer, Dordrecht, p 87–116

DFO (Fisheries and Oceans Canada) (2005) Assessment of finfish cage aquaculture in the marine environment. Science Advisory Report 2005/034, DFO Canadian Science Advisory Secretariat, Ontario

Díaz R, Rabalais NN, Breitburg DL (2012) Agriculture's impact on aquaculture: hypoxia and eutrophication in marine waters. Organisation for Economic Co-operation and Development. www.oecd.org/tad/sustainableagriculture/49841630.pdf (accessed 20 Feb 2014)

ä Douvere F, Ehler CN (2009) New perspectives on sea use management: initial findings from European experience with marine spatial planning. J Environ Manage 90: 77–88

ä Duarte CM, Holmer M, Olsen Y, Soto D and others (2009) Will the oceans help feed humanity? Bioscience 59: 967–976

European Commission (2014) Natura 2000 network. ec.europa.eu/environment/nature/natura2000 (accessed 16 Feb 2014)

FAO (Food and Agriculture Organisation of the United Nations) (2010a) The state of world fisheries and aquaculture 2010. FAO, Rome

FAO (2010b) Aquaculture development. 4. Ecosystem approach to aquaculture. FAO Technical Guidelines for Responsible Fisheries 5, Suppl 4. FAO, Rome

FAO (2012a) The state of world fisheries and aquaculture 2012. FAO, Rome

FAO (2012b) FAO Fishstat Plus database, Fisheries and Aquaculture Information and Statistics Service www.fao.org/fishery/statistics/ (accessed 8 Dec 2013)

FAO (2013a) The FAO Fisheries and Aquaculture Depart-

ment's efforts in implementing the recommendations of the past sessions of the COFI sub-committee on Aquaculture. Seventh session of the Sub-Committee on Aquaculture (SCA) of the FAO Committee on Fisheries (COFI), St. Petersburg, 7–11 Oct 2013

FAO (2013b) Applying spatial planning for promoting future aquaculture growth. In: Seventh session of the Sub-Committee on Aquaculture (SCA) of the FAO Committee on Fisheries (COFI), St. Petersburg, 7–11 October 2013, Discussion document: COFI:AQ/VII/2013/6. www.fao.org/3/a-mk029e.pdf

GFCM (General Fisheries Commission for the Mediterranean) (2011) Site selection and carrying capacity in Mediterranean marine aquaculture: key issues (WGSC-SHoCMed). Unpublished document (GFCM:XXXV/2011/Dma.9). http://gfcmsitestorage.blob.core.windows.net/documents/web/GFCM/35/GFCM_XXXV_2011_Dma.9.pdf (accessed 15 Feb 2014)

GFCM (2012) Resolution GFCM/36/2012/1 on guidelines on allocated zones for aquaculture (AZA). http://bit.ly/Resolution-GFCM-36-2012-1 (accessed 16 Feb 2014)

GMFMC (Gulf of Mexico Fishery Management Council) and NOAA (National Oceanic and Atmospheric Administration) (2009) Fishery management plan for regulating offshore marine aquaculture in the Gulf of Mexico. GMFMC, Tampa, FL

ä Gustafson L, Antognoli M, Fica ML, Ibarra R and others (2014) Risk factors perceived predictive of ISA spread in Chile: applications to decision support. Prev Vet Med 117:276–285

ä Hall CM (2001) Trends in ocean and coastal tourism: the end of the last frontier? Ocean Coast Manage 44:601–618

Hauraki Gulf Forum (2011) Spatial planning for the Gulf: an international review of marine spatial planning initiative and application to the Hauraki Gulf. Hauraki Gulf Forum, Auckland. www.aucklandcouncil.govt.nz/EN/AboutCouncil/representativesbodies/haurakigulfforum/Documents/Spatialplanforthegulf.pdf (accessed 16 Feb 2014)

Hishamunda N, Ridler N, Martone E (2014) Policy and governance in aquaculture: lessons learned and way forward. FAO Fish Aquacult Tech Pap 577, FAO, Rome

ä Hutchings JA, Côté IM, Dodson JJ, Fleming IA and others (2012) Is Canada fulfilling its obligations to sustain marine biodiversity? A summary review, conclusions and recommendations. Environ Rev 20:353–361

IUCN (International Union for the Conservation of Nature) (2009) Guide for the sustainable development of Mediterranean aquaculture. 2. Aquaculture site selection and site management. IUCN, Gland

ä Jessen S (2011) A review of Canada's implementation of the Oceans Act since 1997—from leader to follower? Coast Manage 39:20–56

Karka H, Kyriazopoulos E, Kanellopoulou K (2011) Spatial planning for aquaculture: a special national framework for resolving local conflicts. In: 51st Congress of the European Regional Science Association (ERSA), Barcelona, 30 Aug–3 Sep 2011. www-sre.wu.ac.at/ersa/ersaconfs/ersa11/e110830aFinal01592.pdf (accessed 26 Feb 2014)

Knapp G (2008) Economic potential for US offshore aquaculture: an analytical approach. In: Rubino M (ed) Offshore aquaculture in the United States: economic considerations, implications and opportunities. NOAA Tech Memo NMFS F/SPO-103, US Dept of Commerce, Silver Spring, MD, p 15–50

Larsen J, Roney M (2013) Farmed fish production overtakes beef. Earth Policy Institute, Washington, DC. www.earth-policy.org/plan_b_updates/2013/update114 (accessed 20 Feb 2014)

Longdill PC, Healy TR, Black KP (2008) GIS-based models for sustainable open-coast shellfish aquaculture management area site selection. Ocean Coast Manage 51: 612–624

Lovatelli A, Aguilar-Manjarrez J, Soto D (eds) (2013) Expanding mariculture farther offshore: technical, environmental, spatial and governance challenges. FAO Technical Workshop, 22–25 Mar 2010, Orbetello. FAO Fish Aquacult Proc 24, FAO, Rome

MABF (Ministry of Agriculture Bureau of Fisheries) (2011) China fisheries yearbook. China Agriculture Publishing Company, Beijing

Macías JC, Lozano I, Alamo C (2006) Zonas idóneas para el desarrollo de la acuicultura en espacios marítimo-terrestres de Andalucía. Consejería de Agricultura y Pesca, Junta de Andalucía, Málaga

Marra J (2005) When will we tame the oceans? Nature 436: 175–176

MEF (Ministry of Environment and Forestry) (2007) The notification to identify the closed bay and gulf qualified sensitive where fish farms are not suitable to be established in the seas. Turkish Official Gazette 26413 (in Turkish)

MEF (Ministry of Environment and Forestry) (2009) Monitoring regulations for fish farms. Turkish Official Gazette 27257 (in Turkish)

Niklitschek E, Soto D, Lafon A, Molinet C, Toledo P (2013) Southward expansion of the Chilean salmon industry in the Patagonian fjords: main environmental challenges. Rev Aquacult 5:172–195

NOAA (National Oceanic and Atmospheric Administration) (2011). Marine aquaculture policy. www.nmfs.noaa.gov/ aquaculture/docs/policy/noaa_aquaculture_policy_2011. pdf (accessed 26 Feb 2014)

Pattanaik C, Prasad SN (2011) Assessment of aquaculture impact on mangroves of Mahanadi delta (Orissa), east coast of India using remote sensing and GIS. Ocean Coast Manage 54:789–795

Primavera JH (1997) Socio-economic impacts of shrimp culture. Aquacult Res 28:815–827

Ross LG, Telfer TC, Falconer L, Soto D, Aguilar-Manjarrez J (eds) (2013) Site selection and carrying capacities for inland and coastal aquaculture. FAO/Institute of Aquaculture, University of Stirling, Expert Workshop, 6–8 Dec 2010, Stirling. FAO Fish Aquacult Proc 21, FAO, Rome

Sanchez-Jerez P, Karakassis I (2011) Allowable zone of effect for Mediterranean marine aquaculture (AZE) (WGSC-SHoCMed). Unpublished document (GFCM:CAQ/2012/CMWG-5/Inf.11). http://bit.ly/GFCM-CAQ-AZE-2011 (accessed 15 Feb 2014)

State of Queensland (2011) Great Sandy Regional Marine Aquaculture Plan. Department of Employment, Economic Development and Innovation, Brisbane. www.daff.qld.gov.au/__data/assets/pdf_file/0019/65710/final-GSRMAP.pdf (accessed 16 Feb 2014)

SUBPESCA (2014) Áreas Aptas para la Acuicultura (AAA). Subsecretaría de Pesca y Acuicultura, Gobierno de Chile, Santiago. www.subpesca.cl/institucional/602/w3-article-915.html (accessed 15 Sep 2015)

Toledo-Guedes K, Sanchez-Jerez P, Brito A (2014) Influence of a massive aquaculture escape event on artisanal fisheries. Fish Manag Ecol 21:113–121

VanderZwaag DL, Hutchings JA, Jennings S, Peterman RM (2012) Canada's international and national commitments to sustain marine biodiversity. Environ Rev 20:312–352

Vega-Salinas D (2013) Uso de las concesiones acuícolas de mar en la industria salmonera de Chile. Sustain Agricult Food Environ Res 2:1–35

15

Area use and movement patterns of wild and escaped farmed Atlantic salmon before and during spawning in a large Norwegian river

Karina Moe[1,2], Tor F. Næsje[1,*], Thrond O. Haugen[2], Eva M. Ulvan[1], Tonje Aronsen[1], Tomas Sandnes[3], Eva B. Thorstad[1]

[1]Norwegian Institute for Nature Research (NINA), PO Box 5685 Sluppen, 7485 Trondheim, Norway
[2]Norwegian University of Life Sciences (NMBU), PO Box 5003 NMBU, 1432 Ås, Norway
[3]Aqua Kompetanse AS, 7770 Flatanger, Norway

ABSTRACT: We compared the within-river movements and distribution of wild and escaped farmed Atlantic salmon *Salmo salar* before and during spawning in the Namsen river system of Central Norway. A total of 74 wild and 43 escaped farmed salmon were captured at sea, tagged with radio transmitters and released. Based on our examinations, most, if not all salmon (farmed and wild) entering the River Namsen were sexually mature. Farmed salmon entering the river system had a higher probability than wild individuals of reaching the migration barrier in the upper part of the river, 70 km from the sea. During the pre-spawning and spawning periods, farmed salmon were located mainly in the upper parts (50 to 70 km from the sea), whereas wild salmon were evenly distributed along the entire river during both periods. Consequently, the probability of farmed × wild inter-breeding varied among river sections. Our finding that the distribution of escaped farmed salmon may differ from that of wild salmon and among river sections in the pre-spawning and spawning periods—and that it may also vary over time—must be taken into consideration when (1) designing monitoring programs aimed at estimating the proportion of escaped farmed salmon in rivers and (2) when interpreting monitoring results. Furthermore, targeted fishing in the river aimed at reducing the number of farmed salmon prior to spawning may be more effective in upper rivers sections, and below major migration barriers.

KEY WORDS: Radio telemetry · Farmed escapees · Within-river movements · Pre spawning · Spawning period · Introgression risk · River Namsen

INTRODUCTION

The population of wild Atlantic salmon *Salmo salar* L. has declined over the last decades, despite efforts to reduce fishing pressure (ICES 2014). In areas of intensive salmon aquaculture, populations of wild salmonids may be negatively impacted by increases in the abundance of salmon lice *Lepeophtheirus salmonis* and other infections. In addition, farmed salmon that escape from net pens can migrate to the rivers where they may interbreed with native salmon (Jensen et al. 2010, Taranger et al. 2015, Thorstad et al. 2015). Norway is the world's largest producer of farmed Atlantic salmon, with a total production of 1 220 000 metric tons in 2014. In comparison, the total catches of wild Atlantic salmon in Norway were 490 t in the same year (Anonymous 2015b).

Escaped farmed salmon can contribute to the depletion of wild salmon populations because of their reduced adaptations to environmental conditions. Wild salmon populations differ in genetic composition as a result of local adaptations to different eco-

*Corresponding author: tor.naesje@nina.no

logical conditions (Garcia de Leaniz et al. 2007). The farmed Atlantic salmon in Norway were founded by individuals from a few wild strains in the early 1970s, and have less genetic variation than the wild population due to domestication and selective breeding (Skaala et al. 2005, Karlsson et al. 2010). Consequently, crossbreeding between wild and farmed fish may lead to lower genetic variation and loss of local adaptability in wild Atlantic salmon populations (Ferguson et al. 2007, Glover et al. 2012, 2013). Moreover, the offspring of farmed Atlantic salmon, hybrids and backcrosses have lower survival as juveniles than wild offspring (Fleming et al. 2000, McGinnity et al. 2003). Escaped farmed individuals migrating into rivers may therefore have negative ecological and genetic effects on wild Atlantic salmon populations (Fleming et al. 2000, McGinnity et al. 2003). The average proportion of escaped farmed salmon in samples from Norwegian rivers close to the spawning period varied between 11 and 18 % between 1999 and 2014 (Anonymous 2015b), and genetic hybridization between farmed and wild salmon has been documented (Glover et al. 2012, 2013).

Wild Atlantic salmon return to their natal river to spawn (Hansen et al. 1989, Harden Jones 1968). Imprinting of the environmental characteristics of the river during the smolt and post-smolt migration seems pivotal for precise homing (Hansen et al. 1989). Due to a lack of river imprinting and river experience, farmed salmon may migrate to the uppermost river stretches that are accessible for anadromous salmonids, or to other major migration barriers (Butler et al. 2005, Heggberget et al. 1996, Thorstad et al. 1998). In addition, they may perform more and longer up- and downstream movements than wild Atlantic salmon during the spawning period (Heggberget et al. 1996, Thorstad et al. 1998). The within-river migration of wild Atlantic salmon has been well studied, whereas there are few studies comparing the migration patterns of wild and escaped farmed Atlantic salmon before and during the spawning period (Thorstad et al. 2008). Within-river movements and distribution before and during spawning may be important for evaluating the risk of genetic hybridization between farmed and wild salmon, for developing methods for monitoring the incidence of escaped farmed fish, and for developing measures to remove escaped farmed individuals from rivers.

In 1993, Thorstad et al. (1998) studied the behavior and area use of escaped farmed and wild Atlantic salmon during the spawning period in the River Namsen. They found that farmed salmon were distributed higher upstream in the river than wild salmon, and that farmed salmon exhibited more and longer up- and downstream movements, although the study was based on a small sample size of escaped farmed salmon. In the present study, one of our aims was to confirm in a more comprehensive study that these findings are still valid after 19 additional years of intentional and unintentional selection of farmed salmon. Moreover, we aimed to collect more extensive information on salmon behavior during the pre-spawning period.

The main goals of this study were to compare the area use and movement patterns of wild and escaped farmed Atlantic salmon before and during spawning in a large Norwegian river. Specifically, we used radio telemetry to investigate the following questions: (1) Are wild and farmed Atlantic salmon located in the same river stretches during the pre-spawning and spawning period, and (2) do wild and farmed Atlantic salmon exhibit the same movement patterns prior to and during the spawning period?

MATERIALS AND METHODS

Study area

The Namsen river system in Central Norway has a catchment area of 6265 km^2. The stretch of the river system available to anadromous fish consists of the River Namsen (the main river) and the 2 main tributaries, Høylandsvassdraget and Sanddøla (Fig. 1). River Namsen has a mean annual water discharge at the river mouth of 290 m^3 s^{-1}. In the main river, anadromous salmonids could previously migrate 70 km upriver from the sea to the 35 m high waterfall Nedre Fiskumfoss. In 1977, a fish ladder was constructed in this waterfall, adding 10 km to the anadromous stretch of the main river, to Aunfoss. The River Namsen is a slow-flowing river, and there are no major migration barriers for Atlantic salmon below Nedre Fiskumfoss. In 2012, Atlantic salmon spawning redds were counted from a helicopter. Spawning redds were recorded in all parts of the River Namsen above the saltwater-influenced estuary. This distribution of spawning redds indicates that most parts of the river contain suitable spawning areas for Atlantic salmon (Fig. 1). The total length of accessible stretches for anadromous salmonids in the entire river system, including the tributaries, is 200 km. Based on examination of the gonads of farmed and wild salmon caught in the river in 2012, most, if not all fish entering the River Namsen were sexually mature.

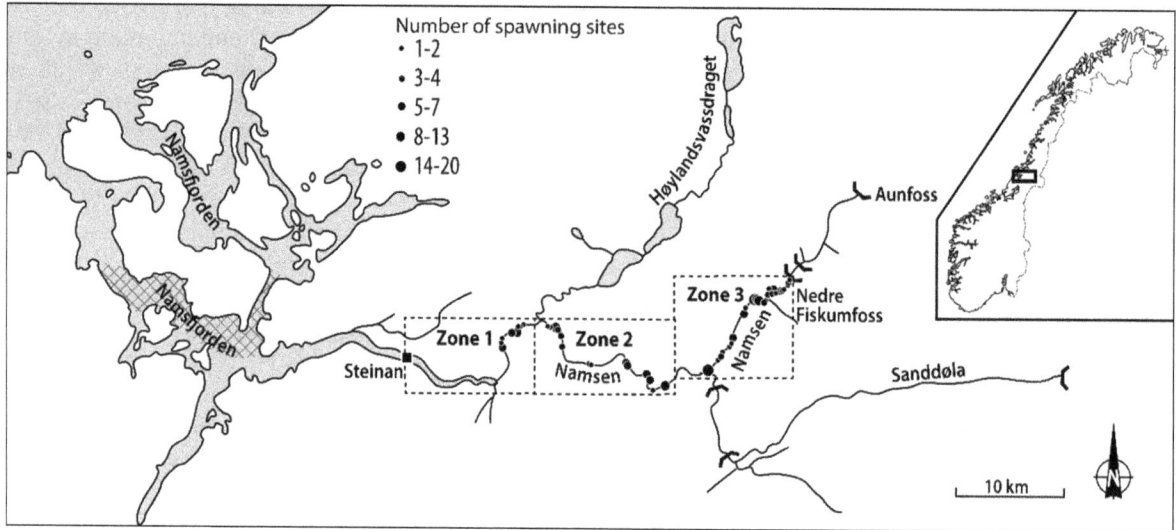

Fig. 1. Namsfjorden and the Namsen river system, showing locations of bag nets where Atlantic salmon *Salmo salar* were captured (hatched area), location of the stationary tracking station at Steinan (■), spawning sites (●) and waterfalls (⟨). Inset shows the location of the study site in Central Norway. River zones are outlined by rectangles with dotted lines. Graphics: Kari Sivertsen, NINA

Fish capture and tagging

A total of 74 wild (24 males, 49 females, 1 unknown sex), and 43 escaped farmed Atlantic salmon (15 males, 15 females, 13 unknown sex) were captured in bag nets in the sea (Namsfjorden, 5.5 to 21.8 km from the river mouth) and tagged with radio transmitters between 10 June and 28 August 2012 (Fig. 1). Farmed Atlantic salmon entered the fjord later than wild Atlantic salmon, hence 29 (67 %) of the farmed fish were tagged between 31 July and 29 August, while 71 (96 %) of the wild salmon were tagged between 15 June and 27 July. Mean (±SD) total body length was 88 ± 9 cm (range: 67 to 109 cm) for the wild and 78 ± 8 cm (range: 64 to 93 cm) for the farmed salmon (for further details, see Table S1 in Supplement 1 at www. int-res.com/articles/suppl/q008p077_supp.pdf).

Each fish was tagged with a radio transmitter (model F2120, Advanced Telemetry Systems; outline dimensions: 21 × 52 × 11 mm, mass in air: 15 g, guaranteed battery life: 149 to 269 d) according to the method described by Thorstad et al. (1998) and Økland et al. (2001). The fish were anesthetized before tagging (2-phenoxyethanol, EEC No 204-589-7, 1 ml l^{-1} of water). Similar transmitters did not reduce the swimming performance of similarly-sized Atlantic salmon in a swim speed chamber (Thorstad et al. 2000). Each individual fish was recognized based on a unique combination of pulse rate and frequency (within the 142.000 to 142.600 MHz range). Identifi-

cation of escaped farmed individuals was based on morphological characteristics and controlled with analysis of growth patterns in 5 to 8 scales collected from each fish during tagging (Fiske et al. 2005). Sex was determined based on external characteristics, if possible. The fish were released at the catch site after tagging and recovery from anesthetization.

Tracking of tagged salmon

The tagged fish were manually tracked from a car using a radio receiver (R4500S ATS) and a whip antennae mounted on the car roof (142 MHz, Laird Technologies). Between 4 July and 15 November 2012, the entire river system accessible to Atlantic salmon was tracked every second week to determine the position of all radio-tagged fish that had entered the watercourse. The large size of the watershed and limited accessibility of certain parts made manual tracking of the entire system too time consuming; hence, the study focused on the main river, River Namsen. Of the 74 wild and 43 farmed Atlantic salmon tagged in the fjord, 59 (78 %) and 32 (74 %) entered the watercourse, respectively. Of these, 2 wild and 2 farmed salmon were registered on only one occasion, and 18 wild and 6 farmed salmon migrated into the tributaries River Høylandsvassdraget and River Sanddøla (Fig. 1). In addition, 1 farmed Atlantic salmon passed the fish ladder at

Nedre Fiskumfoss (Fig. 1), and 20 wild and 5 farmed salmon were caught in the recreational fisheries in the main river. These individuals were therefore excluded from the study, leaving a sample size of 19 wild and 18 farmed Atlantic salmon.

The location of the tagged fish was determined every second day in the main river between Steinan (saltwater influence and no spawning areas downstream of this site) and the Nedre Fiskumfoss waterfall (a 57 km stretch of river; Fig. 1), using 75 permanent tracking stations. The distance between tracking stations was on average 800 m (range: 167 to 1673 m). Tagged fish were assigned to the tracking station at which the strongest signal was received. The detection range of the receiver varied between stations (between approximately 0.5 and 2.0 km). Hence, in a few areas the detection range may have been shorter than the distance between tracking stations due to topography and vegetation.

Tagged fish were tracked every second day during the pre-spawning (4 September to 4 October 2012) and spawning period (5 October to 10 November 2012) of wild Atlantic salmon. Timing of the spawning period was based on a previous study (Thorstad et al. 1998), and confirmed by personal communication with local fishers.

To investigate how the tagged fish used different areas of River Namsen, the river was divided into 3 equally sized zones (19 to 20 km long) from Steinan to Nedre Fiskumfoss (Fig. 1). Movement distances were calculated using the 'locate features along routes' and 'make route event layer' tools in ArcGIS. This was done by using a centreline of the river to calculate the distance from the river mouth to each tracking position.

Statistical analyses

Our analyses were based on the tagged Atlantic salmon that were recorded in the River Namsen during the pre-spawning and spawning periods (excluding those recaptured by anglers after entering the river). In the pre-spawning period, 18 farmed and 19 wild salmon were included in the analysis of area use, total migration distance and daily movements. The analysis of daily downstream migration distances were based on 17 farmed and 18 wild salmon in the pre-spawning period, because 2 individuals (1 farmed and 1 wild salmon) did not exhibit any downstream movements. Details on sample sizes in the different analyses are provided in Table S2 of Supplement 1. During the spawning period, 17 farmed

and 17 wild Atlantic salmon were included in the analysis on area use, total migration distances and daily movements. One farmed salmon was recaptured on 6 October, and was therefore included in the pre-spawning, but not the spawning period analyses. Two wild salmon moved out of the main river after the pre-spawning period, and were consequently removed from the spawning period analysis. The calculations of daily downstream migration distances were conducted on 15 farmed and 16 wild Atlantic salmon in the spawning period, due to 3 individuals (2 farmed and 1 wild) that did not exhibit any downstream movements (details on sample sizes in the different analyses are given in Table S2 of Supplement 1). One farmed Atlantic salmon passed the fish ladder in Nedre Fiskumfoss and migrated further upstream. This individual was not regularly tracked because of a restricted capacity to extend the tracking area required to cover its movements, and was therefore excluded from the analyses. If the most supported model included sex as a factor, individuals with unknown sex were removed from the analysis.

Data on area use and the probability of migrating to the barrier was analyzed using generalized linear mixed effect models (GLMM) (Pinheiro & Bates 2000, Zuur et al. 2009). The probability of migrating to the Nedre Fiskumfoss waterfall barrier was modelled by fitting a generalized linear model (GLM) with logit link function fitted to the binomial response '1' for those fish that reached the barrier, and '0' for those that did not. The GLMMs were fitted using the 'lme4' package in R (Bates et al. 2014). In order to account for within-individual dependency of the observations, the individual fish identification (ID) was *a priori* included as random factor. ANCOVAs including origin, sex and length as explanatory variables were used to test for differences in total daily movement distance, daily downstream movement distance and number of daily movements (i.e. the number of observations where the fish was located in a different tracking station from one tracking event to the next, divided by the number of days in the observation period).

In order to estimate how the salmon's origin, body length and sex and day of year affected the probability of remaining in a certain river section during given time periods, we fitted multinomial logit models to individual-specific mean positions data for both pre-spawning and spawning periods (Hosmer & Lemeshow 1989). All 2-way interactions were included in the global model. The mean position was assigned to 1 of 3 equally sized river zones (Fig. 1). The models were fitted using the 'multinom' function of the 'nnet' package in R. Effect tests for the multi-nomial modes were performed using type III likeli-hood ratio (LR) tests running the ANOVA procedure available from the R package 'car'.

Model selection for fixed effects followed the combined Akaike information criterion (AICc) and backward-selection procedures available in Zuur et al. (2009). For the ran-dom structure, ID was *a priori* included in the models and thus not subjected to model selection (see Supplement 2 at www.int-res.com/articles/suppl/q008p077_supp.pdf for further details on model selection).

RESULTS

Area use in the pre-spawning period

The distribution of wild and escaped farmed Atlantic salmon differed throughout the pre-spawning period (Fig. 2). A total of 9 farmed (50%) and 2 wild (11%) fish mi-grated to the barrier at the Nedre Fiskumfoss waterfall. The most supported GLM to pre-dict the probability of migrating to the bar-rier included only origin (see Table S6 in Supplement 2). The predicted probability of farmed Atlantic salmon migrating to the barrier was 0.5 (95% CI: 0.28–0.72), while that of wild Atlantic salmon was 0.1 (95% CI: 0.03–0.34) (LR $\chi^2 = 37.7$, df = 35, p < 0.01).

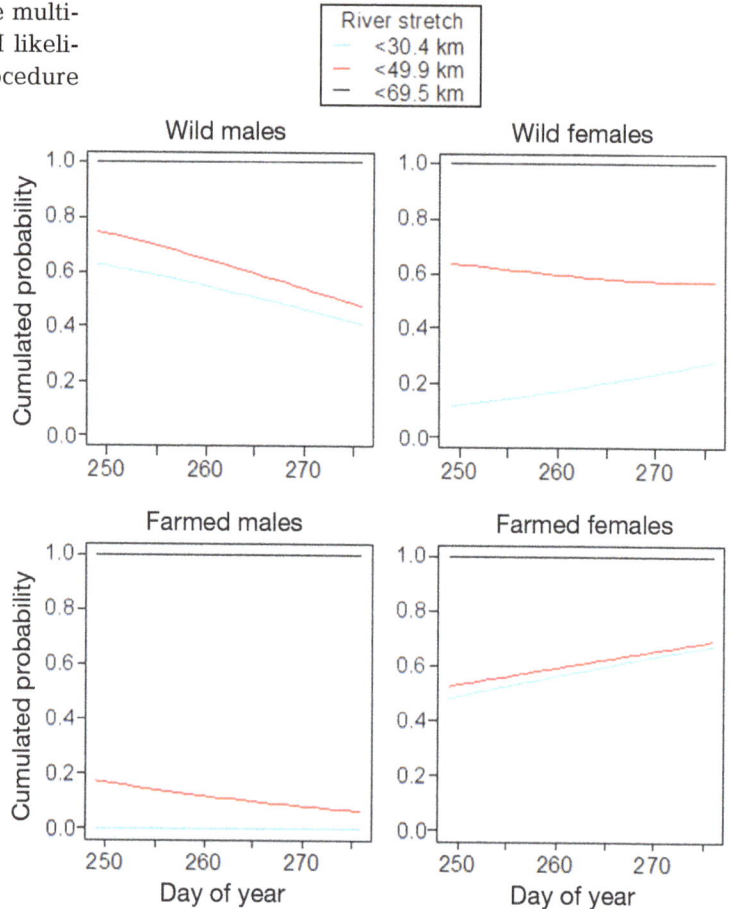

Fig. 3. Predicted probability of use of river stretches (zones) by Atlantic salmon *Salmo salar* in the pre-spawning period as a function of origin, sex and day of year. N = 11 wild females, 8 wild males, 5 farmed females, and 7 farmed males. Individuals of unknown sex were excluded

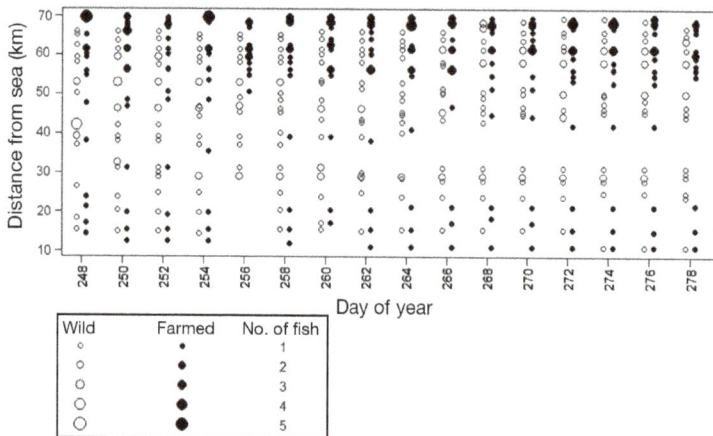

Fig. 2. Location (distance from the river mouth) of farmed and wild Atlantic salmon *Salmo salar* on day of the year 248 to 278 (4 September to 4 October 2012). On Day 256 (12 September), the lower 28 km were not manually tracked. For clarity in the figure, minor temporal separation of wild and farmed salmon observations have been made

Fourteen of the 18 (78%) farmed Atlantic salmon moved to, and stayed within the upper 30 km of the 70 km river stretch, while 3 individuals remained in the lower 20 km (Fig. 2). In contrast, the wild Atlantic salmon were distributed evenly over the entire river stretch.

Males and females had different probabil-ities of using the 3 river zones (lower, middle and upper) during the pre-spawning period (Fig. 3). The most supported multinomial zone-use model included the predictors ori-gin, sex and day of year, and all 2-way inter-actions. There was a significant interaction between origin and sex (multinomial GLM: LR $\chi^2 = 128.4$, df = 2, p < 0.001; Fig. 3). None of the other factors or their interactions were significant (all p > 0.09). Farmed males had a higher predicted probability of using the upper section (50 to 70 km) than both sexes of wild Atlantic salmon (Fig. 3). Farmed females used the lower (10 to 30 km) and upper (50 to 70 km) sections but not the middle section. In contrast, wild females used all river sections while wild males had a similar pattern to farmed females, mainly using the lower and upper sections during the pre-spawning period (Fig. 3).

Area use in the spawning period

The distribution of wild and farmed Atlantic salmon differed throughout the spawning period (Fig. 4). All but 3 farmed salmon remained in the upper 20 km during the spawning period; no farmed salmon were found in the middle 30 km. In contrast, wild salmon were distributed evenly over the whole river (Fig. 4).

Males and females had different probabilities of using the 3 river zones (lower, middle and upper) during the spawning period (Fig. 5). The river section used was best explained by the multinomial GLM model, including the predictors origin, sex and day of year, without including the 3-way interaction between these predictors (origin × sex: LR χ^2 = 59.4, df = 2, p << 0.001; day of year × origin: LR χ^2 = 17.0, df = 2, p < 0.001; day of year × sex: LR χ^2 = 14.4, df = 2, p < 0.001; effect of origin: LR χ^2 = 13.6, df = 2, p = 0.001;

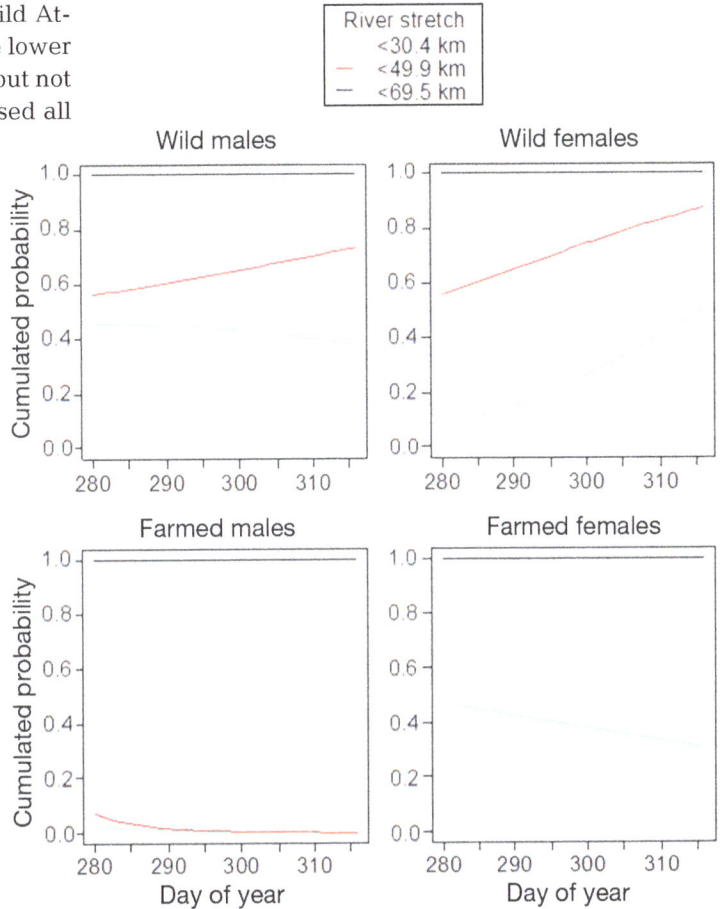

Fig. 5. Predicted probability of use of river stretches (zones) by Atlantic salmon *Salmo salar* during the spawning period as function of origin, sex and day of year. N = 11 wild females, 8 wild males, 5 farmed females, and 7 farmed males. Individuals of unknown sex were excluded

effect of sex: LR χ^2 = 5.1, df = 2, p = 0.078; effect of day of year: LR χ^2 = 3.1, df = 2, p = 0.211). Farmed males had a higher predicted probability of using the upper section (50 to 70 km from the river mouth) than both sexes of wild salmon (Fig. 5). None of the radio-tagged farmed females used the middle section (30 to 50 km) during the spawning period; they were found either in the lower (10 to 30 km) or upper section (50 to 70 km). Wild females used all sections, but the probability of using the middle and lower sections increased with the day of year, indicating that some individuals moved downstream after spawning. The probability of wild males using the middle section increased with the day of year, while the probability of using the lower section decreased (Fig. 5).

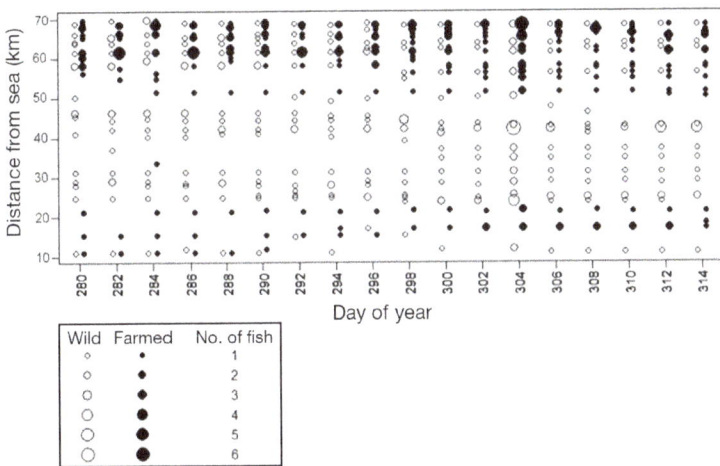

Fig. 4. Location (distance from the river mouth) of farmed and wild Atlantic salmon *Salmo salar* on day of the year 280 to 314 (6 October to 10 November 2012). For figure clarity, minor temporal separation of wild and farmed salmon observations have been made

Movement patterns in the pre-spawning period

Farmed Atlantic salmon performed longer movement distances per day than wild salmon (Table 1). Wild salmon (n = 19) moved on average (±SD) 528 ± 459 m d^{-1} (median: 436, range: 118 to 1785 m d^{-1}), while farmed individuals (n = 18) moved on average 871 ± 510 m d^{-1} (median: 730, range: 293 to 2295 m d^{-1}) based on cumulative movements. Body length affected movement distance differently in wild and farmed salmon (ANCOVA: R^2 = 0.26, F = 3.79, p = 0.02). The movement distance per day increased with body length in farmed individuals, while it decreased with body length in wild individuals. There was no difference in total movement distance per day between males and females (Table 1, Fig. 6). Two farmed Atlantic salmon had a large effect on the linear model parameter estimates, but the results were significant even when these individuals were excluded from the analyses.

The average downstream movements for wild (n = 18) and farmed salmon (n = 17) were 304 ± 232 m d^{-1} (median: 286, range: 74 to 782 m d^{-1}) and 448 ± 305 m d^{-1} (median: 367, range: 11 to 1177 m d^{-1}), respectively. Body length affected downstream movement distance differently in wild and farmed fish

Table 1. Parameter estimates for the most supported linear models to predict total daily movement distance, downstream movement distance and number of daily movements of wild and escaped farmed Atlantic salmon *Salmo salar* during the pre-spawning period. All estimates are reported as contrasts

Model	Explanatory variable	Estimate (±SE)	t	p
Total distance d^{-1} ~Origin × Length	Intercept [Farmed]	−1187 (1074)	−1.11	0.28
	Origin [Wild]	3635 (1600)	2.27	0.03
	Length	26 (13)	1.93	0.06
	Origin [Wild] × Length	−47 (19)	−2.51	0.02
Downstream distance d^{-1} ~Origin × Length	Intercept [Farmed]	−1467 (620)	−2.37	0.024
	Origin [Wild]	1998 (887)	2.25	0.032
	Length	24 (8)	3.10	0.004
	Origin [Wild] × Length	−26 (10)	−2.52	0.017
Movements d^{-1} ~Origin + Sex	Intercept [Farmed, Female]	0.48 (0.05)	10.20	<0.001
	Origin [Wild]	−0.19 (0.06)	−3.39	0.002
	Sex [Male]	−0.20 (0.06)	−3.31	0.003
	Origin [Wild] × Sex	0.14 (0.08)	1.74	0.094

(ANCOVA: R^2 = 0.29, F = 4.3, p = 0.01). Daily downstream movement increased with body length in farmed individuals, while body length had no effect on downstream movement distance in wild individuals. There was no difference in downstream movement distance per day between males and females (Table 1, Fig. 7). There were no significant differences between wild and farmed Atlantic salmon in the average number of downstream movements, since origin alone did not have a significant effect on

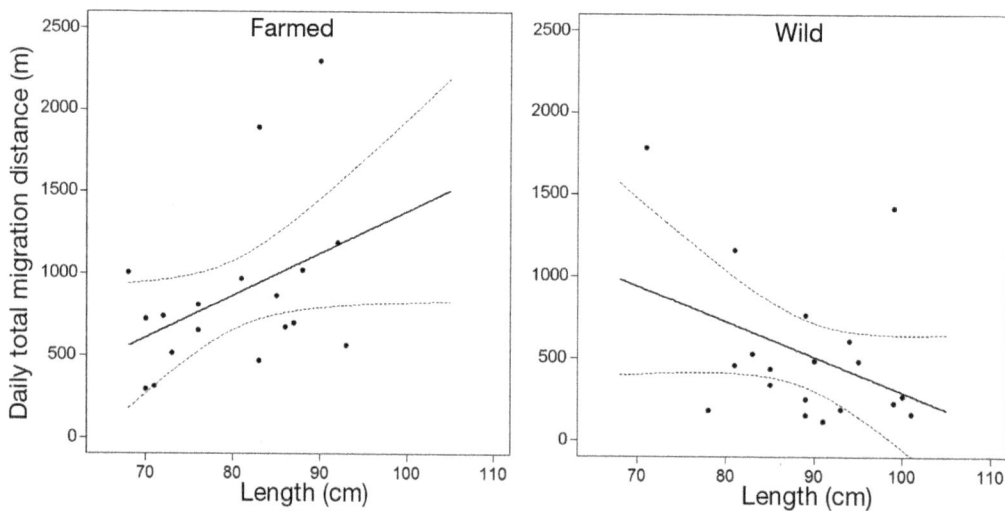

Fig. 6. Predicted (solid line) daily total movement distance of Atlantic salmon *Salmo salar* plotted as a function of body length and origin, showing upper and lower 95% CI (dotted lines). Dots: observed values. The predictions were retrieved from the linear model provided in Table 1. N = 19 wild and 18 farmed Atlantic salmon

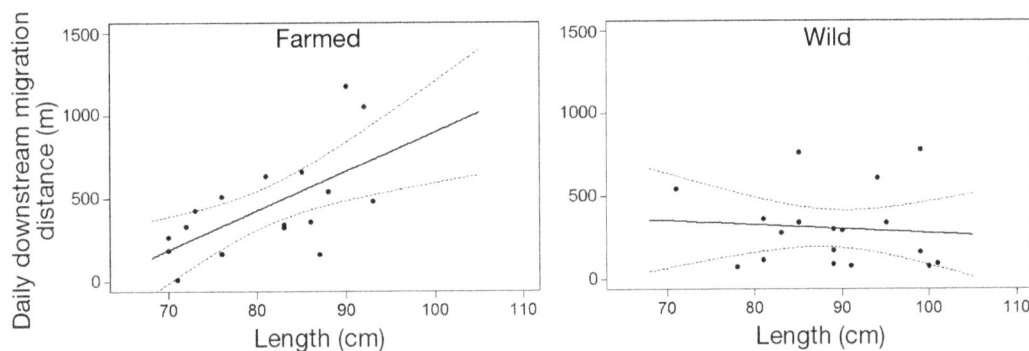

Fig. 7. Predicted (solid line) daily downstream movement distance of Atlantic salmon *Salmo salar* plotted as a function of body length and origin, showing upper and lower 95 % CI (dotted lines). Dots: observed values. The predictions were retrieved from the linear model provided in Table 1. N = 18 wild and 17 farmed Atlantic salmon

downstream movements (ANOVA: $R^2 = 0.07$, $F = 2.5$, p = 0.12).

Wild Atlantic salmon moved on average 0.26 ± 0.12 times d^{-1} (median: 0.23, range: 0.07 to 0.53 times d^{-1}) while farmed salmon moved on average 0.36 ± 0.13 times d^{-1} (median: 0.35, range: 0.13 to 0.63 times d^{-1}). Sex had the same effect on number of daily move-ments in both wild and farmed individuals (ANOVA: $R^2 = 0.42$, $F = 6.6$, p = 0.002). Females exhibited more daily movements than males in both groups, but less so among the wild than the farmed salmon (as indi-cated by a trend towards an interaction between sex and origin; see Table 1). Farmed individuals exhib-ited more daily movements than wild fish for both sexes (Table 1, Fig. 8).

Movement patterns in the spawning period

Wild and farmed Atlantic salmon did not differ in behavior during the spawning period. Farmed salmon moved a mean distance of 435 ± 622 m d^{-1} (median: 252, range: 27 to 2680 m d^{-1}), while wild Atlantic salmon moved a mean distance of 492 ± 357 m d^{-1} (median: 369, range: 68 to 1273 m d^{-1}). Daily movement distance was not affected by origin, sex or body length (ANCOVA: all p ≥ 0.1).

Three individuals (2 farmed, 1 wild) did not regis-ter any downstream movements during the spawn-ing period. Of the fish that did move, farmed fish (n = 15) moved a mean downstream distance of 289 ± 363 m d^{-1} (median: 196, range: 27 to 1499 m d^{-1}), while wild salmon (n = 16) moved a mean downstream dis-tance of 387 ± 300 m d^{-1} (median: 312, range: 41 to 1053 m d^{-1}). Downstream movement distance was not dependent on origin, sex or body length (ANCOVA: all p ≥ 0.15).

Farmed salmon moved on average 0.2 ± 0.1 times d^{-1} (median: 0.2, range: 0.3 to 0.5 times d^{-1}), while wild individuals moved an average of 0.2 ± 0.2 times d^{-1} (median: 0.2, range: 0 to 0.7 times d^{-1}). The num-ber of movements was not dependent on origin, sex or body length (ANCOVA: all p ≥ 0.3).

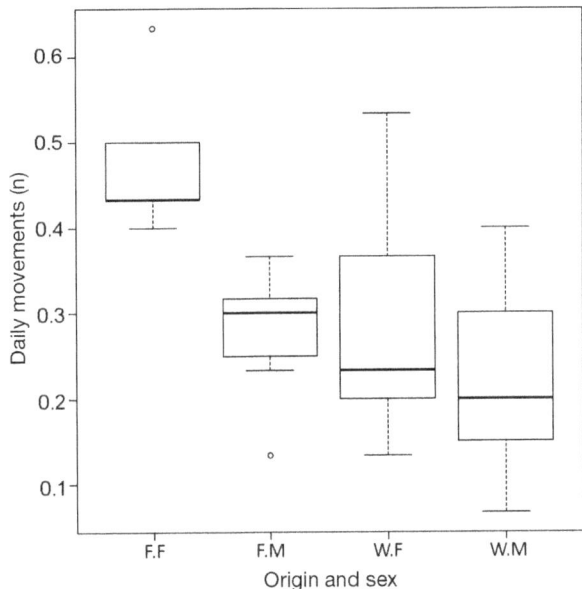

Fig. 8. Observed number of daily movements of Atlantic salmon *Salmo salar* as a function of origin and sex. The bot-tom and top of the box delineate the 25th and 75th percentiles (i.e. the boxes include the middle 50 % of the observations). The whiskers span to the most extreme data point, which is no more than 1.5 times the interquartile rage, and the bold horizontal line represents the median value. W.F: wild females (n = 11); W.F: wild males (n = 8); F.F: farmed females (n = 5); F.M: farmed males (n = 7). Individuals of unknown sex were excluded

DISCUSSION

Area use in the pre-spawning period

In the pre-spawning period, escaped farmed Atlantic salmon mainly stayed in the upper part of the river, while wild salmon were evenly distributed from the river mouth to the upper river section. The predicted probability of migrating to the migration barrier 70 km upstream from the river mouth was 4 times higher for farmed than for wild salmon. These findings support results of earlier studies of upstream migration in escaped farmed Atlantic salmon (Butler et al. 2005, Heggberget et al. 1996, Thorstad et al. 1998), which indicate that farmed salmon, as opposed to native wild individuals, do not have a 'stop signal' when migrating upstream, probably due to lack of imprinting to a certain home site in the river. Consequently, farmed individuals may aggregate below migration barriers in the upper parts of the rivers. Being immature or in poor physical condition may be potential reasons that a few farmed individuals did not migrate this far.

Differences in distribution of wild and escaped farmed Atlantic salmon should be taken into consideration when estimating the proportion of farmed salmon in rivers during the pre-spawning period. The proportion of escaped individuals in rivers is typically monitored by collecting scale samples from angling catches during the regular fishing season, and targeted angling or net fishing in different areas in the pre-spawning period, often close to the spawning period after the regular angling season has ended (Fiske et al. 2005). The proportions of escaped farmed salmon are determined based on morphology and scale analyses of the captured fish (Fiske et al. 2005). Those individuals that are identified as wild based on morphology are released back into the river alive after a scale sample is collected for later verification of wild versus farmed origin. Fish that are identified as escaped farmed salmon based on morphology are killed. Monitoring of escaped farmed salmon is also based on visual identification of wild and farmed fish during drift snorkeling observations in clear-water rivers (Vollset et al. 2014), monitoring of fish ladders by video recordings, and in some instances capturing fish in the spawning areas by paralyzing them with strong light, taking scale samples and releasing them again (Anonymous 2015a). Our results indicate that the proportion of farmed Atlantic salmon captured or observed during such monitoring activities would be sensitive to both timing and location of sampling. Hence, a non-biased

sampling programme for assessing the proportion of escaped farmed Atlantic salmon in the River Namsen and other similar rivers should be accomplished after most of the escaped individuals have entered the river, and include samples from all river stretches equally, covering the lower, middle, and uppermost spawning areas.

Area use in the spawning period

During the spawning period, all except 3 farmed Atlantic salmon remained in the upper 20 km of the river below the migration barrier. In contrast, wild Atlantic salmon were distributed evenly from the lower river sections to the migration barrier. Farmed Atlantic salmon being distributed higher up the river than wild fish is in accordance with results from previous studies in the River Namsen and River Alta (Heggberget et al. 1996, Thorstad et al. 1998). However, both wild and farmed salmon stayed together in the upper reaches of the river, which hold important spawning grounds for wild Atlantic salmon. It is likely that spawning between wild and escaped farmed salmon takes place in these areas. Hybridization between farmed and wild Atlantic salmon in the River Namsen has recently been documented by genetic methods (S. Karlsson et al. unpubl. data). From our findings of spatial variation in the degree of overlapping area use between wild and farmed individuals, a varying degree of hybridization can be expected.

Movement patterns in the pre-spawning period

The escaped farmed salmon exhibited longer total movement distances and more movements per day than did wild individuals during the pre-spawning period. However, when downstream movements were analyzed separately, there was no significant difference between the 2 groups. This may be because a proportion of farmed Atlantic salmon had not completed their upstream migration in the river when the study began, while the wild salmon had entered the river earlier and may have finished their searching phase and started their holding phase close to their spawning area (Økland et al. 2001, Finstad et al. 2005).

The daily total and downstream movement distances of the tagged individuals indicated that movement distances increased with body length for farmed salmon, while body length had little or no effect on the wild individuals' movements. To the best of our knowledge, no previous studies have found that movement distances of farmed Atlantic salmon are related to body length. These within-river movements were short compared to total migration distances and likely not related to restricted energy reserves in the smaller fish, but rather to some other intrinsic factors affecting their behavior.

Females exhibited more daily movements than males in both wild and farmed salmon. The reason may be that females actively searched for a suitable spawning site at different localities prior to spawning (Fleming 1996). Similar to our study, Karppinen et al. (2004) found that wild females tended to exhibit a more erratic migration pattern than wild males, while Økland et al. (2001) and Finstad et al. (2005) did not find differences in movement patterns between wild males and females.

Movement patterns in the spawning period

During the spawning period, there were no differences in daily total and downstream movement distances between wild and escaped farmed Atlantic salmon, or in the number of daily movements. This is in contrast to earlier studies, which documented more extensive up- and downstream movements in farmed than in wild salmon during the spawning period (Thorstad et al. 1998, Økland et al. 1995). Both in the present and previous studies, there were substantial among-individual variations within both groups. A possible reason for this lack of difference may be that the accuracy of the manual tracking used in our study was too low to detect possible small-scale movements. For instance, movements between spawning grounds located 200 m apart in the same area of the river would not have been documented. Furthermore, the individuals may have moved and returned between manual tracking days. Hence, the estimated movement distances and the number of daily movements reported here are most likely underestimated. Another factor that may have affected our results is that not all tagged individuals were located during every manual-tracking day. However, most of the fish had a high number of detections in the pre-spawning (14 to 16 out of 16 tracking days) and spawning (17 to 19 of 19 tracking days) periods.

Farmed Atlantic salmon are subject to selection regimes geared toward optimizing their life history characteristics (such as growth and size at sexual maturation), which now may differ considerably from populations of wild Atlantic salmon and from earlier generations of farmed salmon (Thodesen et al. 1999, Gjedrem & Baranski 2009). Hence, in our study we also aimed to investigate if 19 yr of both intentional and unintentional selection had changed the behavior of the farmed salmon. Based on Thorstad et al. (1998) and the present study, the general area of use of escaped farmed Atlantic salmon in the River Namsen was still the same during the spawning period, since the farmed salmon were mainly in the upper part of the river in both studies. However, selection may have changed the behavior of the escaped farmed Atlantic salmon, as their movements during the spawning period in the present study were more similar to the wild Atlantic salmon than 19 yr ago.

CONCLUSIONS

Results from this and previous studies (Butler et al. 2005, Heggberget et al. 1996, Thorstad et al. 1998) indicate that escaped farmed Atlantic salmon tend to migrate far upstream in the rivers until they reach major migration barriers, which results in a larger proportion of farmed individuals in upper compared to lower parts of rivers. Before the spawning period, these escaped individuals migrated back downstream from the migration barrier pool to nearby downstream spawning areas.

The within-river difference in distribution of farmed versus wild salmon, along with differences in the timing of pre-spawning and spawning, have implications for monitoring the incidence of escaped farmed Atlantic salmon in rivers. Monitoring of farmed and wild salmon should cover river sections in a standardized way to provide a representative sample from the river system.

In many Norwegian rivers, targeted angling or other capture methods are used to reduce the number of escaped farmed Atlantic salmon before spawning. The results of this and previous studies indicate that fishing in the upper parts of the rivers may be most effective in reducing the impact of escaped individuals. Results of the present study also indicate that hybridization between wild and escaped salmon is more likely to occur in the upper rather than the lower parts of the rivers due to the higher incidence of escaped farmed salmon in these areas.

The conclusion that escaped farmed Atlantic salmon tend to migrate far up in the rivers may not be valid for rivers with migration obstacles in lower parts. There are indications that farmed salmon are less capable than wild individuals of passing large and difficult waterfalls (Johnsen et al. 1998). Hence, in river systems with major migration barriers in lower river stretches, there may be an accumulation of escaped farmed salmon below the barriers. However, the ability of escaped individuals to pass large waterfalls has not been well studied.

Acknowledgements. We thank O. Diserud for comments on the statistical analysis, G. Østborg for scale analyses and S. Elden, L. Skorstad, J. A. Lanstad, R. Holm and F. Staldvik for technical assistance. Financial support was provided by the Norwegian Seafood Federation, the Directorate of Fisheries, the County Administrator at the Nord-Trøndelag County, and the County Authority of Nord-Trøndelag.

LITERATURE CITED

Anonymous (2015a) Rømt oppdrettslaks i vassdrag. Rapport fra det nasjonale overvåkningsprogrammet i 2014. Fisken og havet, særn. 2b-2015 (Report from the National Monitoring program in 2014)

Anonymous (2015b) Status for norske laksebestander i 2015 (Status of Norwegian salmon populations in 2015). Rapport fra Vitenskapelig råd for lakseforvaltning, NR8 (Report from the Norwegian Scientific Advisory Committee for Atlantic Salmon Management, in Norwegian)

Bates D, Maechler M, Bolker B, Walker S (2014) lme4: linear mixed-effects models using 'Eigen' and S4. R package version 1.1-7. http://CRAN.R-project.org/package=lme4

ä Butler JRA, Cunningham PD, Starr K (2005) The prevalence of escaped farmed salmon, *Salmo salar* L., in the River Ewe, western Scotland, with notes on their ages, weights and spawning distribution. Fish Manag Ecol 12:149–159

Ferguson A, Fleming IA, Hindar K, Skaala Ø, McGinnity P, Cross T, Prodohl P (2007) Farm escapes. In: Verspoor E, Strameyer L, Nielsen JL (eds) The Atlantic salmon: genetics, conservation and management. Blackwell Publishing, Oxford, p 357–398

ä Finstad AG, Økland F, Thorstad EB, Heggberget TG (2005) Comparing upriver spawning migration of Atlantic salmon *Salmo salar* and sea trout *Salmo trutta*. J Fish Biol 67:919–930

Fiske P, Lund RA, Hansen LP (2005) Identifying fish farm escapees. In: Cadrin SX, Friedland KD, Waldman JD (eds) Stock identification methods. Elsevier Academic Press, Amsterdam, p 659–680

ä Fleming IA (1996) Reproductive strategies of Atlantic salmon: ecology and evolution. Rev Fish Biol Fish 6: 379–416

ä Fleming IA, Hindar K, Mjolnerod IB, Jonsson B, Balstad T, Lamberg A (2000) Lifetime success and interactions of farm salmon invading a native population. Proc R Soc B 267:1517–1523

ä Garcia de Leaniz C, Fleming IA, Einum S, Verspoor E and others (2007) A critical review of adaptive genetic variation in Atlantic salmon: implications for conservation. Biol Rev Camb Philos Soc 82:173–211

Gjedrem T, Baranski M (eds) (2009) Selective breeding in aquaculture: an introduction. Springer, London

ä Glover KA, Quintela M, Wennevik V, Besnier F, Sørvik AGE, Skaala Ø (2012) Three decades of farmed escapees in the wild: a spatio-temporal analysis of Atlantic salmon population genetic structure throughout Norway. PLoS ONE 7:e43129

ä Glover KA, Pertoldi C, Besnier F, Wennevik V, Kent M, Skaala Ø (2013) Atlantic salmon populations invaded by farmed escapees: quantifying genetic introgression with a Bayesian approach and SNPs. BMC Genet 14:74

Hansen LP, Jonsson B, Andersen R (1989) Salmon ranching experiments in the River Imsa: Is homing dependent on sequential imprinting of the smolts? In: Brannon E, Johsson B (eds) Migration and distribution of salmonids: proceedings of the international salmoid migration and distribution symposium. University of Washington, Seattle, p 19–29

Harden Jones FR (1968) Fish migration. Edward Arnold Publishers, London

ä Heggberget TG, Økland F, Ugedal O (1996) Prespawning migratory behaviour of wild and farmed Atlantic salmon, *Salmo salar* L., in a north Norwegian river. Aquacult Res 27:313–322

Hosmer DW, Lemeshow S (1989) Applied logistic regression. Wiley Interscience, New York, NY

ICES (2014) Report of the working group on North Atlantic salmon (WGNAS). ICES CM 2014/ACOM:09, 19–28 March 2014, Copenhagen

ä Jensen Ø, Dempster T, Thorstad EB, Uglem I, Fredheim A (2010) Escapes of fish from Norwegian sea-cage aquaculture: causes, consequences, prevention. Aquacult Environ Interact 1:71–83

Johnsen BO, Jensen AJ, Økland F, Lamberg A, Thorstad EB (1998) The use of radiotelemetry for identifying migratory behaviour in wild and farmed Atlantic salmon ascending the Suldalslågen river in Southern Norway. In: Jungwirth M, Schmutz S, Weiss S (eds) Fish migration and fish bypasses. Fishing New Books, Oxford, p 55–68

ä Karlsson S, Moen T, Hindar K (2010) Contrasting patterns of gene diversity between microsatellites and mitochondrial SNPs in farm and wild Atlantic salmon. Conserv Genet 11:571–582

ä Karppinen P, Erkinaro J, Niemela E, Moen K, Økland F (2004) Return migration of one-sea-winter Atlantic salmon in the River Tana. J Fish Biol 64:1179–1192

ä McGinnity P, Prodohl P, Ferguson A, Hynes R and others (2003) Fitness reduction and potential extinction of wild populations of Atlantic salmon, *Salmo salar*, as a result of interactions with escaped farm salmon. Proc R Soc B 270: 2443–2450

ä Økland F, Heggberget TG, Jonsson B (1995) Migratory behaviour of wild and farmed Atlantic salmon (*Salmo salar*) during spawning. J Fish Biol 46:1–7

▶ Økland F, Erkinaro J, Moen K, Niemela E, Fiske P, McKinley RS, Thorstad EB (2001) Return migration of Atlantic salmon in the River Tana: phases of migratory behaviour. J Fish Biol 59:862–874

Pinheiro JC, Bates DM (2000) Mixed-effects models in S and S-PLUS. Springer, New York, NY

ä Skaala Ø, Taggart J, Gunnes K (2005) Genetic difference between 5 major domesticated strains of Atlantic salmon and wild salmon. J Fish Biol 67:118–128

ä Taranger GL, Karlsen Ø, Bannister RJ, Glover KA and others (2015) Risk assessment of the environmental impact of Norwegian Atlantic salmon farming. ICES J Mar Sci 72: 997–1021

ä Thodesen J, Grisdale-Helland B, Helland SJ, Gjerde B (1999) Feed intake, growth and feed utilization of offspring from wild and selected Atlantic salmon (*Salmo salar*). Aquaculture 180:237–246

▶ Thorstad EB, Heggberget TG, Økland F (1998) Migratory behaviour of adult wild and escaped farmed Atlantic salmon, *Salmo salar* L., before, during and after spawning in a Norwegian river. Aquacult Res 29: 419–428

▶ Thorstad EB, Økland F, Finstad B (2000) Effects of telemetry transmitters on swimming performance of adult Atlantic salmon. J Fish Biol 57:531–535

▶ Thorstad EB, Økland F, Aarestrup K, Heggberget TG (2008) Factors affecting the within-river spawning migration of Atlantic salmon, with emphasis on human impacts. Rev Fish Biol Fish 18:345–371

▶ Thorstad EB, Todd CD, Uglem I, Bjørn PA and others (2015) Effects of salmon lice *Lepeophtheirus salmonis* on wild sea trout *Salmo trutta*—a literature review. Aquacult Environ Interact 7:91–113

▶ Vollset KW, Skoglund H, Barlaup BT, Pulg U and others (2014) Can the river location within a fjord explain the density of Atlantic salmon and sea trout? Mar Biol Res 10: 268–278

Zuur AF, Ieno EN, Walker N, Saveliev AA, Smith GM (2009) Mixed effects models and extensions in ecology with R. Springer, New York, NY

16

Characterising biofouling communities on mussel farms along an environmental gradient: a step towards improved risk management

A. M. Watts[1,2,3,*], S. J. Goldstien[1], G. A. Hopkins[2]

[1]School of Biological Sciences, University of Canterbury, Christchurch 8041, New Zealand

[2]Coastal and Freshwater Group, Cawthron Institute, Nelson 7010, New Zealand

[3]*Present address:* National Institute of Water and Atmosphere (NIWA), 217 Akersten Street, Port Nelson, Nelson 7010, New Zealand

ABSTRACT: Biofouling pests can have significant economic impacts on aquaculture operations, including increased processing and production costs. An important first step towards improved biofouling management is understanding the density and distribution of the biofouling species within a growing region. In this study, biofouling communities were sampled from 73 commercial mussel farms within New Zealand's main mussel growing region, Pelorus Sound. At each farm, photoquadrats (0.08 m^2, n = 6) of biofouling organisms were obtained at 2 depth ranges (3 per range) from suspended long-line droppers, both at the surface (0 to 3 m of the dropper) and bottom (9 to 24 m, depending on dropper length and water depth). Biomass samples and visual estimates of biofouling biomass were also obtained. Strong spatial variation in the structure of biofouling communities was evident, with increasing dissimilarity between communities along Pelorus Sound. Problematic taxa (e.g. the brown alga *Undaria pinnatifida* and calcareous tubeworm *Pomatoceros* sp.) were dominant near the entrance to the Sound, where annual temperature cycles are often reduced and salinity concentrations are higher. Generally, biofouling cover decreased with increasing water depth. A large proportion (48%) of biofouling biomass scores were categorised as high, equating to 121.2 ± 20 g m^{-2} (or 16%) of long-line for a heavily fouled farm, or 10 t for a typical 3 ha farm. Distributional patterns, such as those identified in this study, could be used by aquaculture industries to better inform the timing and placement of susceptible crop species and production stages (e.g. mussel spat). Refined monitoring methods may also facilitate industry participation in collecting long-term biofouling records.

KEY WORDS: Biofouling · Artificial substrate · Community structure · Aquaculture · Mussel · *Perna canaliculus*

INTRODUCTION

Marine farm infrastructure is composed of a diverse range of artificial components, including ropes, floats, anchors, cages, nets and rafts (Fitridge et al. 2012). These surfaces, which intercept water flow and consequently larvae in the water column, provide extensive habitat for colonisation by biofouling organisms (Metri et al. 2002, McKindsey et al. 2007, Dürr & Watson 2010, Adams et al. 2011, Fitridge et al. 2012, Antoniadou et al. 2013, Sievers et al. 2013). The accumulation of biofouling is predominantly detrimental for aquaculture industries, reducing revenue and crop growth, and increasing processing and production costs, disease risk, structural fatigue, farm load, and the mechanical handling and maintenance of equip-

*Corresponding author: ashleigh.watts@niwa.co.nz

ment (Claereboudt et al. 1994, Grant et al. 1998, McKindsey et al. 2009, Fitridge et al. 2012, Fitridge & Keough 2013, Sievers et al. 2013, Lacoste & Gaertner-Mazouni 2014). For example, dominant growth of the ringed tubularian *Ectopleura larynx* in the Norwegian fish farming industry increases the frequency and duration of infrastructure cleaning during peak fouling seasons (Guenther et al. 2009, 2010).

In the mussel aquaculture industry, culture stock provides an ideal and accessible 3-dimensional biofouling surface. The major biofouling groups that colonise mussel shells include sponges, barnacles, spirorids/serpulids and ascidians (Dürr & Watson 2010), many of which are non-indigenous and have detrimental impacts on the appearance, marketability, growth and condition of crop species. For instance, shell-boring polychaete worms such as *Polydora* spp. devalue shellfish appearance, reduce hinge stability, disrupt shell formation and increase the vulnerability of shellfish to predation (Che et al. 1996, Lleonart et al. 2003, Silina 2006, Simon et al. 2006, Fitridge et al. 2012, Fitridge & Keough 2013, Sievers et al. 2014). In addition, increased biomass of the invasive colonial ascidian *Didemnum vexillum* has been found to displace small, cultured New Zealand green-lipped mussels *Perna canaliculus* (Fletcher et al. 2013b).

A number of techniques have been employed and are continually being developed to mitigate and control biofouling in the aquaculture sector (Fitridge et al. 2012), including cleaning and replacing equipment, biocidal paints, chemical treatments, heat treatments, physical removal of pest species or exposure of artificial infrastructure to periods of air-drying, high-pressure power washing and fresh or hot water baths (Enright 1993, Chambers et al. 2006, Forrest & Blakemore 2006, López-Galindo et al. 2010, Carl et al. 2012, Fitridge et al. 2012). Novel antifouling approaches are also being trialled, including 'eco-friendly' antifouling formulations based on natural compounds and novel surface characteristics (Cahill et al. 2013), as well as the development of marine pest biocontrol tools (Atalah et al. 2013). However, the control of biofouling is industry-specific, and effective management requires the development of technologies and methods of application specific to the culture environment. Therefore, control methods in mussel aquaculture predominantly focus on maintaining clean shells through the use of avoidance strategies (Fitridge et al. 2012, Sievers et al. 2014).

While many of the described techniques may be successful in removing soft-bodied biofoulers, some fail to remove or prevent the settlement of several species of barnacles and calcareous tubeworms (Carver et al. 2003, Forrest & Blakemore 2006, LeBlanc et al. 2007). Their implementation can also have undesirable effects, such as the fragmentation of colonial organisms, which may contribute to their localised spread (Hopkins & Forrest 2010, Paetzold & Davidson 2011). In addition, biofouling treatments can increase the intensity of stock stress and mortality (LeBlanc et al. 2007, Antoniadou et al. 2013).

The occurrence and impacts of specific biofouling species on marine farm structures varies spatially (Ceccherelli & Campo 2002, Thomsen et al. 2006) and temporally (Stæhr et al. 2000, Forrest & Taylor 2002), including variation in community structure among years (Underwood & Anderson 1994), locations (Lutz-Collins et al. 2009, Sievers et al. 2014) and depths (Hanson & Bell 1976, Woods et al. 2012). For example, Woods et al. (2012) and Cronin et al. (1999) found reduced biofouling biomass and less diversity in biofouling communities with increased depth on marine farm structures (mussel long-lines and tuna sea cages). Some biofouling populations also proliferate rapidly and then gradually retreat. This is especially true for taxa such as ascidians, including invasive pest species such as *D. vexillum*, *Ciona intestinalis* and *Styela clava* (Valentine et al. 2007, Forrest et al. 2011, Fletcher et al. 2013a). Consequently, it is imperative that more studies quantify variability in these patterns to improve risk management.

An alternative approach for biofouling management in mussel aquaculture could therefore be to incorporate a more information-based approach, as suggested by Sievers et al. (2014) and Fitridge et al. (2012). Such an approach would involve linking knowledge about site-specific patterns of biofouling development (or predictions of their occurrence) with strategies to avoid specific locations during times of heavy biofouling, or strategies to remove biofouling species. This knowledge could also inform decisions regarding the placement of susceptible crop species and stages of production (e.g. mussel seed stock) within a specific region or water depth.

In New Zealand, Pelorus Sound (located in the Marlborough Sounds) is the major growing region for the mussel farming industry, with approximately 645 farms spread across 5000 ha of farming area (Woods et al. 2012). Despite the value and extent of this area, aside from a few site-specific studies (Woods et al. 2012), our knowledge about the spatial and temporal variation of biofouling organisms associated with these aquaculture farms is limited. Hence, this study had 3 specific objectives: (1) to characterise biofouling on mussel farm long-lines in Pelorus Sound, with the expectation that community structure and the relative abundance of pest species would vary across the study region; (2) to investigate the influence of depth on biofouling community structure and species relative abundance, with the expectation that biofouling cover would decrease with increasing depth; and (3) to assess potential mechanisms contributing to biofouling structure by investigating the relationship between biofouling community similarity and distance between marine farms, with the expectation that community similarity would decrease as geographical distance between farms increased.

MATERIALS AND METHODS

Study region

Pelorus Sound is located within the Marlborough Sounds, at the northern end of New Zealand's South Island (Fig. 1). It is a 56 km long, relatively deep (average water depth 40 m) and highly indented estuarine system with variable freshwater input from the Kaituna and Pelorus rivers entering at the head of the Sound, as well as oceanic exchange from upwelling waters in Cook Strait (Heath 1974, Woods et al. 2012). There are complex tidal, estuarine and

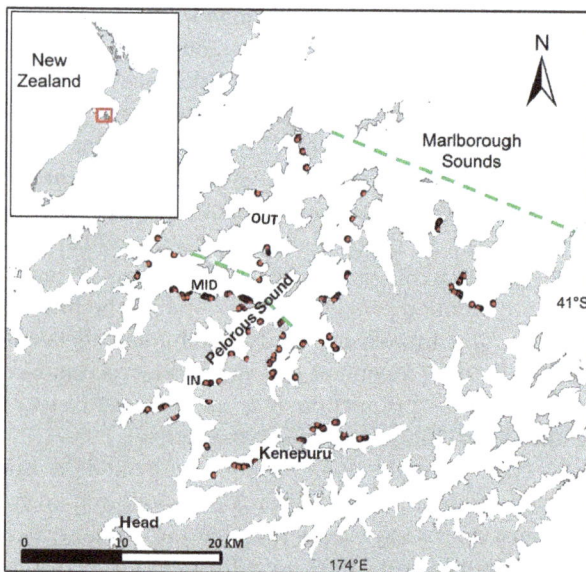

Fig. 1. Marlborough Sounds region, showing the location of commercial green-lipped mussel *Perna canaliculus* farms sampled in Pelorus Sound (filled red circles), from the inner (IN), middle (MID), and outer (OUT) areas. Dashed green lines: delineation of the 3 areas. The 'head' of the Pelorus Sound and the Kenepuru Sound are indicated. Inset shows the location of the Marlborough Sounds at the northern tip of New Zealand's South Island

wind-driven circulation systems in the Pelorus, with mean residence times varying from 21 d in Pelorus channel to 6 d in the Kenepuru. Reduced flushing rates and strong stratification have been recorded within the side arms and embayments near the head of Pelorus Sound. Stratification is generally driven by salinity (Heath 1974, Gibbs et al. 1991), with a gradual decrease in mean salinity towards the confluence with Kenepuru Sound (Heath 1982, Broekhuizen et al. 2015) (Fig. 2a,b). Annual mean temperatures are also warmer near the head of Pelorus Sound, although Broekhuizen et al. (2015) recorded seasonal variation in this pattern with cooler winter temperatures and warmer summer temperatures near the head of Pelorus Sound compared to the outer Sound (Fig. 2c,d).

Biofouling characterisation

In January 2013, a total of 73 commercial greenlipped mussel *Perna canaliculus* farms were sampled within the inner (n = 30), middle (n = 26) and outer (n = 17) Sound (Fig. 1). The 3 area boundaries were set in accordance with previously conducted dispersal studies (Knight et al. 2010), as well as the known environmental gradients and hydrodynamic characteristics in Pelorus Sound (Heath 1974, Gibbs et al. 1991, Broekhuizen et al. 2015). Each mussel farm was considered to be a single replicate within the larger area of interest (inner, middle, outer). Therefore, each mussel farm was randomly selected, and within each farm a random selection of long-line droppers (1–4 per farm; inner n = 50, middle n = 46, outer n = 29) were sampled. Mussel farm long-lines were situated between 50–250 m of the shoreline. Therefore, the inner- and outermost long-lines were excluded to standardise the influence of sampling position. The droppers were lifted from the water column using a winch onboard an industry product-sourcing vessel. Each dropper was systematically photographed (Nikon Coolpix AW100, 16 megapixels) using a 0.4 × 0.2 m photoquadrat; 3 images were taken within the surface 3 m of the dropper (0–3 m of the long-line depth) and 3 within the bottom 3 m (9–12 or 21–24 m depth, depending on the area's water depth and subsequent length of the looped long-line). To ensure photoquadrats were independent, a 50 cm gap was left between photoquadrats, and a 6–18 m gap was left between photoquadrats taken within the 2 depth ranges. Video footage (Sony HDR-XR350VE) was also taken from each mussel long-line dropper that incorporated the areas where photoquadrats were

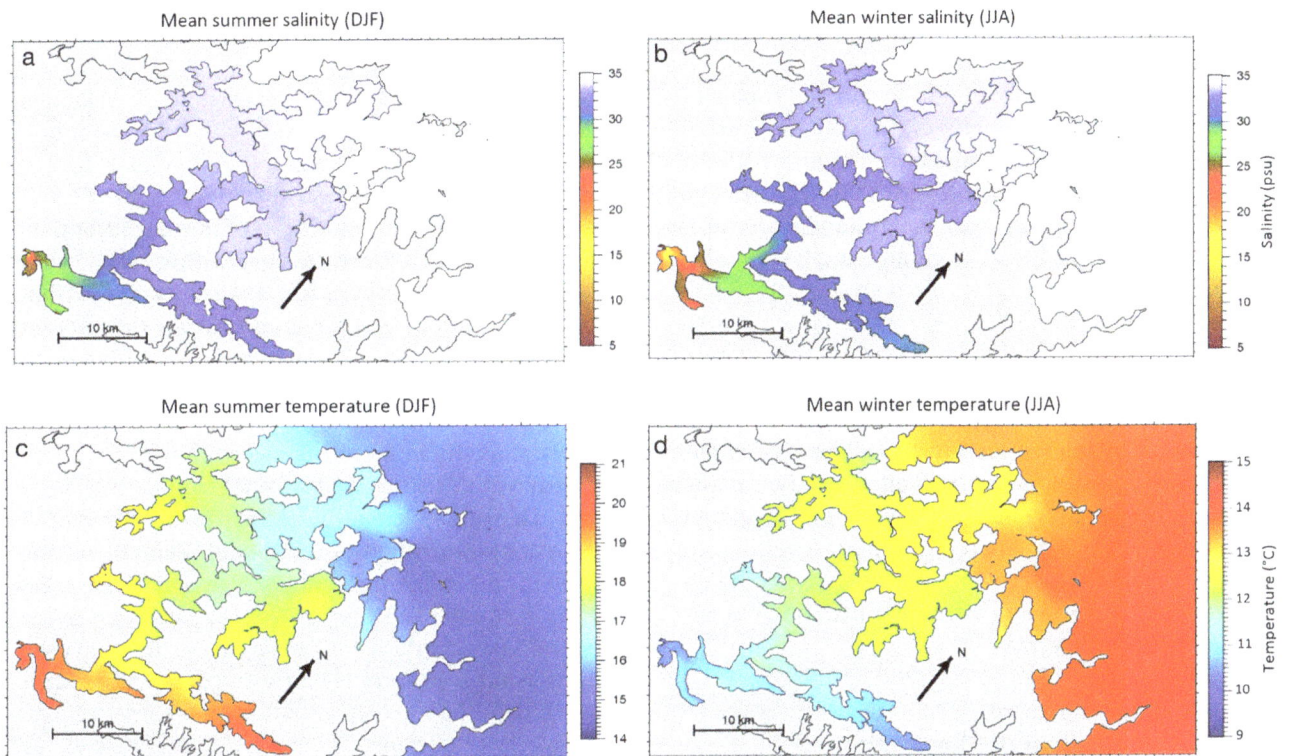

Fig. 2. Summer (DJF: Dec 2012–Jan–Feb 2013) and winter (JJA: Jun–Jul–Aug 2013) mean (a,b) salinity and (c,d) temperatures from a Pelorus Sound model (figure by Mark Hadfield, NIWA). This model was compared with, and matched monthly CTD surveys conducted by NIWA in Pelorus Sound from 2012 to 2014 (see Broekhuizen et al. 2015). Graphs are on a rotated map projection; colour scales represent temperature and salinity ranges and units

taken as well as the remaining long-line dropper length.

The presence of conspicuous biofouling pests (i.e. those with previously documented impacts to aquaculture in New Zealand and overseas) on other farm infrastructure (i.e. backbone ropes, floats, anchor warps) was recorded, and voucher specimens were collected for identification and future reference. In addition, crop line age (at the time of sampling) was obtained from an industry database.

Visual assessments of biofouling biomass and percentage cover of bare rope space (i.e. no crop present) were also recorded from all mussel long-lines across the 3 sampling locations (inner, middle and outer), along with observations of dominant biofouling species (including non-indigenous species). Video footage was used as an extra resource for confirmation of these visual assessments.

During crop-condition assessments, mussel-sourcing staff often estimate and record the presence and extent of biofouling biomass on mussels. Following discussions with these staff, 3 biomass categories were selected and assessed to determine how to reli-

ably estimate biomass from visual assessments, and to determine how these contributed to long-line weight. The 3 biomass categories represented low (2-dimensional, or small patches of 3-dimensional), medium (largely 3-dimensional, but patchy) and high (extensive cover by 3-dimensional) biofouling. Prior to sampling, the reliability of these visual assessments was verified by removing crop and fouling biomass from five 0.4 m sections of crop rope for each of the 3 biomass categories. These sections were spread across the 3 sampling locations and included 2 biomass sections from the inner area, 2 from the middle and 1 from the outer area for the low and high biofouling biomass categories, as well as 1 sample from the inner, 3 from the middle and 1 from the outer area for the medium biofouling biomass category. Samples were individually labelled and transported, cooled (<10°C), to the Cawthron Institute laboratory (Nelson, New Zealand) within 12 h of collection. In the laboratory, the total wet weight of bivalves with biofouling was measured for each sample. Also, the wet weight of biofouling from a subset of 20 green-lipped mussels was obtained from each sample.

Photoquadrat images were cropped in Google Picasa v.3.9, and poor quality images were removed from analyses. The percent cover of biofouling taxa (identifiable and unidentifiable), crop and rope longline beneath 75 randomly stratified points were determined using the Coral Point Count with excel extensions (CPCe) (Kohler & Gill 2006). The selection of 75 points per image was determined during preliminary precision and accuracy analyses, which identified whether point count analyses reflected the 'true' percent cover of taxa, represented by area–length analyses. Variation in precision, accuracy and time efficiency using a variety of points per image in CPCe were also assessed (Watts 2014). Biofouling organisms were identified to the highest taxonomic resolution possible from photographs. For certain groups, such as filamentous algae and ascidians, identification was restricted to higher level descriptions. Voucher species were used to assist with identification, and in some cases verification was provided by taxonomists.

Statistical analyses

Comparison of visual and photographic assessments

To determine the efficiency of using photographs to capture biofouling taxa compared to visual estimates of presence/absence for dominant biofouling taxa, the 2 datasets from these methods were compared. Percent cover data from photoquadrats were transformed into presence/absence data using PRIMER v.6 with PERMANOVA add-on (Anderson 2001a, Clarke & Gorley 2006, Anderson et al. 2007). The percent of records where photoquadrats recorded taxa as present were then compared to the percent of records where visual estimates did not observe these taxa (and vice versa), using R v.3.0.2 (R Core Team 2013). The mean biomass measured from the 3 biomass categories was also compared using a univariate permutational analysis of variance.

Multivariate analysis

Changes in community structure among areas and depths were investigated using distance-based permutational multivariate analysis of variance (PERMANOVA; Anderson 2001b) based on Bray-Curtis dissimilarities of the square-root transformed data (Bray & Curtis 1957). A square-root transformation was selected over a more severe transformation (e.g. fourth

root or presence/absence transformation) where rare species contributed disproportionately more to the analyses (Anderson 2001a).

The experimental design consisted of 4 factors and 1 covariate: (1) area (fixed with 3 levels: inner, middle and outer), (2) depth (fixed with 2 levels: surface and bottom), (3) farm (random with 73 levels, nested within area), and (4) long-line droppers (random with 125 levels, nested within farm), with crop age (i.e. months since long-lines were first seeded) included as a covariate. Crop age ranged from 1 to 67 mo and varied across farms within each area. Interactions between the covariate (age) and the fixed and random variables were excluded from analyses, as the influence of crop age was not a key objective of this study and was not consistently sampled. Therefore, excluding these terms prevented over-parameterization of the model. Each term was tested using 4999 permutations and a Type I SS (sums of squares) (Anderson 2001b, Anderson & Ter Braak 2003). Significant terms were then investigated using a posteriori pairwise comparisons with the PERMANOVA t-statistic and 999 permutations. Multivariate variance components expressed as square-root variance components (i.e. converting values to percentages of Bray-Curtis dissimilarity) were calculated and compared for each term in the analysis (Underwood & Petraitis 1993, Anderson et al. 2007). The distribution of the covariate (crop age) was skewed, and as covariates used in PERMANOVA are expected to show approximately symmetric distributions that are roughly normal, was square-root transformed (Anderson et al. 2007). Bare space and green-lipped mussel percent cover were removed from analyses to ensure similarity was not driven by substrate. In addition, sub-canopy species (i.e. those not visible beneath biofouling) could not be included in the analyses due to the sampling method used in this study (photoquadrats).

A permutational analysis of multivariate dispersions (PERMDISP) was also used, followed by pairwise comparisons, to test for differences in dispersion among areas and between depths using 4999 permutations. In addition, a similarity percentage analysis (SIMPER; Clarke 1993) was conducted to identify the contribution of each species (or taxon) to observed community dissimilarities (cut-off set to 80%) between significant factors. Taxa were considered important if their contribution to percentage dissimilarity was ≥6%. The ratio of the average dissimilarity and standard deviation (Diss/SD) was used to indicate the consistency with which a given species contributed to dissimilarity (Clarke 1993, Clarke & Gorley 2006); values ≥1.5 indicated a high degree of consistency. Differences in community structure among areas were visualised through a principal coordinates analysis (PCO; Gower 1966). Taxa that consistently discriminated between significant terms and had a correlation >0.3 with the PCO axes were displayed as vectors in the PCO plot. Variability in the cover of contributing taxa were assessed using separate 2-way univariate permutational analysis of variance tests (ANOVAs) on the square-root transformed variables (Anderson 2001b). To account for multiple comparisons and control alpha inflation, a False Discovery Rate (FDR) correction was applied (n = 8, p < 0.02 represented significance) (Benjamini & Yekutieli 2001, Narum 2006). Statistical analyses were done using PRIMER v.6 with PERMANOVA add-on (Anderson 2001a, Clarke & Gorley 2006, Anderson et al. 2007).

Spatial correlation of biofouling communities

Alongside information on the spatial structure of ecological communities, distance approaches have proven effective and informative for gauging the spatial turnover of communities (Soininen et al. 2007, Morlon et al. 2008). Species spatial turnover, or beta diversity, often induces reduced community similarity with increasing geographic distance, known as the distance–decay relationship (Morlon et al. 2008). Distance–decay relationships were investigated using Partial Mantel tests calculated in the 'vegan' community ecology package (Oksanen et al. 2011) in R (R Core Team 2013).

We determined the relationship between community structure and geographic distance among farms using a distance–decay analysis between the Bray-Curtis dissimilarity matrix of the community structure data (square-root transformed) and a Euclidean distance matrix of the geographic distances between all farms. The Wisconsin double standardisation method, which improves the gradient detection of dissimilarity measures (Oksanen 2011), was used alongside Bray-Curtis dissimilarity to generate the community structure matrix (Oksanen 1983, Legendre & Gallagher 2001, Anderson 2006). As pairwise similarity values and distances were not truly independent in a statistical sense, partial Mantel statistics were estimated using the 'matrix permutation' method with 9999 permutations. Linear regression was used to describe the relationship between dissimilarity values and geographical distance.

Comparison of visual and photographic assessments

When comparing species cover attained from photoquadrats with that from visual estimates, it was evident that visual estimation alone was not a reliable method for determining biofouling species on mussel long-line droppers. Fifty percent of the records attained from photoquadrats detected taxa as being present when visual estimates recorded them as absent, while only 2 % of the records gained by visual estimates recorded taxa as present when photoquadrats did not. Photoquadrats therefore provide a more reliable representation of biofouling cover on long-line droppers.

Community structure and patterns in the cover of problematic biofoulers

A total of 86 biofouling taxa were identified occupying vacant space on crop long-lines, or occurring epibiotically on mussel shells and other biofouling organisms (Table 1). Overall, communities were dominated by red filamentous algae and the Asian kelp *Undaria pinnatifida* (macroalgae), the blue mussel *Mytilus galloprovincialis*, hydroids, bryozoans and ascidians (Fig. 3). Less abundant taxa included sponges, anemones and mobile taxa, such as amphipods, isopods, sea cucumbers and crabs (Fig. 3). Biofouling cover was significantly greater near the surface of mussel long-lines ($t_1 = 5.75$, p < 0.01).

Community composition

Patterns of variation in community structure are depicted in the PCO plot (Fig. 4a), showing a clustering and separation of communities located in different areas (specifically the inner area from the middle and outer areas; Fig. 4a), and at different depths (Fig. 4b). These patterns were reflected in PERMANOVA analyses, which indicated a significant interactive effect of area and depth on the structure of biofouling communities (depth × area effect, p < 0.001; Table 2). Pairwise comparisons revealed that community structure differed significantly between the 2 depths, but this effect varied among all areas (for all pairwise comparisons, p < 0.001; Table 2). Average percent similarity in community structure was lowest between inner and outer areas, across

Table 1. Taxa and species occurring in biofouling communities on green-lipped mussel *Perna canaliculus* long-lines across Pelorus Sound, New Zealand

Taxon	Group	Genus and species	Taxon	Group	Genus and species
Macroalgae	Cladophora	*Cladophora* sp.	Mollusa	Bivalvia	*Mytilus galloprovincialis*
		Ulva sp.	Bryozoa	Cheilostomata	
		Codium fragile		*Erect*	*Bugula* sp.
		Unidentified green filamentous algae			*Bugula stolonifera*
	Phaeophyta	*Colpomenia* sp.			*Bugula flabellata*
		Undaria pinnatifida			*Bugula neritina*
		Macrocystis pyrifera			Unidentified erect bryozoan
		Spatoglossum sp.		*Encrusting*	*Watersipora* sp.
	Rhodophyta	*Porphyra* sp.			Unidentified encrusting bryozoan
		Ceramium sp.	Echino-dermata	Aspido-chirotida	*Australostichopus mollis*
		Gracilaria sp.			
		Echinothamnion sp.		Ophiuroidea	*Ophionereis fasciata*
		Asparagopsis sp.		Asteroidea	*Coscinasterias calamaria*
		Unidentified red filamentous alga			*Patiriella* sp.
		Unidentified red filamentous alga 1			
		Unidentified red filamentous alga 2	Teleostei	Triptery-giidae	*Fosterygion varium*
		Unidentified red filamentous alga 3			
		Unidentified red filamentous alga 4	Ascidiacea	*Colonial*	*Didemnum* sp.
Porifera		*Halichondria* sp.			*Didemnum incanum*
		Sycon sp.			*Didemnum vexillum*
		Haliclona sp.			*Didemnum lambitum*
		Unidentified sponge			*Diplosoma listerianum*
					Diplosoma sp.
Cnidaria	Hydrozoa	*Amphisbetia bispinosa*			*Leptoclinides novaezelandiae*
		Sertularella sp.			*Lissoclinum notti*
		Unidentified hydroid sp. 1			*Aplidium phortax*
		Unidentified hydroid sp. 2			*Botrylloides leachii*
	Anthozoa	*Culicia rubeola*			*Botryllus schlosseri*
		Actinothoe albocincta			Unidentified colonial ascidian sp. 1
		Diadumenidae			Unidentified colonial ascidian sp. 2
		Bunodeopsis sp.			Unidentified colonial ascidian sp. 3
Annelida	Sabellida	*Galeolaria hystrix*			Unidentified colonial ascidian sp. 4
					Unidentified colonial ascidians
		Pomatoceros sp.		*Solitary*	*Ascidiella aspersa*
		Spirorbidae			*Asterocarpa humilis*
		Serpulidae			*Ciona intestinalis*
Crustacea	Amphipoda	*Caprella* sp.			*Corella eumyota*
	Isopoda	*Paridotea ungulata*			*Molgula* sp.
	Brachyura	*Halicarcinus* sp.			*Pyura pachydermatina*
		Hemigrapsus sp.			*Cnemidocarpa bicornuta*
		Notomithrax minor			Unidentified solitary ascidian sp. 1
		Notomithrax peronei			Unidentified solitary ascidian sp. 2
		Unidentified crab			Unidentified solitary ascidian sp. 3
					Unidentified solitary ascidian sp. 4
Arthropoda	Sessilia	*Elminius modestus*			Unidentified solitary ascidian sp. 5
		Balanus trigonus			Unidentified solitary ascidian sp. 6
					Unidentified solitary ascidians

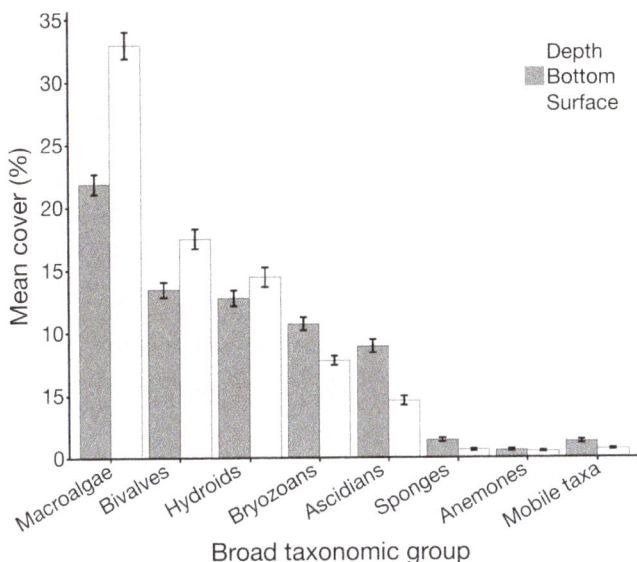

Fig. 3. Mean (±SE) percentage cover of broad taxonomic groups of biofouling organisms found on green-lipped mussel *Perna canaliculus* long-line droppers. Results are shown across all areas sampled in Pelorus Sound, in accordance with depth

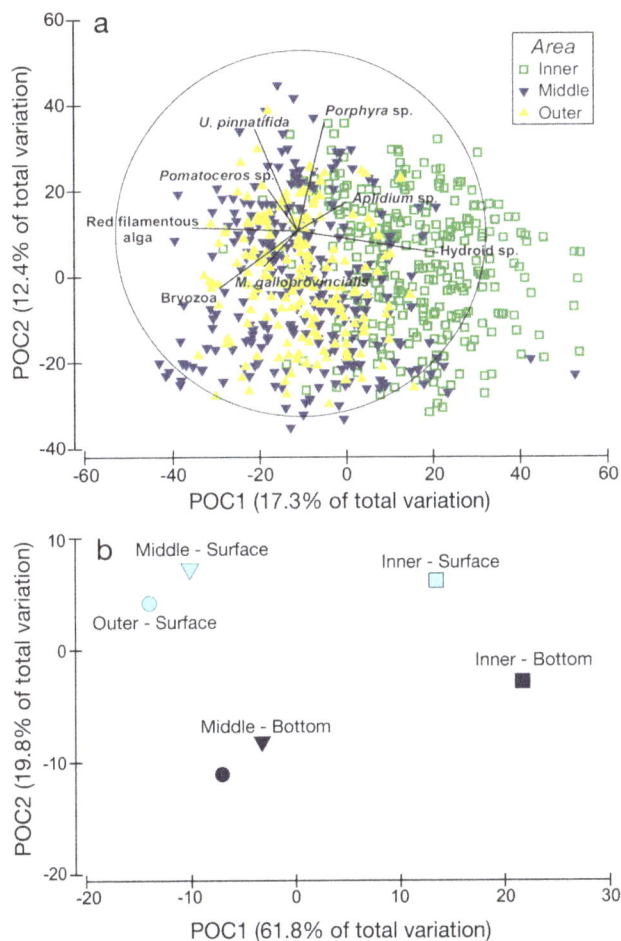

Fig. 4. Principal coordinate analysis (PCO) plots on the basis of Bray-Curtis dissimilarities of the square-root transformed (a) percent cover data of biofouling communities across the inner (square), middle (upside-down triangle) and outer (upwards triangle) Sound, along with the species contributing to differences in biofouling assemblages (correlation > 0.3), and (b) across the inner (square), middle (upside-down triangle) and outer (circle) Sound, in association with depth. Distances among centroids for all photoquadrat results across the inner, middle and outer Sound (with depth, represented by symbol shading: light blue symbols = surface 3 m and dark blue = bottom 3 m) are presented in (b)

depths (Table 2). Furthermore, the effect of depth on community structure significantly varied between long-lines: depth × long-line (Farm[Area]), p < 0.001 (Table 2). There was also a significant positive relationship between crop age and community structure (age, p < 0.001; Table 2). The greatest component of variation for biofouling community structure occurred at the lowest spatial scale (the residual, 28 %), followed by between farms within areas (Table 2). Moderate variability was observed among areas, and the lowest variance component was estimated across crop ages (Table 2).

The percentage cover of several pest species varied with area and depth across Pelorus Sound. There was a trend for high cover of hydroid species *Amphisbetia bispinosa* and macroalgae, including *Cladophora* sp., *U. pinnatifida* and *Colpomenia* sp. near the entrance of Pelorus Sound, within the surface 3 m of sampled long-line droppers. There was also a trend for high cover of the calcareous tubeworm *Pomatoceros* sp. near the entrance of Pelorus Sound, but within the bottom 3 m. The cover of *M. galloprovincialis* tended to be highest in the surface 3 m of long-line droppers in the middle of Pelorus Sound, and the colonial and solitary ascidians *Didemnum vexillum* and *Ciona intestinalis* (respectively) had a tendency for high cover near the head of Pelorus Sound, also within the surface 3 m.

SIMPER analysis revealed that 6 taxa (red filamentous alga, an unidentified hydroid, encrusting bryozoans, *U. pinnatifida*, *Porphyra* sp. and *M. galloprovincialis*) contributed consistently to the community dissimilarities between areas and depths (Table 3); however, their contributions were relatively low (6 to 11 %). The largest average dissimilarity was between fouling in the bottom 3 m of long-line droppers in the inner Sound and fouling in the surface 3 m in the outer Sound (70.17 %; Table 3). Patterns revealed by the SIMPER analysis were confirmed by univariate permutational ANOVAs on the cover of the most

Table 2. Multivariate PERMANOVA results based on Bray-Curtis dissimilarities for spatial differences in community structure (square-root transformed) at the scales of area and depth, with age as a co-variate. Pairwise comparisons for the depth × area interaction effect, within factor area, are included. Estimates of multivariate variation (variation %), the estimated sizes of average similarities between areas (AS), mean sums of squares (MS), F-statistics (F) and pairwise t-statistics (t) are also included. *$p < 0.05$; **$p < 0.01$; ***$p < 0.001$.

Source of variation	df	MS	F	Variation (%)
Age	1	24058	3.95***	5
Depth	1	45716	19.35***	11
Area	2	74469	9.58***	16
Farm(Area)	70	6933	2.70***	21
Depth × Area	2	7411	3.19***	6
Long-line(Farm[Area])	52	2531	3.31***	17
Depth × Farm(Area)	70	2108	1.57***	12
Depth × Long-line(Farm[Area])	52	1346	1.76***	14
Residuals	511	766		28
Pairwise comparisons				
Groups			t	AS (%)
Surface depth				
Inner–Middle			3.19***	41
Inner–Outer			3.06***	36
Middle–Outer			1.56***	43
Bottom depth				
Inner–Middle			3.48***	37
Inner–Outer			3.13***	33
Middle–Outer			1.82***	43

prominent taxa contributing to dissimilarities, and on taxa with a high correlation (>0.3) with the PERMANOVA PCO axis. Depth had a significant effect on the cover of encrusting bryozoans, an unidentified hydroid and *M. galloprovincialis*, which was area dependent (area × depth effect, $p < 0.01$). Encrusting bryozoans tended to have high cover within the bottom 3 m of mussel long-line droppers, in the middle and outer areas (Fig. 5). The unidentified hydroid and *M. galloprovincialis* typically had high cover at the surface of mussel long-line droppers in the inner and middle areas, respectively (Fig. 5). Depth and area also had a significant effect on the cover of red filamentous alga and *U. pinnatifida*, but there was no interaction between these factors (depth effect, $p < 0.01$; area effect, $p < 0.01$). Red filamentous alga and *U. pinnatifida* had higher cover within the outer area (Fig. 5). Area alone had a significant effect on the cover of *Porphyra* sp. (area effect, $p < 0.001$), with a tendency for high cover in the inner area (Fig. 5). Univariate PERMANOVA tests also revealed that the cover of *Aplidium* sp. and *Pomatoceros* sp., which had a high correlation (>0.3) with the PERMANOVA PCO axis (Fig. 4a), significantly differed with depth

and across areas (area × depth effect, $p < 0.001$). The cover of *Aplidium* sp. and *Pomatoceros* sp. tended to be highest at the bottom depth, in the inner and outer areas, respectively (Fig. 5).

PERMDISP analyses showed significant differences in multivariate dispersion between areas, which varied with depth (area × depth effect, $p < 0.001$; Table 4). Pairwise comparisons revealed that this was due to significantly greater dispersion near the bottom of the mussel long-lines within the inner area of Pelorus Sound compared to the surface of the mussel long-lines ($t_1 = 4.444$, $p < 0.001$; Table 4).

Spatial correlation between fouling communities

A relatively weak but significant positive relationship was detected between the geographical distance of sampled mussel farms across Pelorus Sound and the dissimilarity in biofouling community structure ($b = 0.35$, $r = 0.40$, $p = 0.03$). Specifically, dissimilarity in the structure of biofouling assemblages increased, with a steep slope, with geographical distance (Fig. 6).

Biofouling levels and biomass

There were no significant differences between the 3 categories used to assess biofouling biomass ($p > 0.05$). Although there was an increase in biomass weights (kg) associated with an increase in the category of biomass cover (low, medium to high) (Fig. 7a), high variation and overlap in the biomass categories indicate the need to refine such a method before it could be used reliably to accurately predict biomass levels. However, for the purpose of this study it was considered appropriate to provide course estimates of fouling biomass levels within the study system.

A large proportion of observed biomass scores were high (16 % of the total wet weight across sampled sections) and predominantly occurred in the outer Pelorus Sound (Fig. 7b), while the proportion of low and medium biomass scores were lower (5 and 12 % respectively), and primarily occurred in the middle and inner areas (Fig. 7b). Sections of culture rope (0.4 m) visually assessed as having 'low' biofouling biomass had on average 57.8 ± 14.5 g m^{-2} (43–72 g m^{-2}) of biofouling (wet weight), whereas 'medium'

Table 3. Similarity percentage analysis (SIMPER) summary showing taxa with the highest percent contribution (reflected in the taxa order) towards the dissimilarity between areas, within different depths across Pelorus Sound (IB: inner bottom; IS: inner surface; MB: middle bottom; MS: middle surface; OB: outer bottom; OS: outer surface). Contributing taxa were red alga (red filamentous alga), Hydroid (hydroid species-1), Bryozoa (encrusting bryozoans), Porphyra (*Porphyra* sp.), Mytilus (*Mytilus galloprovincialis*) and Undaria (*Undaria pinnatifida*). Taxa % contribution was ≥6%. All taxa listed consistently contributed (%) to group dissimilarity; (−) no data

	IB	IS	MS	MB	OS
OB	Avg. diss.: 67.01 Red alga Hydroid Bryozoa Porphyra	Avg. diss.: 62.63 Hydroid Bryozoa Red alga Porphyra	Avg. diss.: 59.26 Mytilus Red alga Bryozoa Hydroid	Avg. diss. : 57.47 Bryozoa Red alga Hydroid Porphyra	Avg. diss.: 60.91 Red alga Bryozoa Hydroid −
OS	Avg. diss.: 70.17 Red alga Hydroid Porphyra Bryozoa Undaria	Avg. diss.: 63.63 Hydroid Red alga Porphyra Bryozoa Undaria	Avg. diss.: 56.98 Mytilus Red alga Hydroid Bryozoa −	Avg. diss.: 60.02 Red alga Hydroid Bryozoa − −	
MB	Avg. diss.: 63.48 Hydroid Bryozoa Red alga Porphyra	Avg. diss.: 59.21 Hydroid Bryozoa Porphyra −	Avg. diss.: 55.57 Mytilus Porphyra Hydroid Bryozoa		
MS	Avg. diss.: 66.59 Red alga Hydroid Mytilus Porphyra Bryozoa	Avg. diss.: 59.10 Hydroid Mytilus Porphyra Bryozoa −			
IS	Avg. diss.: 60.14 Porphyra Bryozoa Red alga				

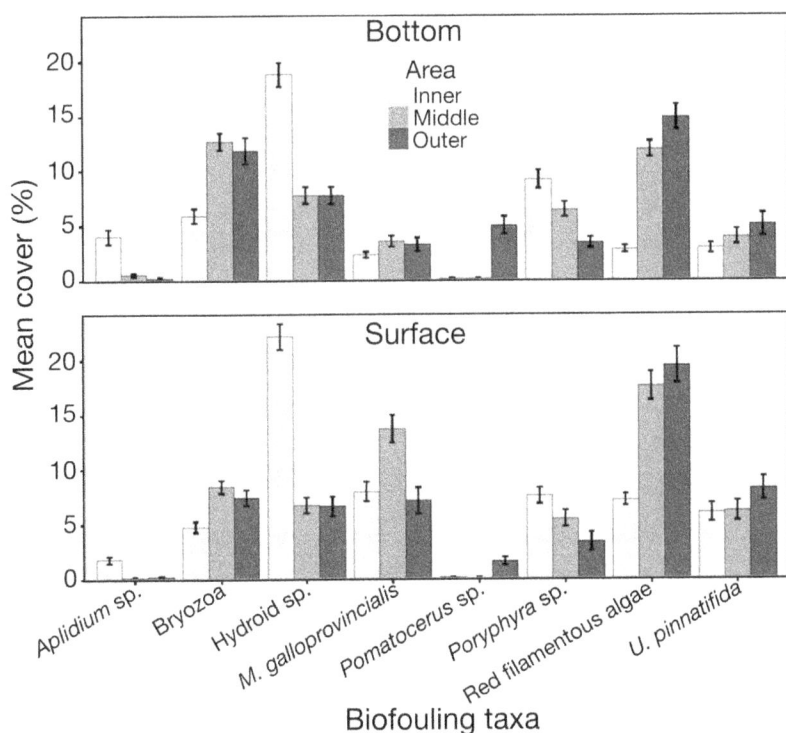

Fig. 5. Similarity percentage (SIMPER) analysis results for the mean (±SE) percent cover of taxa contributing to dissimilarities between areas, at different depths across Pelorus Sound. The highest contributing taxa were red filamentous algae, *Hydroid* sp. (an unidentified hydroid species), Bryozoa (encrusting bryozoans), *Porphyra* sp., *Mytilus galloprovincialis* and *Undaria pinnatifida*. The average abundance of taxa with a correlation >0.3 included *Aplidium* sp. and *Pomatoceros* sp., and sp. refers to a single species

Table 4. Differences in multivariate dispersion (PERMDISP) of the depth × area interaction effect, with the associated pairwise comparisons for dispersion among and within areas, across depths. Analyses were based on Bray-Curtis dissimilarities and data were square-root transformed. *p < 0.05; **p < 0.01; ***p < 0.001

Source of variation	F
Depth × Area	11.587***
Pairwise comparisons	
Groups	t
Bottom Inner–Surface Inner	4.44***
Bottom Inner–Surface Middle	6.26***
Bottom Inner–Bottom Middle	6.28***
Bottom Inner–Bottom Outer	2.24*
Bottom Inner–Surface Outer	1.75
Surface Inner–Surface Middle	1.84
Surface Inner–Bottom Middle	1.70
Surface Inner–Bottom Outer	1.74
Surface Inner–Surface Outer	1.94
Surface Middle–Bottom Middle	0.21
Surface Middle–Bottom Outer	3.39***
Surface Middle–Surface Outer	3.47***
Bottom Middle–Bottom Outer	3.36***
Bottom Middle–Surface Outer	3.44***
Bottom Outer–Surface Outer	0.3

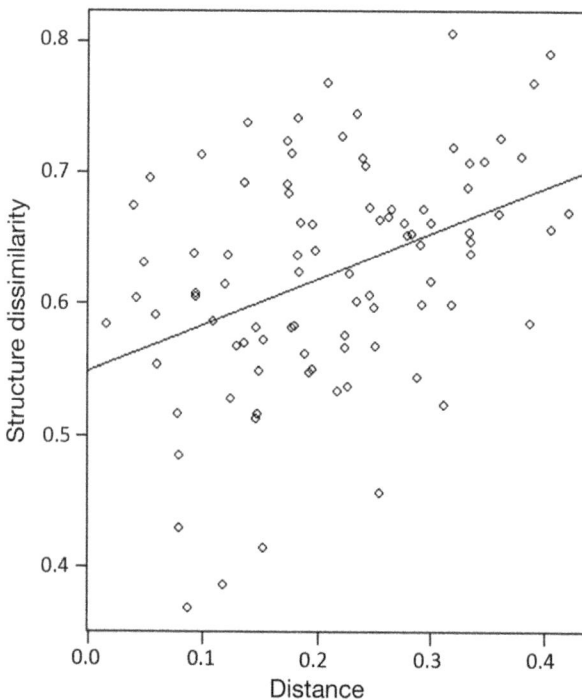

Fig. 7. Mean (±SE) biofouling biomass (g 0.4 m⁻¹) of samples taken to represent (a) pre-set biofouling biomass categories (low, medium and high), and (b) the mean percent occurrence of low, medium and high biofouling biomass categories across the 3 study areas (inner, middle and outer)

and 'high' biomass samples had on average 83.8 ± 15.5 g m⁻² (68–99 g m⁻²) and 121.2 ± 20 g m⁻² (100–140 g m⁻²) of biofouling, respectively.

DISCUSSION

Patterns of spatial variation in biofouling communities

This study demonstrated that mussel farm longline droppers in Pelorus Sound support diverse biofouling communities that are spatially variable. A total of 86 distinct taxa were identified, with biofouling cover dominated by macroalgae and suspension-feeders (>60% of the total cover) such as other bivalves (specifically blue mussels), hydroids, bryozoans and ascidians. Epibenthic communities are generally composed of suspension-feeders and macroalgae, al-

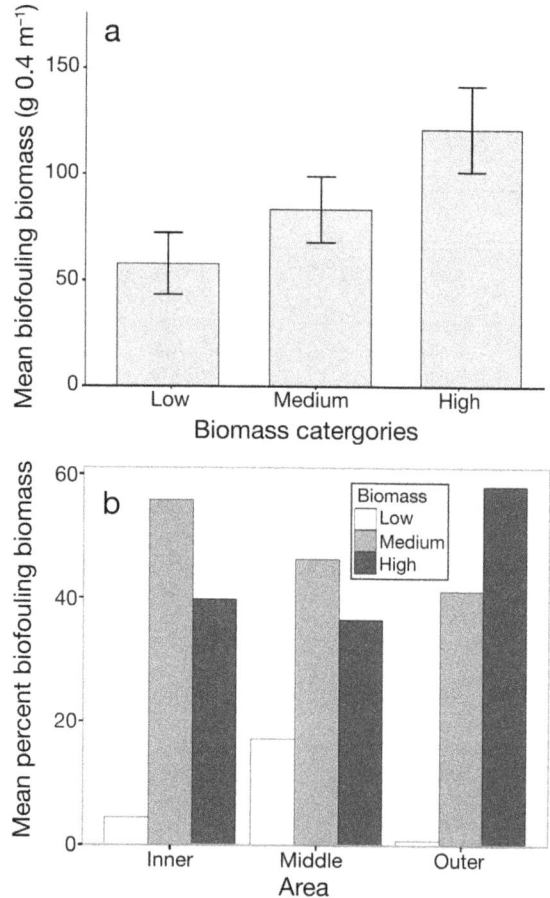

Fig. 6. Distance–decay relationship for the Bray-Curtis matrix of biofouling community dissimilarities against the Euclidean matrix of the geographic distances between sampled marine farms

though this is dependent on water depth and location (Cook et al. 2006). Suspension-feeders are recognised for substantially contributing to the overall biomass of biofouling communities on artificial structures (Lesser et al. 1992, Cronin et al. 1999, Howes et al. 2007, Fitridge et al. 2012). Therefore, these findings are not unexpected, and a dominance of macroalgae and suspension-feeders within biofouling communities have been found on other marine farms, including oyster cultures and tuna farms (Cronin et al. 1999, Mazouni et al. 2001).

The structure of biofouling communities became less similar with increasing geographical distance between sampled mussel farms, making this study the first known, documented account of distance–decay in biofouling communities associated with artificial structures. The slope of the relationship between distance and community similarity in this study was similar to those recorded for intertidal sessile assemblages, but greater than those documented for marine fish communities (Oliva & Teresa 2005, Tsujino et al. 2010). This contrast makes sense given that the rate of decline of similarity would be greater for organisms with a lower dispersal potential (e.g. ascidian species with a short-lived larval stage or intertidal algae with limited dispersal ranges) in open and continuous marine systems, compared to mobile organisms (Soininen et al. 2007, Tsujino et al. 2010). Furthermore, smaller organisms, which respond more intensively to fine scale environmental variation due to their shorter generation times, may have lower similarity at small distances (Gillooly et al. 2002, Tsujino et al. 2010).

Variation in biofouling communities along Pelorus Sound showed a prominence of known problematic species within the surface depths near the Sound entrance. Biofouling cover was greater within the surface 3 m of mussel long-line droppers, with macroalgae and sessile filter-feeders dominating these communities; a pattern that was expected given that light, nutrients and oxygen are not a limiting resource in surface waters. This finding is also in agreement with other reports of decreased biofouling biomass and diversity with increasing water depths (Cronin et al. 1999, Braithwaite et al. 2007, Guenther et al. 2010, Fitridge et al. 2012).

Mussel farms near the entrance of the Sound are located within deep, exposed areas, with strong winds from the north and north-west generating high energy storm waves and promoting extensive coastal erosion (Mcintosh 1958). In contrast, mussel farms near the head of Pelorus Sound are in shallower, sheltered areas, experiencing low wave action and

periods of reduced or stagnant water flow, which may induce warmer summer temperatures (Gibbs et al. 1991). Furthermore, episodic storm events can generate high freshwater inputs from the Pelorus and Kaituna rivers into areas near the head of Pelorus Sound (Gibbs et al. 1991, Broekhuizen et al. 2015). Differential cover of certain biofouling taxa and associated differences in community structure found in this study could therefore be a reflection of tolerance to local environmental conditions. For example, *Undaria pinnatifida* is well adapted for life on exposed coastlines, maintaining firm attachment from its holdfast (Curiel et al. 1998, Russell et al. 2008, Nelson 2013). Nanba et al. (2011) found that *U. pinnatifida* sporophytes grown in exposed sites attained larger sizes and had faster growth rates compared to those grown in sheltered sites in Japan. Similarly, the firm attachment of calcareous tube-worms to hard substrate by cementational adhesion makes them well adapted for high wave energy environments (Moate 1985, Callow & Callow 2002, Bromley & Heinberg 2006). In contrast, some ascidians such as *Ciona intestinalis* are more commonly found fouling sheltered habitats, as hydrodynamic processes in exposed areas preclude successful larval settlement (Howes et al. 2007). Furthermore, while it has been documented that ascidians perform best at salinities above 25 PSU and are rarely tolerant of brackish conditions (Lambert 2005), some may be well adapted to fluctuations in ion concentrations. For example, in laboratory experiments *Didemnum vexillum* showed higher growth rates and survival under low salinities, ranging from 10 to 20 PSU (Gröner et al. 2011).

A number of important problematic pests were identified on mussel long-line droppers in this study, including ascidians (*C. intestinalis* and *D. vexillum*), macroalgae (*U. pinnatifida*, *Cladophora* sp. and *Colpomenia* sp.), tube-building polychaetes (*Pomatoceros* sp.), blue mussels (*Mytilus galloprovincialis*), and the hydroid *Amphisbetia bispinosa*. These species have had, or currently have, detrimental impacts on commercial industries in New Zealand and overseas. For example, *A. bispinosa* has been problematic for mussel cultures in the south Hauraki Gulf, adding weight that can enhance crop loss, and damaging shells, reducing their aesthetic value (Heasman & de Zwart 2004). Similarly, *C. intestinalis* has become a serious biofouling problem for many shellfish operations in Scotland (Karayucel 1997), South Africa (Hecht & Heasman 1999), Chile (Uribe & Etchepare 2002) and eastern Canada (Cayer et al. 1999).

In support of previous work (Woods et al. 2012), biofouling biomass associated with communities on mussel farms in this study appear to substantially contribute to long-line wet weight. In New Zealand, a typical 3 ha mussel farm would have 9 backbone ropes, each measuring 110 m, and supporting a 3750 m crop long-line (Marine Farming Association pers. comm.). Based on our limited assessment, if biofouling was categorised as low across the entire farm, this would extrapolate to approximately 5 t of biofouling biomass along crop long-lines. Similarly, biofouling biomass would be expected to reach 7 and 10 t for farms categorised as having medium and high biomass, respectively. This weight has important implications for the service lifetime, maintenance and costs of aquaculture buoyancy and anchoring systems associated with these marine farms, and is especially important for farms located in the outer Pelorus Sound, where high biofouling biomass categories were predominantly recorded.

Improved biofouling management opportunities

Distributional, site-specific patterns of biofouling observed in Pelorus Sound could be used by aquaculture managers to identify areas where dominant biofouling pests are either absent or less prolific. This is particularly relevant for culture species or production stages that are more susceptible to biofouling (Fitridge & Keough 2013, Sievers et al. 2014). For example, if spat holding areas were placed away from the entrance of Pelorus Sound in deeper waters, this may avoid high levels of biofouling biomass and heavy settlement by notoriously detrimental species such as the brown alga *Colpomenia* sp., which create problems for spat retention in the Marlborough Sounds, or the hydroid species *A. bispinosa*, which renders mussels unsuitable for half-shell trade (Heasman & de Zwart 2004). Similarly, higher cover by blue mussels *M. galloprovincialis*, arguably the most problematic pest for mussel farming in the region, within the surface 3 m of long-line droppers may be reduced by lowering mussel long-lines deeper into the water column. However, while moving stock or lowering long-lines into the water column may reduce biofouling, losses in production due to less submerged culture rope or reduced food levels and associated alterations in crop growth would need to be considered. The effects each biofouling species has on mussel growth/economic value would also need to be quantified to informatively assess the feasibility of translocating stock to avoid these species.

Furthermore, in some locations within Pelorus Sound, biofouling cover, specifically blue mussel 'over-settlement', has been observed to occur across the entire length of mussel long-lines (i.e. is not depth dependent), and attempts to avoid this species have to date proven futile (authors' pers. obs). Therefore, an improved understanding of the reproductive seasonality of important pest species may identify 'windows' during which susceptible production stages may be less prone to impacts.

Through mathematical modelling, scientists could utilise site-specific biofouling patterns to inform marine farmers of the geographical distance between farms that might reduce the localised natural spread of problematic biofoulers. This could be achieved by modelling the potential connectivity of biofouling populations between farms and within farming areas, based on their pelagic larval duration and area-specific environmental conditions (Watts 2014). Such information would be particularly beneficial for mussel farmers and local biosecurity efforts, given that biofouling communities are often dominated by non-indigenous species (Tyrrell & Byers 2007).

Potential steps forward for biofouling management

This study has identified potential opportunities to improve the management of biofouling risks in Pelorus Sound, such as avoiding locations where known problematic marine pests proliferate. However, avoidance may not always be possible given that marine farmers are often constrained by permitted space allocations. Anecdotal evidence suggests that seasonal patterns could also be exploited. However, for biofouling management through avoidance to be adopted by industry and applied successfully, finer-scale sampling and seasonal patterns would be required to identify 'windows of opportunity' in both space and time. A cost-effective approach would be for industry members to frequently document marine pest presence/absence and abundance during routine farm visits (e.g. during product sourcing). The refinement of the biofouling biomass categories (low, medium and high) used in this study would be one way to achieve this goal, although participants would need to be trained to identify key pests, and there would need to be consistency in how abundance estimates were made. For the latter, it would be worthwhile to involve industry staff (e.g. through a workshop) in developing criteria for the different levels of abundance used, and these levels would ideally relate to likely levels of impact (i.e. high abundance of blue mussels = likely high impact). Consideration would also need to be given to the stages of industry production, as biofouling impacts may occur at lower densities for different stages. Data could then be uploaded to an online database, as suggested by Sievers et al. (2014), and used to better inform overall industry practices, as is already undertaken for the detection of harmful phytoplankton (Trainer et al. 2003). Identifying areas containing problematic bio-fouling species would also be useful to industry members for the development of biofouling management plans that address the spread of biofouling pests via industry vessel and infrastructure/crop movements; i.e. improved vector and pathway management.

CONCLUSIONS

Realistically, avoidance of all biofouling pests within a growing region is highly unlikely, and farms will continually be exposed to species with a large dispersal potential (e.g. blue mussels) or to those that are widely distributed and reproduce year-round (e.g. *Didemnum vexillum*). For such species, an im-proved understanding of actual impacts would underpin management decisions about whether additional risk mitigation measures (e.g. mechanical removal or treatment) are justified. An improved understanding of the environmental drivers of bio-fouling proliferation would also be advantageous; particularly to understand areas at risk from newly introduced pest species or those pests that are ex-pected to arrive in a region through domestic spread (i.e. those species already present in the country). Also, given that artificial substrates can enhance the recruitment of epibiota, particularly early succes-sional species (Glasby et al. 2007), and to gain insight into local community connectivity, a study of the regional pool of biofouling species available from surrounding native habitats would be required.

Acknowledgements. Sanford Ltd. and Marlborough Under-water Ltd. are acknowledged for their logistical support and field assistance. Thanks also to Rod Asher, Fiona Gower, Lauren Fletcher, Javier Atalah and Aaron Quarterman (Cawthron Institute) for technical support, as well as Niall Broekhuizen and Mark Hadfield (NIWA) for assistance with the Pelorus Sound model graphs, and Mike Page (NIWA) for ascidian identifications. This research was funded by the New Zealand's Ministry of Business, Innovation and Employment (Adding Value to New Zealand's Cultured Shellfish Industry: Maximising Profitability, Minimising Risk). Financial assistance was also provided by the Cawthron Institute Master's Scholarship and the University of Canterbury Master's Scholarship.

LITERATURE CITED

ä Adams CM, Shumway SE, Whitlatch RB, Getchis T (2011) Biofouling in marine molluscan shellfish aquaculture: a survey assessing the business and economic implications of mitigation. J World Aquac Soc 42:242–252

Anderson MJ (2001a) A new method for non-parametric multivariate analysis of variance. Austral Ecol 26:32–46

ä Anderson MJ (2001b) Permutation tests for univariate or multivariate analysis of variance and regression. Can J Fish Aquat Sci 58:626–639

ä Anderson MJ (2006) Distance-based tests for homogeneity of multivariate dispersions. Biometrics 62:245–253

ä Anderson MJ, Ter Braak CJF (2003) Permutation tests for multi-factorial analysis of variance. J Stat Comput Simul 73:85–113

Anderson MJ, Gorley RN, Clarke KR (2007) PERMANOVA+ for PRIMER: guide to software and statistical methods. PRIMER-E, Plymouth

ä Antoniadou C, Voultsiadou E, Rayann A, Chintiroglou C (2013) Sessile biota fouling farmed mussels: diversity, spatio-temporal patterns, and implications for the basi-biont. J Mar Biol Assoc UK 93:1593–1607

ä Atalah J, Bennett H, Hopkins GA, Forrest BM (2013) Evalu-ation of the sea anemone *Anthothoe albocincta* as an augmentative biocontrol agent for biofouling on artificial structures. Biofouling 29:559–571

ä Benjamini Y, Yekutieli D (2001) The control of the false dis-covery rate in multiple testing under dependency. Ann Stat 29:1165–1188

ä Braithwaite RA, Carrascosa MCC, McEvoy LA (2007) Bio-fouling of salmon cage netting and the efficacy of a typi-cal copper-based antifoulant. Aquaculture 262:219–226

ä Bray JR, Curtis JT (1957) An ordination of the upland forest communities of southern Wisconsin. Ecol Monogr 27: 325–349

Broekhuizen N, Hadfield M, Plew D (2015) A biophysical model for the Marlborough Sounds. Part 2: Pelorus Sound. NIWA report CHC2014-130, Marlborough Dis-trict Council, Christchurch

ä Bromley RG, Heinberg C (2006) Attachment strategies of organisms on hard substrates: a palaeontological view. Palaeogeogr Palaeoclimatol Palaeoecol 232:429–453

ä Cahill PL, Heasman K, Jeffs A, Kuhajek J (2013) Laboratory assessment of the antifouling potential of a soluble-matrix paint laced with the natural compound polygo-dial. Biofouling 29:967–975

ä Callow ME, Callow JA (2002) Marine biofouling: a sticky problem. Biologist 49:10–14

ä Carl C, Poole AJ, Vucko MJ, Williams M, Whalan S, de Nys R (2012) Enhancing the efficacy of fouling-release coat-ings against fouling by *Mytilus galloprovincialis* using nanofillers. Biofouling 28:1077–1091

Carver CE, Chisholm A, Mallet AL (2003) Strategies to miti-gate the impact of *Ciona intestinalis* (L.) biofouling on shellfish production. J Shellfish Res 22:621–631

Cayer D, MacNeil M, Bagnall A (1999) Tunicate fouling in Nova Scotia aquaculture: a new development. J Shellfish Res 18:1–327

ä Ceccherelli G, Campo D (2002) Different effects of *Caulerpa racemosa* on two co-occurring seagrasses in the Mediter-ranean. Bot Mar 45:71–76

ä Chambers LD, Stokes KR, Walsh FC, Wood RJ (2006) Mod-ern approaches to marine antifouling coatings. Surf Coat Tech 201:3642–3652

ä Che LM, Le Campion-Alsumard T, Boury-Esnault N, Payri C, Golubic S, Bezac C (1996) Biodegradation of shells of the black pearl oyster, *Pinctada margaritifera* var. *cumingii*, by microborers and sponges of French Polynesia. Mar Biol 126:509–519

ä Claereboudt MR, Bureau D, Côté J, Himmelman JH (1994) Fouling development and its effect on the growth of juvenile giant scallops (*Placopecten magellanicus*) in suspended culture. Aquaculture 121:327–342

ä Clarke KR (1993) Non-parametric multivariate analyses of changes in community structure. Aust J Ecol 18:117–143

Clarke K, Gorley R (2006) PRIMER v6: user manual/tutorial. PRIMER-E, Plymouth

ä Cook EJ, Black KD, Sayer MDJ, Cromey CJ and others (2006) The influence of caged mariculture on the early development of sublittoral fouling communities: a pan-European study. ICES J Mar Sci 63:637–649

► Cronin ER, Cheshire AC, Clarke SM, Melville AJ (1999) An investigation into the composition, biomass and oxygen budget of the fouling community on a tuna aquaculture farm. Biofouling 13:279–299

ä Curiel D, Bellemo G, Marzocchi M, Scattolin M, Parisi G (1998) Distribution of introduced Japanese macroalgae *Undaria pinnatifida*, *Sargassum muticum* (Phaeophyta) and *Antithamnion pectinatum* (Rhodophyta) in the Lagoon of Venice. Hydrobiologia 385:17–22

Dürr S, Watson DI (2010) Biofouling and antifouling in aquaculture. In: Dürr S, Thomason JC (eds) Biofouling. Wiley-Blackwell, Oxford, p 267–287

Enright C (1993) Control of fouling in bivalve aquaculture: biofouling has been identified as a major impediment to the development of a commercial oyster culture industry. World Aquac 24:44–46

ä Fitridge I, Keough MJ (2013) Ruinous resident: the hydroid *Ectopleura crocea* negatively affects suspended culture of the mussel *Mytilus galloprovincialis*. Biofouling 29: 119–131

ä Fitridge I, Dempster T, Guenther J, de Nys R (2012) The impact and control of biofouling in marine aquaculture: a review. Biofouling 28:649–669

ä Fletcher LM, Forrest BM, Atalah J, Bell JJ (2013a) Reproductive seasonality of the invasive ascidian *Didemnum vexillum* in New Zealand and implications for shellfish aquaculture. Aquacult Environ Interact 3:197–211

ä Fletcher LM, Forrest BM, Bell JJ (2013b) Impacts of the invasive ascidian *Didemnum vexillum* on green-lipped mussel *Perna canaliculus* aquaculture in New Zealand. Aquacult Environ Interact 4:17–30

ä Forrest BM, Blakemore KA (2006) Evaluation of treatments to reduce the spread of a marine plant pest with aquaculture transfers. Aquaculture 257:333–345

► Forrest BM, Taylor MD (2002) Assessing invasion impact: survey design considerations and implications for management of an invasive marine plant. Biol Invasions 4: 375–386

Forrest B, Hopkins G, Webb S, Tremblay L (2011) Overview of marine biosecurity risks from finfish aquaculture development in the Waikato Region. Cawthron Report No. 1871, Cawthron Institute, Nelson

ä Gibbs M, James M, Pickmere S, Woods P, Shakespeare B, Hickman R, Illingworth J (1991) Hydrodynamic and water column properties at six stations associated with mussel farming in Pelorus Sound, 1984–85. NZ J Mar Freshw Res 25:239–254

ä Gillooly JF, Charnov EL, West GB, Savage VM, Brown JH (2002) Effects of size and temperature on developmental time. Nature 417:70–73

► Glasby TM, Connell SD, Holloway MG, Hewitt CL (2007) Nonindigenous biota on artificial structures: Could habi-

tat creation facilitate biological invasions? Mar Biol 151: 887–895

ä Gower JC (1966) Some distance properties of latent root and vector methods used in multivariate analysis. Biometrika 53:325–338

Grant J, Stenton-Dozey J, Monteiro P, Pitcher G, Heasman K (1998) Shellfish culture in the Benguela System: a carbon budget of Saldanha Bay for raft culture of *Mytilus galloprovincialis*. J Shellfish Res 17:41–49

ä Gröner F, Lenz M, Wahl M, Jenkins SR (2011) Stress resistance in two colonial ascidians from the Irish Sea: the recent invader *Didemnum vexillum* is more tolerant to low salinity than the cosmopolitan *Diplosoma listerianum*. J Exp Mar Biol Ecol 409:48–52

ä Guenther J, Carl C, Sunde LM (2009) The effects of colour and copper on the settlement of the hydroid *Ectopleura larynx* on aquaculture nets in Norway. Aquaculture 292: 252–255

ä Guenther J, Misimi E, Sunde LM (2010) The development of biofouling, particularly the hydroid *Ectopleura larynx* on commercial salmon cage nets in Mid-Norway. Aquaculture 300:120–127

Hanson CH, Bell J (1976) Subtidal and intertidal marine fouling on artificial substrate in northern Puget Sound. Fish Bull 74:377–385

Heasman K, de Zwart E (2004) Preliminary investigation on *Amphisbetia bispinosa* colonisation on mussel farms in the Coromandel. Cawthron Report No. 928, Cawthron Institute, Nelson

ä Heath R (1974) Physical oceanographic observations in Marlborough Sounds. NZ J Mar Freshw Res 8:691–708

ä Heath R (1982) Temporal variability of the waters of Pelorus Sound, South Island, New Zealand. NZ J Mar Freshw Res 16:95–110

Hecht T, Heasman K (1999) The culture of *Mytilus galloprovincialis* in South Africa and the carrying capacity of mussel farming in Saldanha Bay. World Aquac 30:50–55

ä Hopkins GA, Forrest BM (2010) Challenges associated with pre-border management of biofouling on oil rigs. Mar Pollut Bull 60:1924–1929

► Howes S, Herbinger CM, Darnell P, Vercaemer B (2007) Spatial and temporal patterns of recruitment of the tunicate *Ciona intestinalis* on a mussel farm in Nova Scotia, Canada. J Exp Mar Biol Ecol 342:85–92

Karayucel S (1997) Mussel culture in Scotland. World Aquac 28:4–11

Knight BR, Goodwin EO, Jiang WM, Carbines G (2010) Development of settlement model for Marlborough Sounds blue cod (*Parapercis colias*). Cawthron Report No. 1707, Cawthron Institute, Nelson

ä Kohler KE, Gill SM (2006) Coral point count with Excel extensions (CPCe): a Visual Basic program for the determination of coral and substrate coverage using random point count methodology. Comput Geosci 32:1259–1269

ä Lacoste E, Gaertner-Mazouni N (2014) Biofouling impact on production and ecosystem functioning: a review for bivalve aquaculture. Rev Aquac 7:187–196

ä Lambert G (2005) Ecology and natural history of the protochordates. Can J Zool 83:34–50

ä LeBlanc N, Davidson J, Tremblay R, McNiven M, Landry T (2007) The effect of anti-fouling treatments for the clubbed tunicate on the blue mussel, *Mytilus edulis*. Aquaculture 264:205–213

► Legendre P, Gallagher ED (2001) Ecologically meaningful transformations for ordination of species data. Oecologia 129:271–280

► Lesser MP, Shumway SE, Cucci T, Smith J (1992) Impact of

fouling organisms on mussel rope culture: interspecific competition for food among suspension-feeding invertebrates. J Exp Mar Biol Ecol 165:91–102

Lleonart M, Handlinger J, Powell M (2003) Spionid mudworm infestation of farmed abalone (*Haliotis* spp.). Aquaculture 221:85–96

López-Galindo C, Casanueva JF, Nebot E (2010) Efficacy of different antifouling treatments for seawater cooling systems. Biofouling 26:923–930

Lutz-Collins V, Ramsay A, Quijón PA, Davidson J (2009) Invasive tunicates fouling mussel lines: evidence of their impact on native tunicates and other epifaunal invertebrates. Aquat Invasions 4:213–220

Mazouni N, Gaertner JC, Deslous-Paoli JM (2001) Composition of biofouling communities on suspended oyster cultures: an *in situ* study of their interactions with the water column. Mar Ecol Prog Ser 214:93–102

Mcintosh CB (1958) Maps of surface winds in New Zealand. NZ Geog 14:75–81

McKindsey CW, Landry T, O'Beirn FX, Davies IM (2007) Bivalve aquaculture and exotic species: a review of ecological considerations and management issues. J Shellfish Res 26:281–294

McKindsey CW, Lecuona M, Huot M, Weise AM (2009) Biodeposit production and benthic loading by farmed mussels and associated tunicate epifauna in Prince Edward Island. Aquaculture 295:44–51

Metri R, da Rocha RM, Marenzi A (2002) Epibiosis reduction on productivity in a mussel culture of *Perna perna* (Linné, 1758). Braz Arch Biol Technol 45:325–331

Moate R (1985) Offshore fouling communities and settlement and early growth in *Tubularia larynx* and *Pomatoceros triqueter*. PhD thesis, Plymouth University

Morlon H, Chuyong G, Condit R, Hubbell S and others (2008) A general framework for the distance-decay of similarity in ecological communities. Ecol Lett 11:904–917

Nanba N, Fujiwara T, Kuwano K, Ishikawa Y, Ogawa H, Kado R (2011) Effect of water flow velocity on growth and morphology of cultured *Undaria pinnatifida* sporophytes (Laminariales, Phaeophyceae) in Okirai Bay on the Sanriku coast, Northeast Japan. J Appl Phycol 23: 1023–1030

Narum SR (2006) Beyond Bonferroni: less conservative analyses for conservation genetics. Conserv Genet 7:783–787

Nelson W (2013) New Zealand seaweeds, an illustrated guide. Te Papa Press, Wellington

Oksanen J (1983) Ordination of boreal heath-like vegetation with principal component analysis, correspondence analysis and multidimensional scaling. Vegetatio 52: 181–189

Oksanen J (2011) Multivariate analysis of ecological communities in R: vegan tutorial. R package version 2.0-1, R Foundation for Statistical Computing, Vienna

Oksanen J, Blanchet F, Kindt R, Legendre P and others (2011) Vegan: community ecology package. R package version 2.0-2, R Foundation for Statistical Computing, Vienna

Oliva ME, Teresa GM (2005) The decay of similarity over geographical distance in parasite communities of marine fishes. J Biogeogr 32:1327–1332

Paetzold SC, Davidson J (2011) Aquaculture fouling: efficacy of potassium monopersulphonate triple salt based disinfectant (Virkon® Aquatic) against *Ciona intestinalis*. Biofouling 27:655–665

R Core Team (2013) R: a language and environment for statistical computing. R Foundation for Statistical Computing, Vienna

Russell LK, Hepburn CD, Hurd CL, Stuart MD (2008) The expanding range of *Undaria pinnatifida* in southern New Zealand: distribution, dispersal mechanisms and the invasion of wave-exposed environments. Biol Invasions 10:103–115

Sievers M, Fitridge I, Dempster T, Keough MJ (2013) Biofouling leads to reduced shell growth and flesh weight in the cultured mussel *Mytilus galloprovincialis*. Biofouling 29:97–107

Sievers M, Dempster T, Fitridge I, Keough MJ (2014) Monitoring biofouling communities could reduce impacts to mussel aquaculture by allowing synchronisation of husbandry techniques with peaks in settlement. Biofouling 30:203–212

Silina AV (2006) Tumor-like formations on the shells of Japanese scallops *Patinopecten yessoensis* (Jay). Mar Biol 148:833–840

Simon CA, Ludford A, Wynne S (2006) Spionid polychaetes infesting cultured abalone *Haliotis midae* in South Africa. Afr J Mar Sci 28:167–171

Soininen J, McDonald R, Hillebrand H (2007) The distance decay of similarity in ecological communities. Ecography 30:3–12

Stæhr PA, Pedersen MF, Thomsen MS, Wernberg T, Krause-Jensen D (2000) Invasion of *Sargassum muticum* in Limfjorden (Denmark) and its possible impact on the indigenous macroalgal community. Mar Ecol Prog Ser 207:79–88

Thomsen MS, Wernberg T, Stæhr PA, Pedersen MF (2006) Spatio-temporal distribution patterns of the invasive macroalga *Sargassum muticum* within a Danish *Sargassum*-bed. Helgol Mar Res 60:50–58

Trainer VL, Eberhart BTL, Wekell JC, Adams NG, Hanson L, Cox F, Dowell J (2003) Paralytic shellfish toxins in Puget Sound, Washington state. J Shellfish Res 22:213–223

Tsujino M, Hori M, Okuda T, Nakaoka M, Yamamoto T, Noda T (2010) Distance decay of community dynamics in rocky intertidal sessile assemblages evaluated by transition matrix models. Popul Ecol 52:171–180

Tyrrell MC, Byers JE (2007) Do artificial substrates favor nonindigenous fouling species over native species? J Exp Mar Biol Ecol 342:54–60

Underwood AJ, Anderson MJ (1994) Seasonal temporal aspects of recruitment and succession in an intertidal estuarine fouling assemblages. J Mar Biol Assoc UK 74: 563–584

Underwood AJ, Petraitis PS (1993) Structure of intertidal assemblages in different locations: How can local processes be compared? In: Ricklefs RE, Schluter D (eds) Species diversity in ecological communities: historical and geographical perspectives. University of Chicago Press, Chicago, IL, p 39–51

Uribe E, Etchepare I (2002) Effects of biofouling by *Ciona intestinalis* on suspended culture of *Argopecten purpuratus* in Bahia Inglesa, Chile. Bull Aquacult Assoc Can 102:93–95

Valentine PC, Carman MR, Blackwood DS, Heffron EJ (2007) Ecological observations on the colonial ascidian *Didemnum* sp. in a New England tide pool habitat. J Exp Mar Biol Ecol 342:109–121

Watts AM (2014) Biofouling patterns and local dispersal in an aquaculture system in the Marlborough Sounds, New Zealand. MSc thesis, University of Canterbury, Christchurch

Woods C, Floerl O, Hayden BJ (2012) Biofouling on Greenshell™ mussel (*Perna canaliculus*) farms: a preliminary assessment and potential implications for sustainable aquaculture practices. Aquacult Int 20:537–557

Permissions

List of Contributors

Gary S. Caldwell and Selina M. Stead
School of Marine Science and Technology, Newcastle University, Newcastle NE1 7RU, UK

Georgina Robinson
School of Marine Science and Technology, Newcastle University, Newcastle NE1 7RU, UK
Department of Ichthyology and Fisheries Science, Rhodes University, Grahamstown 6140, South Africa

Clifford L. W. Jones
Department of Ichthyology and Fisheries Science, Rhodes University, Grahamstown 6140, South Africa

Matthew J. Slater
School of Marine Science and Technology, Newcastle University, Newcastle NE1 7RU, UK
Alfred-Wegener-Institute, Helmholtz Center for Polar and Marine Research, Am Handelshafen 12, 27570 Bremerhaven, Germany

Antonio Medina, Guillermo Aranda, Silvia Gherardi and Agustín Santos
University of Cádiz, Department of Biology (Zoology), Faculty of Marine and Environmental Sciences, Campus de Excelencia Internacional del Mar (CEI•MAR), 11510 Puerto Real, Cádiz, Spain

Begonya Mèlich and Manuel Lara
Grup Balfegó, Pol. Ind. edifici 'Balfegó', 43860 L'Ametlla de Mar, Tarragona, Spain

I. A. Johnsen, L. C. Asplin, A. D. Sandvik and R. M. Serra-Llinares
Institute of Marine Research, Nordnes, 5817 Bergen, Norway

Eveline Diopere and Filip A. M. Volckaert
Laboratory of Biodiversity and Evolutionary Genomics, University of Leuven, Ch. Deberiotstraat 32, 3000 Leuven, Belgium

Jonas Bylemans
Laboratory of Biodiversity and Evolutionary Genomics, University of Leuven, Ch. Deberiotstraat 32, 3000 Leuven, Belgium
Institute for Applied Ecology, University of Canberra, Canberra, ACT 2612, Australia

Gregory E. Maes
Laboratory of Biodiversity and Evolutionary Genomics, University of Leuven, Ch. Deberiotstraat 32, 3000 Leuven, Belgium
Centre for Sustainable Tropical Fisheries and Aquaculture, Comparative Genomics Centre, College of Marine and Environmental Sciences, Faculty of Science and Engineering, James Cook University, Townsville, 4811 QLD, Australia

Alessia Cariani and Fausto Tinti
Department of Biological, Geological and Environmental Sciences, University of Bologna, Ravenna 48123, Italy

Helen Senn
WildGenes Laboratory, Royal Zoological Society of Scotland, Edinburgh EH12 6TS, UK

Martin I. Taylor
School of Biological Sciences, University of East Anglia, Norwich NR4 7TJ, UK

Sarah Helyar
Institute for Global Food Security, Queen's University Belfast, Belfast BT9 5BN, UK

Luca Bargelloni
Department of Public Health, Comparative Pathology, and Veterinary Hygiene, University of Padova, Viale dell'Università 16, 35020 Legnaro, Italy

Alessio Bonaldo and Ilaria Guarniero
Department of Veterinary Medical Sciences DIMEVET, University of Bologna, 40064 Bologna, Italy

Gary Carvalho
Molecular Ecology and Fisheries Genetics Laboratory, School of Biological Sciences, Environment Centre Wales, Bangor University, Bangor, Gwynedd LL57 2UW, UK

Hans Komen
Animal Breeding and Genomics Centre, Wageningen Univer 338, 6700AH Wageningen, The Netherlands

Jann Th. Martinsohn
JRC.G.4 – Maritime Affairs, Institute for the Protection and Security of the Citizen (IPSC), Joint Research Centre (JRC), European Commission, Via Enrico Fermi 2749, 21027 Ispra VA, Italy

Einar E. Nielsen
Section for Marine Living Resources, National Institute of Aquatic Resources, Technical University of Denmark, Vejlsøvej 39, 8600 Silkeborg, Denmark

Rob Ogden
TRACE Wildlife Forensics Network, Royal Zoological Society of Scotland, Edinburgh EH12 6TS, UK

Adam D. Hughes and Kenneth D. Black
Scottish Association for Marine Sciences, Dunbeg, Oan, Argyll PA37 1QA, UK

Ruihuan Li
Key Laboratory of Marine Chemistry Theory and Technology, MOE, Ocean University of China/ Qingdao Collaborative Innovation Center of Marine Science and Technology, Qingdao 266100, PR China
State Key Laboratory of Tropical Oceanography, South China Sea Institute of Oceanology, Chinese Academy of Sciences, 164 West Xingang Road, Guangzhou 510301, PR China

Sumei Liu
Key Laboratory of Marine Chemistry Theory and Technology, MOE, Ocean University of China/ Qingdao Collaborative Innovation Center of Marine Science and Technology, Qingdao 266100, PR China
Laboratory for Marine Ecology and Environmental Science, Qingdao National Laboratory for Marine Science and Technology, Qingdao, PR China

Jing Zhang
State Key Laboratory of Estuarine and Coastal Research, East China Normal University, Shanghai 200062, PR China

Zengjie Jiang and Jianguang Fang
Carbon Sink Fisheries Laboratory, Key Laboratory of Sustainable Utilization of Marine Fisheries Resources, Ministry of Agriculture, Yellow Sea Fisheries Research Institute, Chinese Academy of Fishery Sciences, 106 Nanjing Road, Qingdao 266071, PR China

Lindsay M. Brager and Peter J. Cranford
Department of Fisheries and Oceans, Bedford Institute of Oceanography, 1 Challenger Drive, Dartmouth, Nova Scotia B2Y 4A2, Canada

Henrice Jansen
Wageningen IMARES–Institute for Marine Resources and Ecosystem Studies, Korringaweg 5, 4401 NT Yerseke, The Netherlands

Øivind Strand
Institute of Marine Research, Nordnesgaten 50, 5005 Bergen, Norway

Zhenlong Sun, Qin-Feng Gao, Shuanglin Dong and Fang Wang
Key Laboratory of Mariculture, Ministry of Education, Ocean University of China, Qingdao, Shandong 266003, PR China

Bin Xia
Key Laboratory of Mariculture, Ministry of Education, Ocean University of China, Qingdao, Shandong 266003, PR China
Marine Science and Engineering College, Qingdao Agricultural University, Qingdao, Shandong 266109, PR China

Mingshuang Sun, Wangwang Ye and Da Song
Key Laboratory of Marine Chemistry Theory and Technology, Ministry of Education, Ocean University of China, Qingdao 266100, PR China

Jing Hou and Guiling Zhang
Key Laboratory of Marine Chemistry Theory and Technology, Ministry of Education, Ocean University of China, Qingdao 266100, PR China
Qingdao Collaborative Innovation Center of Marine Science and Technology, Ocean University of China, Qingdao 266100, PR China

Jeffrey S. Ren and Jeanie Stenton-Dozey
National Institute of Water and Atmospheric Research, 10 Kyle Street, Christchurch 8440, New Zealand

Jihong Zhang
Yellow Sea Fisheries Research Institute, Chinese Academy of Fishery Sciences, 106 Nanjing Road, Qingdao 266071, PR China

Function Laboratory for Marine Fisheries Science and Food Production Processes, Qingdao National Laboratory for Marine Science and Technology, 1 Wenhai Road, Aoshanwei, Jimo, Qingdao 266200, PR China

Javier Atalah and Barrie M. Forrest
Cawthron Institute, Private Bag 2, Nelson 7010, New Zealand

Hayden Rabel
Statistics New Zealand, Private Bag 4741, Christchurch 8140, New Zealand

Sandra Casas and Jerome La Peyre
School of Renewable Natural Resources, Louisiana State University Agricultural Center, Baton Rouge, LA 70803, USA

William Walton, Glen Chaplin and Scott Rikard
Auburn University Shellfish Laboratory, Dauphin Island, AL 36528, USA

John Supan
School of Renewable Natural Resources, Louisiana State University Agricultural Center, Baton Rouge, LA 70803, USA
Louisiana Sea Grant College Program, Louisiana State University, Baton Rouge, LA 70803, USA

Sarantis Sofianos
Department of Environmental Physics, University of Athens, 15784 Athens, Greece

Yannis Hatzonikolakis
Department of Environmental Physics, University of Athens, 15784 Athens, Greece
Hellenic Centre for Marine Research (HCMR), Athens-Sounio Avenue, Mavro Lithari, 19013 Anavyssos, Greece

Kostas Tsiaras and George Triantafyllou
Hellenic Centre for Marine Research (HCMR), Athens-Sounio Avenue, Mavro Lithari, 19013 Anavyssos, Greece

John A. Theodorou
Department of Fisheries and Aquaculture Technology, Technological Educational Institute of Western Greece, Nea Ktiria, Mesolonghi 30200, Greece

George Petihakis
Hellenic Centre for Marine Research (HCMR), 71003 Heraklion, Greece

P. Sanchez-Jerez and A. Forcada
Department of Marine Science and Applied Biology, University of Alicante, 03080 Alicante, Spain

I. Karakassis
Institute of Marine Biology of Crete, 71003 Heraklion, Greece

F. Massa and D. Fezzardi
General Fisheries Commission for the Mediterranean, Food and Agriculture Organisation of the United Nations, Via Vittoria Colonna 1, 00193 Rome, Italy

J. Aguilar-Manjarrez
Fisheries and Aquaculture Department, Food and Agriculture Organization of the United Nations, Viale delle Terme di Caracalla, 00153 Rome, Italy

D. Soto and R. Chapela
Fisheries and Aquaculture Department, Food and Agriculture Organization of the United Nations, Viale delle Terme di Caracalla, 00153 Rome, Italy
CETMAR, C/ Eduardo Cabello s/n, 36208 Bouzas-Vigo Spain

P. Avila
Junta de Andalucía, Agencia de Gestión Agraria y Pesquera de Andalucía, C/ Severo Ochoa 38, Parque Tecnológico de Andalucía, 29590 Campanillas, Málaga, Spain

J. C. Macias
Aquaculture Consultant, C/ Crucero 2^{a}F n1, Sanlúcar de Barrameda, 11540 Cadiz, Spain

P. Tomassetti and G. Marino
Institute for Environmental Protection and Research (ISPRA), Via Vitaliano Brancati 48, 00144 Rome, Italy

J. A. Borg
9Department of Biology, Faculty of Science, University of Malta, 20810 Msida, Malta

V. Franičević
Ministry of Agriculture, Department of Fisheries, I. Mažurani a 30, 23000 Zadar, Croatia

G. Yucel-Gier
Institute of Marine Sciences and Technology, Dokuz
Eylül University, 35340 Izmir, Turkey

I. A. Fleming
Department of Ocean Science, Memorial University
of Newfoundland, St. John's, NL A1C 5S7, Canada

X. Biao
School of Geography Sciences, Nanjing Normal
University, 1 Wenyuan Road, Nanjing 210023, PR
China

H. Nhhala
Centre Aquacole Institut National de Recherche
Halieutique, BP n°31, M'diq, Morocco

H. Hamza
Direction Générale des Pêches et de l'Aquaculture
Ministère de l'Agriculture, 30 Rue Alain Savary,
1002 Tunis, Tunisia

T. Dempster
Sustainable Aquaculture Laboratory – Temperate
and Tropical (SALTT), Department of Zoology,
University of Melbourne, Victoria 3010, Australia

Karina Moe
Norwegian Institute for Nature Research (NINA),
Sluppen, 7485 Trondheim, Norway Norwegian
University of Life Sciences (NMBU), NMBU, 1432
Ås, Norway

**Tor F. Næsje, Eva M. Ulvan, Tonje Aronsen and
Eva B. Thorstad**
Norwegian Institute for Nature Research
(NINA), Sluppen, 7485 Trondheim, Norway

Thrond O. Haugen
Norwegian University of Life Sciences (NMBU),
NMBU, 1432 Ås, Norway

Tomas Sandnes
Aqua Kompetanse AS, 7770 Flatanger, Norway

A. M. Watts
School of Biological Sciences, University of
Canterbury, Christchurch 8041, New Zealand
Coastal and Freshwater Group, Cawthron Institute,
Nelson 7010, New Zealand
National Institute of Water and Atmosphere
(NIWA), 217 Akersten Street, Port Nelson, Nelson
7010, New Zealand

S. J. Goldstien
School of Biological Sciences, University of
Canterbury, Christchurch 8041, New Zealand

G. A. Hopkins
Coastal and Freshwater Group, Cawthron Institute,
Nelson 7010, New Zealand

Index